Smart Wearables, Nanofibers, and Textiles

This book provides an in-depth exploration of the development and application of smart wearables, nanofibers, and sustainable textiles. It covers advancements in energy harvesting, health monitoring, and sustainable design, offering readers a holistic understanding of the intersection between wearable technology, advanced materials, and environmental responsibility.

- Explores the challenges and prospects of integrating energy harvesting and storage into smart wearables.
- Discusses the role of artificial intelligence and machine learning in enhancing the functionality of wearable devices.
- Highlights innovations in textile-based sensors, conducting yarns, and eco-friendly photonic and fluorescent materials for wearables.
- Examines the use of synthetic and biopolymer nanofiber films in food packaging and their implications for sustainability.
- Provides insights into sustainable textile processing and the future of eco-carriers in green diffusion.

By presenting state-of-the-art research and technologies, this book serves as a valuable resource for researchers, engineers, and professionals, including materials scientists, textile engineers, and biomedical engineers.

Majid Montazer is a prominent expert in textile engineering, particularly recognized for his innovative research in nanofinishing techniques. He serves as a full professor at Amirkabir University of Technology in Iran, where he has made significant contributions to the advancement of antibacterial, self-cleaning, conductive, technical, and multifunctional textiles. He has authored more than 400 scientific publications and several chapter books.

Ali Mianehro specializes in advanced nanofibers and textiles, nanocomposites, electrospinning, and surface modification of nanofibers. He is affiliated with the KTH Royal Institute of Technology in Stockholm, Sweden. His work integrates nanotechnology and surface engineering, with a significant focus on sustainable nanofibers and biomass valorization.

Emerging Materials and Technologies

Series Editor: Boris I. Kharissov

The *Emerging Materials and Technologies* series is devoted to highlighting publications centered on emerging advanced materials and novel technologies. Attention is paid to those newly discovered or applied materials with potential to solve pressing societal problems and improve quality of life, corresponding to environmental protection, medicine, communications, energy, transportation, advanced manufacturing, and related areas.

The series takes into account that, under present strong demands for energy, material, and cost savings, as well as heavy contamination problems and worldwide pandemic conditions, the area of emerging materials and related scalable technologies is a highly interdisciplinary field, with the need for researchers, professionals, and academics across the spectrum of engineering and technological disciplines. The main objective of this book series is to attract more attention to these materials and technologies and invite conversation among the international R&D community.

Boron Nitride Nanostructures
Applications in Energy and Healthcare
Indranil Lahiri, Amrita De Adhikari, Souvik Ghosh, Rita Joshi, Satish Jaiswal, and Debrupa Lahiri

Smart Polymers
An Emerging Vista for Nanotheranostics
Dipankar Chattopadhyay, Adrija Ghosh, Jonathan Tersur Orasugh and Suprakas Sinha Ray

Lightweight Materials for Electric Vehicles
Sustainable Materials, Production Process, and Modeling Techniques
Edited by R. Suresh, C. Durga Prasad, Satish Kumar, Bharath K.N. and Ajith G. Joshi

Graphitic Carbon Nitrides
Michio Inagaki, Masahiro Toyoda and Yasushi Soneda

The Science and Applications of 3D Bioprinting
Edited by Thiago Domingues Stocco and Anderson de Oliveira Lobo

Metal-Organic Frameworks for Anticorrosive Coatings
Edited by Ashish Kumar, Ambrish Singh and Humira Assad

Evaluation of Electrochemical Corrosion
Experimental Procedures and Characterization
Nam Nguyen-Dang

Smart Functional Wearables
From Nanofibers to Textiles
Majid Montazer and Ali Mianehro

For more information about this series, please visit:
www.routledge.com/Emerging-Materials-and-Technologies/book-series/CRCEMT

Smart Wearables, Nanofibers, and Textiles

Edited by
Majid Montazer
and
Ali Mianehro

CRC Press
Taylor & Francis Group
Boca Raton London New York

CRC Press is an imprint of the
Taylor & Francis Group, an **informa** business

Designed cover image: © 2026 Shutterstock

First edition published 2027
by CRC Press
2385 NW Executive Center Drive, Suite 320, Boca Raton FL 33431

and by CRC Press
4 Park Square, Milton Park, Abingdon, Oxon, OX14 4RN

CRC Press is an imprint of Taylor & Francis Group, LLC

ISBN: 978-1-032-82992-0 (hbk)
ISBN: 978-1-032-82991-3 (pbk)
ISBN: 978-1-003-50720-8 (ebk)

DOI: 10.1201/9781003507208

Typeset in Times
by SPi Technologies India Pvt Ltd (Straive)

Contents

Contributors

Ajay Agarwal
Department of Electronics Engineering
Indian Institute of Technology
Jodhpur, India

Ozan Avinc
Textile Engineering Department
Pamukkale University
Denizli, Turkiye

Md. Nazif Hasan Chowdhury
Department of Textile Engineering
Port City International University
Chittagong, Bangladesh

G M Faysal
Department of Textile Engineering
Northern University Bangladesh
Dhaka, Bangladesh

Seyed Majid Ghoreishian
The Center for Energy and Environmental Solutions (CEES)
College of STEM-T, South Carolina State University
Orangeburg, South Carolina, USA

Md. Ariful Islam
Department of Textile Engineering
Northern University Bangladesh
Dhaka, Bangladesh

Lea Sophie Kämper
Hochschule Niederrhein
Faculty of Textile and Clothing Technology
Mönchengladbach, Germany

Ece Kalayci
Textile Engineering Department
Pamukkale University
Denizli, Turkiye

Anupam Kumari
Department of Electronics Engineering
Indian Institute of Technology
Jodhpur, India

Boris Mahltig
Faculty of Textile and Clothing Technology
Hochschule Niederrhein
Mönchengladbach, Germany

Md. Himel Mahmud
Department of Textile Engineering
Northern University Bangladesh
Dhaka, Bangladesh

Vimala Mathew
National Institute of Electronics & Information Technology (NIELIT)
Calicut, India

Steven J. McNeil
Textile Department
AgResearch Ltd.
Christchurch, New Zealand

Kannissery Pramod
College of Pharmaceutical Sciences
Government Medical College
Kozhikode, India

Md. Reazuddin Repon
Department of Bioproducts and Biosystems, School of Chemical Engineering
Aalto University
Espoo, Finland

Hubert Bryan Riley
The Center for Energy and Environmental Solutions (CEES)
College of STEM-T, South Carolina State University
Orangeburg, South Carolina, USA

Manindra Nath Roy
Department of Textile Engineering
Northern University Bangladesh
Dhaka, Bangladesh

Palasseri Ayisha Sana
College of Pharmaceutical Sciences
Government Medical College
Kozhikode, India

Lea Schulz
Faculty of Textile and Clothing Technology
Hochschule Niederrhein
Mönchengladbach, Germany

Angan Sengupta
Department of Chemical Engineering
Interdisciplinary Research Program of Space Science and Technology
Jodhpur, India

Anicka Siegert
Faculty of Textile and Clothing Technology
Hochschule Niederrhein
Mönchengladbach, Germany

Kuriyedath Parambil Khadeeja Thanha
College of Pharmaceutical Sciences
Government Medical College
Kozhikode, India

Preface

We are living through a profound shift in how humans interact with technology. Nowadays, fibers are not just for ordinary clothing; they are also used for advanced applications far beyond traditional textiles including aerospace engineering, biomedicine, telecommunications, and smart wearables. As we demand higher performance and functionality, escalating environmental challenges create an urgent need for resource conservation and a lower impact on our planet. It is no longer enough for technologies to be functional; they must also be sustainable.

This book titled *Smart Wearables, Nanofibers, and Textiles* as part of the *Emerging Materials and Technologies* book series provides a comprehensive guide to developing eco-friendly, smart, functional wearables, bridging the gap between fundamental materials science and practical applications. It offers a deep understanding of eco-friendly approaches in wearable design and provides detailed discussions on advanced textiles, fiber-form electronics, sensors, optical wearables, and smart active nanofibers for environmental and biomedical uses. Furthermore, it illustrates how the field fosters innovation and contributes toward a sustainable future by incorporating artificial intelligence (AI) and machine learning (ML) into wearable technology development.

Ultimately, this book serves as a vital roadmap for researchers, engineers, and practitioners such as materials scientists, textile engineers, and biomedical engineers. By exploring state-of-the-art research, it provides the insights needed to develop innovative, eco-friendly fibrillar structures and wearable devices that successfully harmonize recent technological advances with environmental responsibility.

1 Challenges and Prospects of Smart Wearables for Energy Harvesting and Energy Storage

Md. Himel Mahmud, Manindra Nath Roy, Md. Ariful Islam, G M Faysal, and Md. Reazuddin Repon

INTRODUCTION

In recent years, smart wearables have become one of the most popular issues in the smart product industry, drawing the attention of manufacturers who are expanding their applications beyond everyday usage to medical checkups and other disciplines. By utilizing wearable technologies, smart wearables are cleverly created for everyday use. It is designed to be used right on the body or incorporated into accessories, apparel, etc. [1]. The market for smart wearables is dominated by Apple, Samsung, and Huawei [2]. Since smart wearables require a steady power source, energy-harvesting and storage technologies are vital for the smart wearables market. Figure 1.1 exhibits some common sources used for harvesting and storing energy. Lithium-ion batteries (LIBs), which are required to be periodically recharged or replaced, are being most widely used as an energy-storage solution for wearables, which have a wide variety of power necessities [1]. Environmental energy-harvesting-based self-powered sensor research has gained popularity. Kinetic energy is produced when individuals move and solar energy is limitless [4]. Since the human body is a vast energy reservoir, numerous researchers have looked into ways to produce electricity from breathing, blood pressure, body temperature, and body movement. For some wearables used in military and special field applications, generating energy through movement is significantly more feasible than plugging in for charging [5]. Wearable technology that harnesses human energy should be both environmentally and skin-friendly [6]. Realizing a steady energy supply for wearables has proven to be a significant difficulty thus far. While LIBs' present cycle life can satisfy wearable technology, wearables of the future will need to be lightweight, multifunctional, and small. Due to the restricted

DOI: 10.1201/9781003507208-1

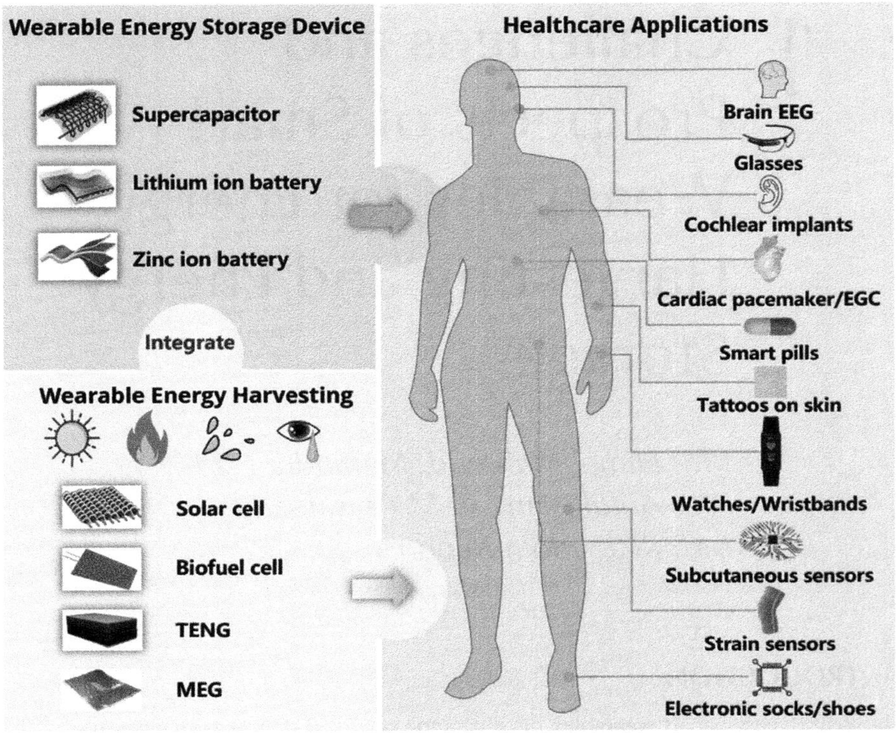

FIGURE 1.1 Typical energy-storage and energy-harvesting devices for wearables [3].

capacity of LIBs, certain wearables are unable to constantly recharge and replace the batteries [7].

Wearable energy-harvesting and storage systems have a long history and have seen numerous revolutionary advancements [8]. The wearability and flexibility required for useful healthcare applications were severely hampered by the large, stiff batteries that initially powered these gadgets. However, cutting-edge advances in materials engineering have resulted in devices that are stretchy, flexible, and lightweight. For example, the development of biofuel cells (BFCs), triboelectric nanogenerators (TENGs), and flexible LIBs and zinc-ion batteries (ZIBs) has enabled the creation of smart wearables with ergonomic comfort and adaptability, which were impossible to think of before [9, 10]. With wearable technology becoming more and more integrated into daily life, systems that provide personalized health information and captivating user experiences are in high demand. This progress requires sophisticated sensor and data processing technology, along with incredibly efficient energy-storage devices [11]. Wearable energy-storage technologies that store energy to be used later and ensure a steady power supply for worn medical equipment include ZIBs [12], LIBs [13], and supercapacitors (SCs) [14, 15]. On the other hand, energy-harvesting

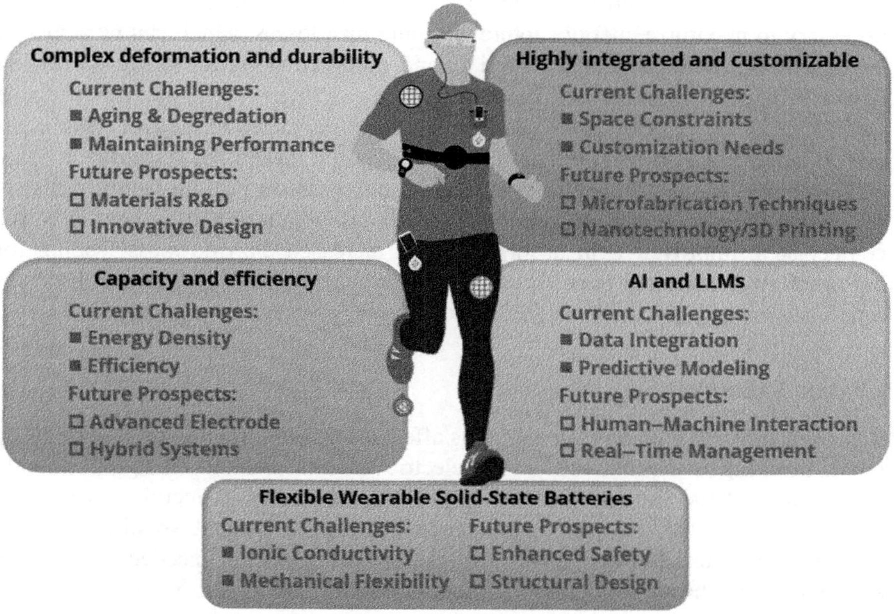

FIGURE 1.2 Challenges and prospects of energy harvesting and storage for wearables [3]. **(Published under a Creative Commons Attribution 4.0.)**

devices, such as TENGs, magnetoelastic generators (MEG) [16], solar cells (SCs) [11], and BFCs [17], not only collect but also transform optimum energy into electric power, thereby offering an uninterrupted energy supply that increases device durability and improves user friendliness [18, 19].

The difficulties and possibilities of energy-harvesting techniques and wearable energy-storage devices, as shown in Figure 1.2, are covered in this chapter.

CHALLENGES AND PROSPECTS OF ENERGY-HARVESTING TECHNOLOGY

In recent years, various energy-harvesting technologies have become popular—piezoelectric nanogenerators (PENGs), TENGs, thermoelectric generators (TEGs), SCs, BFCs, and hybrid generators (HGs)—for smart wearables. However, these energy-harvesting technologies are not immune to their drawbacks.

Efficiency Enhancement/Output Power

The prime challenge of the popular energy-harvesting devices is the efficiency of the technology because it directly influences the quantity of usable energy produced from the source. Moreover, the energy requirement is increasing each day. A typical smart watch, for example, consumes about 30 mW of energy, which is higher than the current yield of the present technologies [20]. To meet the energy demand, it is necessary to maximize energy capture and minimize losses, which can be achieved by developing new functional materials and modified designs for the harvesting device or integrated harvesting. For example, $Mg_{3+\delta}Sb_xBi_{2-x}$ [21] and ultra-flexible $BaTiO_3$ [22] are developed to integrate with textile materials due to their high-performance biomechanical energy harvesting. In the meantime, three-dimensional (3D) printing or omnidirectional weaving are being used to create flexible energy-harvesting devices, taking into account greater temperature gradient and effective contact area, improved light absorption, and increased enzyme mass loading for high-performance energy harvesting [23]. Additionally, porous structure and surface roughness are being considered for improving energy-conversion efficiency.

Output Stability

It is often found that energy harvesting is affected by the surroundings of the energy input, resulting in fluctuations in the electric outputs. Due to changing climatic conditions and the little surface area available for energy collection, small-scale devices in particular frequently experience considerable output volatility. Robust and flexible materials, innovative designs, and strategies are needed to stabilize energy output. The most popular way to solve this problem is using a rectifier or a buffer with the harvesting devices. Another possible approach is integrating energy-harvesting devices with storage devices—supercapacitors [24, 25] and batteries [26]—which will support during the unfavorable conditions.

Mechanical and Environmental Durability

For developing smart wearables, textile materials frequently go through subsequent modifications like physical, chemical, or biological as the textile materials are electrically insulating [27, 28]. However, in the ambient environment, oxygen, water, temperature, as well as mechanical stress can damage energy-harvesting systems during daily use [29]. For example, the adhesive conductive layer can crack or even peel off from wearable substrates, like textile-embedded SCs, due to body movement. Therefore, achieving optimum mechanical strength and durability of the modification or functional coating depends on the adhesion between functional coatings and soft, flexible materials like textiles. These problems can be addressed by increasing the roughness of the flexible materials' surface for better loading of coating materials, and a rough surface helps higher coating functional material via solution process [30]. In addition, polymer-assisted metal deposition can also improve the mechanical robustness of smart wearables. Another cutting-edge technique for solving mechanical degradation could be imparting self-healing properties to the energy harvesters. For example, a TENG harvester with self-healing polydimethylsiloxane (PDMS) elastomers can easily stretch and recover itself [31]. Therefore, integration of smart wear with self-healing materials could open a new door to address this issue.

To add that, the textile-based enzymatic biofuel cell (EBFC) became unstable under body fluid due to ion fouling [32]. Regarding this, encapsulation is a potential option, but traditional encapsulating layers like hermetic plastic reduce the breathability and energy performance of the harvester, for instance, stress transport within TENGs and PENGs [33], heat flow in TEGs [34], body fluid absorption in EBFCs [35], translucency of SCs [36], etc. Therefore, innovative encapsulant materials and technologies—biobased films [37], ultrathin white-ethylene-vinyl acetate (EVA) film [38], 3D printing packages [39]— are highly possible options for conventional encapsulating layers replacement.

Wearing Comfort/Wearability

The modification of the wearable materials for energy harvesting significantly affects the wearing comfort. For example, metal coatings or metal wires used in manufacturing conductive textiles may hamper the textile's softness, breathability, fit, and weight.

To make the energy-harvesting textiles lightweight, metal coatings or metal wires are being replaced by lightweight conductive carbon-based materials like carbon nanotubes (CNTs) and graphene [40]. Besides that, ultralight materials—aerogels [41], aerographite [42], metallic foams [43], polymeric foams [44], and metallic microlattices [45]—are also contributing to producing lightweight smart textiles.

Another primary parameter for wearing comfort is breathability, which may be affected by the clogging of the pores of the textile or flexible materials during modification, which may stop the transmission of moisture and air from the body to the outside. As a result, damp and clammy sensations will be felt by the wearers. To release body heat and absorb perspiration, hydrophilic materials like cotton or viscose [46] can conform to the wearer to increase the ventilation of the skin. Moreover, the appropriate pore size for better ventilation may also be controlled by increasing the volume of the interlaced yarn. Sometimes, on-body energy-harvesting devices like TEGs and EBFCs require tight contact with the body and often are sewn into the clothes, which results in the discomfort of the wearer due to skin irritation, itching, and cold touch [47]. To enhance the softness of the wear surface, manufacturing materials can be altered with soft materials like silk, wool, or cotton [48] and conductive materials like carbon-based materials or conductive polymers [49]. Additionally, woven or knitted fabrics with more porous structures can boost the softness. Besides this, energy harvesters are now themselves a source of threat to the human body because they contain hazardous materials, most commonly synthetic dyes N719 used in dye-sensitized solar cells (DSSCs) [50], PZT and Bi–Te alloys [51, 52], mediators in the electrodes of EBFCs [53], etc. release toxic substances. PEDOT-Cl in TEG may cause allergic reactions by direct contact with the human body. To encounter this issue, toxic-free materials should be used in the energy harvester; for example, natural dyes like blackberries could be used as a safe dye source for DSSCs.

Washability

Still, the washability of the smart wear is a major matter of concern for the manufacturer and users, especially while energy harvesters are attached to the wear as a coating. After each household's laundry, the performance of the harvesters decreases due to the mechanical deformation in water with detergent. Addressing this problem, developing self-cleaning functional coatings or mounting superhydrophobic surfaces may be potential solutions [54]. Regarding self-cleaning, surface washing is not required, which will help to protect the coating during wear, and superhydrophobic surfaces resist the wear being contamination by dirt particles.

Aesthetics Properties

Though aesthetic properties are not the primary concern of the smart wear or textiles, color and designs of garments may affect the wearable technology market [30]. For that, without compromising performance like energy-harvesting efficiency, flexible design, and optional color choice may lead to the commercialization.

Large-Scale Fabrication

Smart wearables are still limited to lab-scale fabrication, and without bulk production, no real impact will be found in society [55]. However, the cost and sophisticated manufacturing technology are the main barriers to robust production. In technological aspects, the production of conductive yarn for manufacturing smart wear becomes challenging because the conductivity decreases with the increase in length [56, 57]. Moreover, the conductive coating on the yarn surface increases the diameter of the yarn, which causes problems for weaving fabric in the loom [58, 59]. In addition, due to the coating on the yarn, stiffness and brittleness increase, which affects the stretchability properties of the yarn. As a result, during weaving or knitting, some faults may appear easily, like crack propagation, delamination of these coatings, and performance degradation [60]. Hence, advancements in conductive yarn production with proper diameter and mechanical strength are required to advance the smart wearables market. To produce suitable yarn or direct smart wear with highly controllable material compositions and device structure, 3D printing may provide a lucrative solution for both industry and academia [61]. Some research has already been explored, like PZT for PENGs [62], BiSbTe alloys for TEGs [63], electrodes for EBFCs [64], organic–inorganic metal halide perovskites for PSCs [65], etc. Though 3D printing has some difficulties—availability of printable materials, scalable fabrication, and the mechanical robustness of 3D-printed textiles—, it has a bright potential considering the fabrication efficiency, as well as design flexibility.

Power Management

Effective power management in the energy harvester is crucial; otherwise, it will yield low output due to its self-consumption [66, 67]. Minimizing power consumption, specific algorithms and hardware, and efficient power point detection are necessary to balance the input and output. For example, sleep–wake operation can substantially extend the operational lifespan of wearable devices [68]. Here, still deeper research is required to accommodate irregular energy inputs, optimizing energy storage and consumption.

Real-Time Service

The energy harvesters are basically dependent on the environment and their surroundings to harvest energy, which runs the electrical devices that work for health and environmental monitoring. However, due to the unfavorable conditions, like a lack of sunlight for DSSCs, the energy-harvesting and the connected devices will stop working [69]. Consequently, real-time service will be hindered. For ensuring real-time service, additional storage devices or self-powered sensors can be integrated with the harvesters [70]. Future research should focus on enhancing real-time services through the energy harvester to provide consistent and reliable data under various conditions.

Overall, the advancement of energy-harvesting technology for smart wear presents a promising pathway for powering next-generation on-body electronic systems. Although significant opportunities exist, a number of challenges must still be addressed. Further research is required to enhance energy-conversion efficiency, output stability, mechanical robustness, wearability, washability, device packaging, aesthetics, scalable and advanced manufacturing techniques, self-healing capabilities, and standardized evaluation systems for smart textile-based energy harvesters. With continued progress in these areas, the widespread adoption and commercialization of smart energy-harvesting textiles across diverse applications are anticipated, ultimately contributing to improved quality of life in the near future.

CHALLENGES AND PROSPECTS OF ENERGY-STORAGE TECHNOLOGY

Key Challenges in Energy-Storage Technology

A power management unit (PMU) is necessary for wearable energy harvester applications in order to control the power that is collected, stored, and supplied to the devices. However, depending on the climate and energy sources, gathered energy varies greatly. PMUs are essential for effectively controlling and converting captured energy for practical applications in a variety of energy-harvesting devices [71]. Wearable energy-storage device design and performance improvement are difficult tasks due to the complex interaction of different parameters that determine their deployment. For wearable energy-storage systems to function better overall and be more useful, it is essential to fully comprehend and optimize power density, cycle life, energy density, and mechanical characteristics like stretchability and flexibility (Figure 1.2) [3]. The following limitations are prevalent in developing wearables more versatile and easier to access for all:

a) Balancing Energy Density and Power Density: These two aspects are critical characteristics that have a direct impact on the device's responsiveness and operating lifetime. Even though obtaining high-power density usually necessitates trade-offs that may compromise energy density, high energy density helps to prolong the operational cycle and shorten the recharging interval. Since flexible materials frequently have lower energy-storage capabilities, finding a compromise between flexibility and high energy and power density is difficult [3].

b) Cycle Life: After multiple cycles of charging and discharging, a battery with good cycling stability can still generate energy and have capacity. This is essential for wearable devices because users rely on a consistent power supply and reliable battery life. Wearable electronics must continually maintain performance throughout multiple electrochemical cycles and adapt to frequent body motions without compromising their ability to store or distribute energy [3]. However, this poses a significant issue since traditional rigid energy systems cannot withstand these strains, which results in aging or performance decline [72]. The materials may become worn down and cracked as a result of frequent mechanical deformation, which could result in a significant loss of performance or complete failure.
c) Ensuring Mechanical Flexibility and Stretchability: Given the various deformation patterns they can encounter in practical applications, flexible batteries must maintain both their structural integrity and electrochemical performance in the face of multiple physical deformations. Excellent stretchability and bendability allow batteries to adapt to different application scenarios and environmental conditions without compromising their operation. The need for rigorous design considerations and optimizations is highlighted by the fact that increasing flexibility and stretchability can sometimes make it challenging to maintain the optimal energy and power density [3].
d) Maintaining Structural and Functional Integrity under Movement: Frequent physical deformation can weaken internal components, reducing the ability to transmit and store energy [3]. The mechanical characteristics and overall performance of flexible power sources will consequently be impacted by the cell structural design and assembly. In order to ensure the omnidirectional flexibility required for irregular and continuous mechanical movements, wire-shaped electronics are better suited than planar power sources. For example, wearable electronics must be fully integrated with the human body or clothing. Furthermore, 3D lights gathering, which efficiently gathers light from any direction, is made possible by a wire-shaped solar cell. Furthermore, the associated power sources can be stretched by embedding each component at the neutral strain point, which will be considerably more useful in future flexible electronics [73].
e) Flexibility and Electrochemical Performance: Improving mechanical properties, like as flexibility, softness, and stretchability, often results in a drop in achievable energy and power density, which impacts the electrochemical performance [3].
f) Capability and Effectiveness: High capacity and efficiency are necessary to ensure wearable energy-storage and harvesting devices continue to function without frequent recharging. The goal is to include effective energy-harvesting technologies that can increase stored energy and lessen the need for frequent external charging, in addition to enhancing energy density while preserving or decreasing the device's size and weight. Minimizing energy losses during harvesting and storage is equally important to increase the device's usable capacity [3].

g) Binders: Because binders are intrinsically electro-insulating, they add dead weight and increase resistance even though they are frequently used in traditional electrodes. One efficient method to get rid of the binders and improve performance is to make freestanding electrodes [73].
h) Electrolytes: Liquid electrolytes must be appropriately encapsulated to prevent leaking, and a separator is required to avoid internal short-circuiting. Although the mechanical and conductivity properties are insufficient, it has been proposed that nonflammable, sustainable, and flexible in shape solid-state electrolytes, that additionally serve as separators, could replace liquid ones. An appropriate electrolyte for flexible sources of power ought to provide good contact with electrodes and safety in addition to being highly conductive and flexible. The challenge of finding appropriate electrolytes is another challenge [73].

Potential Developments in Energy Storage

Although numerous obstacles hinder the progress and usability of smart wearables, researchers continue to address these limitations and develop improved energy-storage solutions (Figure 1.2). The challenges discussed in the previous section have motivated significant research efforts within the scholarly community, leading to the development of the following potential solutions.

Flexible Lithium-Ion Batteries

Because of their high energy density, high output voltage, extended lifespan, and eco-friendliness, LIBs are the primary power source for portable electronics [74]. The ability of these batteries to bend, stretch, and twist is commonly referred to as flexibility. The design and production of flexible LIBs are particularly challenging because of the requirements of wearables, where approximately a hundred days of daily activity can amount to one million mechanical loading–unloading (deformation) cycles [75]. Flexible LIBs must be able to tolerate various deformations while maintaining stable mechanical and electrochemical performance in order to support structural design and the variety of application situations [76]. Based on their dimensional properties, three flexible LIB forms have been developed: stretchable LIBs, planar LIBs, and fiber-shaped LIBs.

Fiber-Shaped LIBs

Kim et al. reported the first fiber Li-ion battery in 2012 [77]. However, the battery's flexibility was limited by the traditional slurry coating process at the cathode, and the use of a liquid electrolyte may be troublesome due to the potential leakage hazards. It is acknowledged that substituting polymer gel electrolytes for liquid electrolytes is a broad and efficient method of resolving safety issues and attaining great flexibility in wearable batteries [78]. Even yet, they can exhibit remarkable endurance and flexibility, continuing to function after 100,000 cycles of deformation [79].

However, stretching is a crucial deformation for flexible LIBs, especially in wearable technology. Because there are currently no active materials that are naturally elastic, making stretchable batteries is more difficult than making bending ones.

This restriction calls for innovative strategies that combine battery performance with flexible structural designs, guaranteeing a smooth transition between flexibility and functionality. One of the first groups to produce stretchable LIBs was John Rogers' team [80]. In this work, segmented active components and interconnected structures were integrated into novel designs using thin silicone elastomers as the basic material. The method maintained a capacity density of roughly 1.1 mA h cm^{-2} while enabling the batteries to expand reversibly up to 300%. Peng et al. created a wavy-structured Li-ion battery by sandwiching a mixture of electrode active materials and uncured PDMS fluid between two CNT sheets to achieve improved stretchability [81]. When this assembly was put on a PDMS substrate that had already been stretched, the electrode showed a wavy pattern after the substrate was cured and released. This design showed that it could withstand 1,000 cycles of stretch-release at 400% strain.

Planar LIBs

Compared to traditional stacked thin-film designs, planar LIBs offer more flexibility due to their structural architecture, which unifies all components onto a flat substrate. A unique bidirectional flexible snake-origami LIB with good flexibility and a high energy density was created by Song et al. With an energy density of 357 W h L^{-1} (133 W h kg^{-1}), the resulting planar LIBs showed remarkable performance and retained high energy-storage capacities even after being continuously bent and mechanically deformed up to 5,000 times [3].

Stretchable LIBs

A unique self-charging method for low-power wearable electronics made of flexible interdigitated LIBs was created by Choi et al. [82]. Even after multiple bending cycles, this battery's charge–discharge curve remained constant, with only a 3.2% decrease in discharge capacity. The wearable device's traditional battery was swapped out with the suggested self-charging technology, making it more portable and cozier. By incorporating a flexible LIB belt, Hu et al. created a wearable power unit [83]. Even after 30 folds at 180°, the LIB belt continued to work exceptionally well electrochemically. The TENG-cloth successfully charged it over three cycles, powering a smartphone-connected heartbeat monitor strap. The potential of this wearable and self-charging power system for upcoming smart electronics is confirmed by this successful demonstration.

Flexible Zinc-Ion Batteries

Flexible zinc-ion batteries (FZIBs) are a practical source of energy for smart wearables due to their high theoretical capacity, superior safety, affordability, and environmental friendliness [84, 85]. To achieve high electrochemical performance and exceptional mechanical deformability of FZIBs used for flexible electronics, excellent deformation robustness, low contact resistance, high space utilization, and straightforward device fabrication are required [86, 87]. Based on their structural architecture, these batteries are currently commonly divided into one-dimensional (1D) and two-dimensional (2D) FZIBs. High flexibility, light weight, and an adaptable structure are among the unique advantages of 1D fiber-type batteries, which enhance both volumetric specific capacity and adaptability. These batteries can also be easily incorporated into textiles, making large-scale and economical manufacture possible. Parallel, helical, and coaxial architectures are commonly used to describe the internal configurations of 1D fiber-type ZIBs.

One-Dimensional Zinc-Ion Batteries (1D ZIBs)

Two fiber electrodes are positioned next to one another in a parallel arrangement within the packaging device of a parallel structure, ZIB [88]. The ease of assembly and scalability of parallel structures make them very beneficial. One or two electrode fibers are wound around a central axis to create helical-structured FZIBs, which are subsequently covered in a solid or gel electrolyte. Compared to parallel fiber-type ZIBs, this design offers a larger electrochemical contact area and better space efficiency, which promotes better electrode–electrolyte interaction and improves electrochemical performance. Better energy density distribution and mechanical endurance are made possible by helical structures' superior design, which is essential for wearable applications. Furthermore, fiber-type ZIBs' mechanical strength is improved by their helical design. However, unequal electric fields frequently cause problems for batteries' parallel and helical structures. For fiber-type ZIBs, the coaxial construction offers a viable solution to this problem. A solid or gel electrolyte, an outside electrode layer, and a protective packing layer surround a center fiber in coaxial fiber ZIBs. By ensuring a consistent electric field in the radial direction, this multilayered arrangement maximizes the electrode surface area. Furthermore, the coaxial design facilitates the large-scale integration of ZIBs by preventing component misalignment during deformation. Additionally, the coaxial form makes it possible to combine multiple functionalities onto a single fiber, increasing the batteries' adaptability [89]. After 3,000 bending cycles, the coaxial-style integrated FZIB exhibits exceptional performance, retaining 97.75% of its capacity. With a high areal capacity (1.24 mA h cm^{-2}) and energy density (1.31 mW h cm^{-2}), it functions well at −20°C [90].

Two-Dimensional Zinc-Ion Batteries (2D ZIBs)

Because of their thin, film-like or paper-like structure, 2D flexible ZIBs are becoming more and more popular for applications in wearable and planar electronics [91, 92]. They are usually divided into planar and sandwich-type designs based on the electrode configuration. Planar ZIBs place microelectrodes, electrolytes, and current collectors on a flat substrate, enabling a compact design with high volumetric capacity, though their performance still depends heavily on precise fabrication methods such as high-loading screen printing [93]. Conversely, sandwich-type structures—which stack the cathode, electrolyte, and anode—are more widely used and simpler to produce, but they frequently have issues with ionic transport and long-term stability [94]. Despite these difficulties, recent advancements have significantly improved energy density, flexibility, and cycling durability [95, 96]. These include multichannel polymer electrolytes, fiber-integrated strain sensors, and stable full-cell designs. Strong prospects for self-powered wearable technology are demonstrated by the successful integration of ZIBs with electronic skins, pressure-sensitive sensors, and even solar-powered wristbands [97, 98].

Flexible Supercapacitors (FSCs)

The energy density of supercapacitors, often referred to as electrochemical capacitors, is between that of regular capacitors and secondary batteries, making them an auspicious family of electrochemical energy-storage systems with high power [99]. Low weight, compression, and excellent flexibility are the main advantages of FSCs over traditional supercapacitors [100]. FSCs are becoming more and more popular because of their potential to revolutionize energy storage in smart wearables and portable electronics. Their unique qualities make it easier to develop energy-storage devices that are more efficient and smaller [101]. Better stretchability, transparency, biocompatibility, and wearability are found in flexible plastic, elastomeric, and textile substrates [3]. Compared to traditional inflexible and bulky power supply units, a textile-based SC guarantees improved wearability of the electronic components, better incorporation with the clothing, and better wearability while worn, in addition to intrinsic flexibility and wearability. Additionally, energy storage can be virtually unnoticeable by integrating textile-based FSCs into regular apparel. Additionally, it guarantees improved active material mass loading, which raises energy, capacitance, and power density [3, 14].

Thus, FSCs based on textiles have a lot of potential for wearable electronic applications since they provide compact, portable, and flexible substitutes for existing energy storage systems. FSCs can be classified as planar-type SCs, all-in-one SCs, sandwich-type SCs, and stacked SCs based on different electrode architectures. For example, Liu et al. used polyvinyl alcohol (PVA)–H_2SO_4 as the gel electrolyte and hybrid CNF/GP/MPCMs as electrodes to create a flexible sandwich-shaped all-solid-state SCs [102]. The produced SCs were able to withstand deformations at various angles and retained 94.1% of their original capacitance after many bending cycles, thanks to their remarkable capacitance (66.34 mF cm^{-2}). Xiao et al. developed a 3D self-sustaining reduced graphene oxide/poly(3,4-ethylenedioxythiophene), also known as rGO/PEDOT hybrid fiber fabric for all-solid-state FSCs [103]. In a 1 M H_2SO_4 electrolyte, the rGO/PEDOT electrode exhibits a high specific capacitance of 314.4 F g^{-1}. The solid-state FSCs made of rGO/PEDOT fiber textiles exhibit a good capacity retention (84%) during 5,000 bending cycles and an outstanding energy density of 22.7 μW h cm^{-2}. For the enormous demands of SCs with mechanical flexibility and high-power density, more intricate sandwich architectures have also been created. In Cheng et al.'s first spatial-interleaving FSCs, graphene microelectrodes are reversely layered by the layer-by-layer method in a 3D region [104]. The remarkable specific areal capacitance of 36.46 mF cm^{-2} and energy density of 5.34 μW h cm^{-2} are attained by the 100 μm-thick device. Each layer's interdigitated microelectrode architecture guarantees remarkable mechanical flexibility, retaining roughly 98.7% performance over a hundred bending cycles.

Because they are smaller, lighter, and thinner, planar SCs have attracted a lot of interest [105]. In order to effectively separate the two electrodes and exclude the possibility of a short circuit, the electrode materials are placed on the same plane with a tiny space between them. Ion diffusion is substantially shortened by the small intervals between electrode arrays, providing a high-power output [106, 107]. PEDOT: PSS and MnO_2-supported interdigitated electrodes were created utilizing a screen-printing method in Lai et al.'s PET-knitted fabric-based FSCs [108]. The FSCs performed exceptionally well, attaining high energy density (12.03 μW h cm^{-2}) and areal capacitance (135.4 mF cm^{-2}) while remaining stable across multiple charge–discharge cycles. They are perfect for small, portable electronic devices because of their streamlined planar architecture, which makes system integration simple. Furthermore, planar micro-supercapacitors have greater volumetric energy density and may be folded to fit into small places, which is very advantageous for downsized applications.

Despite the fact that the two most common FSC configurations in the literature today are sandwiched-type and planar-type SCs [3]. The overall performance of sandwiched-structure SCs is impacted by tension on the top electrodes and compression on the bottom electrodes caused by the curved textile substrates and separator. As a result, under bending deformation, both electrodes in the sandwiched supercapacitor would concurrently experience compression and tensile strain, leading to electrode distortion and unstable electrochemical performance. In order to increase bending stability, researchers suggest an all-in-one SC. A customizable solid-state supercapacitor made by combining the active components of polyaniline nanofibers with a PVA gel electrolyte (PANI–PVA) through interface engineering was described by Liu et al. [109]. The electrode/electrolyte interface in supercapacitor devices is improved by the novel all-in-one PANI–PVA architecture, which results in an exceptional specific capacitance of 1.88 F cm^{-2}, an excellent rate performance of 80%, and a remarkable cycling stability, maintaining 78.3% after 3,000 cycles.

An MXene textile supercapacitor with a high $Ti_3C_2T_x$ areal loading was created by Gogotsi et al. and implemented in series using a stacked design. A prototype with five cells in series, with a loading of 24.2 mg cm^2 that runs at 6 V and generates an exceptional 0.401 mW h cm^{-2} energy density, 0.248 mW cm^{-2} power density, and also an aerial capacitance illustrates its promise for high-performance energy-storage applications [110]. Because FSCs are thin, light, and adaptable energy-storage devices, they may stick to the skin with ease. Gao et al. have created a satisfying skin-friendly incorporated wireless charger for FSC that is skin-like in order to accommodate the local peculiarities of its intended surfaces' unevenness as well as aesthetics [111]. Fan et al. integrated printable SCs onto wearable bracelets as an energy supply intelligent sensing system, achieving an astounding surface capacitance of 12.9 mF cm^{-2} [112]. Additionally, Zhang et al. presented an all-in-one planar micro-supercapacitor array for self-powered stretchable systems and high-performance wearable energy-storage devices [113]. Excellent electrochemical performance is demonstrated by the hybrid electrodes, which maintain 68.5% of their capacity after 5,000 cycles and have a high capacity at both low and high currents.

Lastly, it is clear that the thin FSC skin facilitates wireless charging, storing energy in high-capacitance micro-supercapacitors, and conforms nicely to the curves of the human body. This design is perfect for wearables, portable prosthetic limbs, and health monitoring applications. These FSCs are much more useful and convenient for wearable applications due to their wireless charging capability.

Flexible Electrode Materials

Carbon Nanotubes (CNTs)

CNTs have garnered significant interest as supercapacitor electrodes because of their exceptional mechanical and thermal resilience, large specific surface area (1,240–2,200 m^2 g^{-1}), strong electrical conductivity (104–105 S cm^{-1}), and unusual porous structure [114]. They are also great electrode materials for flexible SCs because of their high aspect ratio and open tubular network. Highly integrated stretchable SCs were created by Niu et al. (2013) using buckled single-walled carbon nanotube (SWCNT) sheets. Directly produced SWNT network films were combined with a PDMS substrate with increased pre-strain to create these buckling SWCNT film electrodes [115].

Graphene

Graphene is made up of only one sheet of hexagonally arranged sp^2-hybridized carbon atoms. Its unique 2D morphology, superior thermal conductivity, electrical conductivity, mechanical and chemical stability, and ultrahigh surface area (2630 m2 g^{-1}) are just a few of its unmatched physical and chemical characteristics that indicate its alluring potential material for flexible SCs [114]. Niu et al. used lateral ultrathin interdigitated microelectrodes made of rGO on a polyethylene terephthalate (PET) substrate [116]. They created all-solid-state flexible ultrathin micro-supercapacitors. Compared to traditional SCs (86 F g^{-1}), the resultant devices attained a significantly higher specific capacitance of 285 F g^{-1}. Additionally, the developed micro-supercapacitor's flawlessly linear charge/discharge curve with a constant current of 1 A g^{-1} results in a columbic efficiency of around 98%.

In contrast, pseudo-capacitors store charge using electronically conducting polymers and transitional metal oxide-based redox-active materials through quick and reversible faradic reactions. Since their specific capacitance is significantly higher than that of EDLCs, pseudo-capacitors have been the subject of much research in recent decades [114]. According to Yang et al. [117], the edge-orientation of MoS_2 thin films made on Mo substrates functions as an elastic electrode having a high overall capacitance of 12.5 mF cm^{-2} [117].

Conducting Polymers

Conducting polymers have emerged as attractive candidates for flexible supercapacitors because of their comparatively high theoretical charge-storage capacities, typically ranging from 100 to 140 mAh g^{-1}. Their popularity also stems from several additional advantages, including environmental friendliness, a wide operational voltage range, notable charge-storage efficiency with good reversibility, and adjustable redox behavior. Widely investigated polymers such as polypyrrole (PPy), PANI, and polythiophene (PTP) continue to stand out as strong contenders for next-generation smart, flexible SC systems. Moreover, many conducting polymers respond sensitively to changes in their surroundings—such as fluctuations in temperature, pH, ionic concentration, solvent composition, or ligand interactions. This stimulus-responsive nature, often accompanied by distinct color changes under different conditions, makes them particularly well-suited for use as electrode materials in supercapacitors and other intelligent energy-storage or energy-conversion devices [114].

For instance, Chen et al. [118] created smart SCs by coating aligned CNT sheet electrodes with the conducting polymer PANI. The SCs can operate at a strain of up to 100% for 200 cycles, and the capacitance attained was 308.4 F g^{-1}. During the stretching and healing process, the hue shift is instantly visible to the unaided eye.

Composite Electrodes

Composite electrodes combining carbon frameworks with pseudocapacitive materials have gained attention for delivering higher energy and power densities in supercapacitors. Choi et al. developed a multi-walled carbon nanotube (MWNT)/ MnO_2 yarn supercapacitor by depositing MnO_2 onto a twisted MWNT yarn, which functioned as a flexible support, conductive channel, and current collector. Efficient charge transfer between MnO_2 and the aligned CNT bundles enabled a specific capacitance of 25.4 F cm^{-3}, with energy and power densities of 3.52 mW h cm^{-3} and 127 mW cm^{-3}, respectively. The device also maintained its capacitance after 1000 bending cycles at 90°, demonstrating strong mechanical durability for flexible energy-storage systems. [114].

Hybrid Composites Electrodes

Hybrid carbon–polymer composites have been actively investigated for flexible SCs. Yu et al. developed a stretchable solid-state device that maintained conductivity at 50% strain. Their PANI/MWCNT/PDMS electrode showed 481 mF cm^{-2} areal capacitance and 1023 F g^{-1} specific capacitance at 5 mV s^{-1}, with 95% retention after 500 stretching cycles. The supercapacitor also delivered an energy density of 0.15 mW h cm^{-3} (11 W h kg^{-1}), demonstrating strong potential for flexible energy-storage applications [119].

Flexible Solid-State Electrolytes

Excellent electrochemical performances are ensured by high conductivity and good physical contact with liquid electrolytes, which are frequently utilized in conventional power systems. However, as was already indicated, there are a number of disadvantages to using liquid electrolytes in flexible LIBs, including safety concerns, the requirement for separators, unstable performance under continuous mechanical deformation, varying operating temperatures, and more. This leads to the proposal of numerous innovative ideas for shape-conformable solid-state electrolytes. Gel polymer electrolytes (GPEs), which have good conductivity, low rates of electrolyte leakage, low flammability, and mechanical flexibility, are the most researched solid-state electrolytes [73].

Kil et al. [120] recently presented a flexible GPE made of an ethoxylated trimethylolpropane triacrylate (ETPTA) polymer matrix that was UV-cured, an electrolyte based on $LiPF_6$ and Al_2O_3 NPs. In order to regulate the rheological characteristics of the GPE, Al_2O_3 NPs were used as a functional filler. In contrast to a prior work [121], the addition of Al_2O_3 NPs allowed for improved mechanical flexibility and strong conductivity, making it more appropriate for 3D electrodes with intricate geometries to preserve good interface contact. In light of this, the GPE was further developed for usage in flexible LIBs. The growth of lithium dendrites was demonstrated to be inhibited by the uniformly distributed and densely packed Al_2O_3 NPs as protective barriers, which can stop the battery from short-circuiting.

Sang-Young Lee's group introduced a flexible sulfide solid electrolyte in 2015 by reinforcing it with a poly(paraphenylene terephthalamide) (PPTA) nonwoven scaffold. Using a doctor-blade coating followed by cold pressing, they formed SE-NW-SE and

NW-SE-NW structures that achieved ionic conductivities of 0.20 and 0.16 mS cm^{-1}. A full solid-state cell with LCO and LTO delivered an energy density of 44 W h kg $cell^{-1}$ and an initial discharge capacity of 89 mA h g_LCO^{-1}. Although its energy density was below that of commercial LIBs, future integration of advanced electrodes such as sulfur/Li metal could significantly boost performance [15].

Application of Artificial Intelligence (AI) and Long Learning Method (LLM)

AI can help maximize the effectiveness of energy-harvesting devices as well as the performance of energy-storage systems by combining enormous volumes of data from these various sources. Furthermore, one major difficulty that AI may help with is creating precise predictive models for battery life, performance under different conditions, and ideal material combinations [122, 123]. Energy-storage system design and operation can be improved using machine learning algorithms, increasing longevity and efficiency. By modifying charging cycles, anticipating consumption trends, and reducing energy losses, AI-driven models can improve storage efficiency and maximize the system's usable capacity [124]. However, by facilitating the effective processing of enormous volumes of data, LLMs play a major role in the advancement of these technologies. Material combinations with potential for high-performance energy applications have been highlighted by the automatic analysis and summarization of experimental results from a large body of literature using LLMs such as GPT models.

CONCLUSION

Smart wearables demand uninterrupted, nonintrusive health monitoring and fitness checkups for a certain period, with breaks in between or at a stretch. That's why the power supply unit, as well as the storage unit, should be on par with these wearables' power demand to perform as per its intended design. Energy-harvesting and energy-storage devices intended for wearables have faced numerous obstacles; many have already been overcome, and many remain to be addressed. In this chapter, the challenges for wearable electronics' energy supply have been presented, with shortcomings in harvesting and storing. At the same time, contemporary approaches made by researchers in this domain, along with the upcoming trajectory of this field, are also discussed. It is evident that, for a long time, LIBs have been the most widely used solution, but they fall behind in terms of the advancement made by smart electronics. They lack flexibility, consistency, longevity, optimization, and overall feasibility required. To address these challenges, different initiatives have been taken so far, for example, PENGs, TENGs, TEGs, SCs, BFs, and HGs are the latest in the power harvesting vicinity. But these technologies are also in a growth stage and are not completely capable of totally supporting the wearables industry right away. For instance, in the case of power harvesting, they fall short as they lack efficiency, stability, mechanical durability, environmental dependency, absence of comfort, absence of washability, large-scale development, etc.

On the other hand, in the power storage technologies, which are also a vital key aspect for harvesting devices, several hindrances can be observed. For instance, they fall short in balancing between the density of energy and power and also lack the mechanical flexibility and stretchability required in wearables; they lose integrity during movement, and the type of binders and electrolytes used can also limit their functionality. But the development of high-capacity supercapacitors, flexible batteries, and advanced battery chemistry seems to be promising. To overcome the persisting limitations, more studies in this domain are expected, where scientists from multidisciplinary fields can work hand in hand to design suitable materials and technology required to match the desired performance with the devices.

In the upcoming days, the use of AI and LLM can play a crucial role in material selection, optimization, and predicting the behavior of these devices in terms of the environment in which they will be used. In combination with the state-of-the-art development in material engineering, AI, large language models, and the electronics field, we can build wearables ready to be a daily companion for the betterment of human health monitoring and assisting treatments.

AI DISCLOSURE STATEMENT

The authors acknowledge the use of AI-based tools for minor language editing and grammatical correction.

REFERENCES

[1] H. Yang, J. Yu, H. Zo, and M. Choi, "User acceptance of wearable devices: An extended perspective of perceived value," *Telema Infor*, vol. 33, no. 2, pp. 256–269, 2016, doi: 10.1016/j.tele.2015.08.007

[2] E. Park, "User acceptance of smart wearable devices: An expectation-confirmation model approach," *Telema Infor*, vol. 47, p. 101318, 2020, doi: 10.1016/j.tele.2019.101318

[3] Q. Zhang, D. Soham, Z. Liang, and J. Wan, "Advances in wearable energy storage and harvesting systems," *Med-X*, vol. 3, no. 1, p. 3, 2025, doi: 10.1007/s44258-024-00048-w

[4] X. Kang, S. Jia, Z. Lin, H. Zhang, L. Wang, and X. Zhou, "Flexible wearable hybrid nanogenerator to harvest solar energy and human kinetic energy," *Nano Energy*, vol. 103, p. 107808, 2022, doi: 10.1016/j.nanoen.2022.107808

[5] T. Starner, "Human-powered wearable computing," *IBM Systems Journal*, vol. 35, no. 3.4, pp. 618–629, 1996, doi: 10.1147/sj.353.0618

[6] Y. Xu et al., "Paper-based wearable electronics," *iScience*, vol. 24, no. 7, p. 102736, Jul 23 2021, doi: 10.1016/j.isci.2021.102736

[7] M. N. Hasan, N. Nayan, M. Nafea, A. G. A. Muthalif, and M. S. Mohamed Ali, "Novel structural design of wearable thermoelectric generator with vertically oriented thermoelements," *Energy*, vol. 259, p. 125032, 2022, doi: 10.1016/j.energy.2022.125032

[8] J.-H. Lee et al., "Flexible solid-state hybrid supercapacitors for the internet of everything (IoE)," *Energy & Environmental Science*, vol. 15, no. 6, pp. 2233–2258, 2022.

[9] L. Han et al., "A flexible, high-voltage and safe zwitterionic natural polymer hydrogel electrolyte for high-energy-density zinc-ion hybrid supercapacitor," *Chemical Engineering Journal*, vol. 392, p. 123733, 2020.

[10] X. Gong, Q. Yang, C. Zhi, and P. S. J. A. E. M. Lee, "Stretchable energy storage devices: From materials and structural design to device assembly," *Advanced Energy Materials*, vol. 11, no. 15, p. 2003308, 2021.

[11] S. A. Hashemi, S. Ramakrishna, A. G. J. E. Aberle, and E. Science, "Recent progress in flexible–wearable solar cells for self-powered electronic devices," *Energy & Environmental Science*, vol. 13, no. 3, pp. 685–743, 2020.

[12] P. Yu, Y. Zeng, H. Zhang, M. Yu, Y. Tong, and X. Lu, "Flexible Zn-Ion batteries: Recent progresses and challenges," *Small*, vol. 15, no. 7, p. e1804760, Feb 2019, doi: 10.1002/smll.201804760

[13] J. Chang, Q. Huang, Y. Gao, and Z. Zheng, "Pathways of developing high-energy-density flexible lithium batteries," *Adv Mater*, vol. 33, no. 46, p. e2004419, 2021, doi: 10.1002/adma.202004419

[14] J. Wen, B. Xu, Y. Gao, M. Li, and H. J. E. S. M. Fu, "Wearable technologies enable high-performance textile supercapacitors with flexible, breathable and wearable characteristics for future energy storage," *Energy Storage Materials*, vol. 37, pp. 94–122, 2021.

[15] W. Liu, M. S. Song, B. Kong, and Y. Cui, "Flexible and stretchable energy storage: Recent advances and future perspectives," *Adv Mater*, vol. 29, no. 1, p. 1603436, Jan 2017, doi: 10.1002/adma.201603436

[16] W. Wang, A. Yu, J. Zhai, and Z. L. J. A. F. M. Wang, "Recent progress of functional fiber and textile triboelectric nanogenerators: Towards electricity power generation and intelligent sensing," *Advanced Fiber Materials*, vol. 3, no. 6, pp. 394–412, 2021.

[17] A. J. Bandodkar et al., "Soft, stretchable, high power density electronic skin-based biofuel cells for scavenging energy from human sweat," *Energy & Environmental Science*,vol. 10, no. 7, pp. 1581–1589, 2017.

[18] A. Ali, H. Shaukat, S. Bibi, W. A. Altabey, M. Noori, and S. A. J. E. S. R. Kouritem, "Recent progress in energy harvesting systems for wearable technology," *Energy Strategy Reviews*,vol. 49, p. 101124, 2023.

[19] K. Dong, X. Peng, and Z. L. Wang, "Fiber/Fabric-based piezoelectric and triboelectric nanogenerators for flexible/stretchable and wearable electronics and artificial intelligence," *Adv Mater*, vol. 32, no. 5, p. e1902549, Feb 2020, doi: 10.1002/adma.201902549

[20] J. Li, J. Zhao, and J. A. Rogers, "Materials and designs for power supply systems in skin-interfaced electronics," *Acc Chem Res*, vol. 52, no. 1, pp. 53–62, 2019, doi: 10.1021/acs.accounts.8b00486

[21] R. Shu et al., "Mg3+δSbxBi2−x Family: A promising substitute for the State-of-the-Art n-Type thermoelectric materials near room temperature," *Adv Funct Mater*, vol. 29, no. 4, p. 1807235, 2019, doi: 10.1002/adfm.201807235

[22] G. Dong et al., "Super-elastic ferroelectric single-crystal membrane with continuous electric dipole rotation," *Science*, vol. 366, no. 6464, pp. 475–479, 2019, doi: 10.1126/science.aay7221

[23] J. Yuan, M. J. Ford, Y. Xu, Y. Zhang, G. C. Bazan, and W. Ma, "Improved tandem all-polymer solar cells performance by using spectrally matched subcells," *Adv Energy Mater*, vol. 8, no. 14, p. 1703291, 2018, doi: 10.1002/aenm.201703291

[24] Y. Zhou, C.-H. Wang, W. Lu, and L. Dai, "Recent advances in fiber-shaped supercapacitors and lithium-ion batteries," *Adv Mater*, vol. 32, no. 5, p. 1902779, 2020, doi: 10.1002/adma.201902779

[25] D. Chen, K. Jiang, T. Huang, and G. Shen, "Recent advances in fiber supercapacitors: Materials, Device configurations, and applications," *Adv Mater*, vol. 32, no. 5, p. 1901806, 2020, doi: 10.1002/adma.201901806

[26] F. Mo, G. Liang, Z. Huang, H. Li, D. Wang, and C. Zhi, "An overview of fiber-shaped batteries with a focus on multifunctionality, scalability, and technical difficulties," *Adv Mater*, vol. 32, no. 5, p. 1902151, 2020, doi: 10.1002/adma.201902151

[27] K. Jost et al., "Carbon coated textiles for flexible energy storage," *Energy & Environ Sci*, vol. 4, no. 12, pp. 5060–5067, 2011, doi: 10.1039/C1EE02421C

[28] M. Stoppa and A. J. S. Chiolerio, "Wearable electronics and smart textiles: A critical review," *Sensors*, vol. 14, no. 7, pp. 11957–11992, 2014.

[29] D. Lee, J. S. Sang, P. J. Yoo, T. J. Shin, K. W. Oh, and J. J. P. Park, "Machine-washable smart textiles with photothermal and antibacterial activities from nanocomposite fibers of conjugated polymer nanoparticles and polyacrylonitrile," *Polymers*,vol. 11, no. 1, p. 16, 2018.
[30] T. L. Andrew et al., "Melding vapor-phase organic chemistry and textile manufacturing to produce wearable electronics," *Accounts of Chemical Research*, vol. 51, no. 4, pp. 850–859, 2018.
[31] G. Chen, Y. Li, M. Bick, and J. J. C. R. Chen, "Smart textiles for electricity generation," *Chemical Reviews*, vol. 120, no. 8, pp. 3668–3720, 2020.
[32] C. Wang, R. Feng, F. J. J. O. C. Yang, and i. science, "Enhancing the hydrophilic and antifouling properties of polypropylene nonwoven fabric membranes by the grafting of poly (N-vinyl-2-pyrrolidone) via the ATRP method," *Journal of Colloid and Interface Science*, vol. 357, no. 2, pp. 273–279, 2011.
[33] Y. Qin, X. Wang, and Z. L. J. N. Wang, "Microfibre–nanowire hybrid structure for energy scavenging," *Nature*, vol. 451, no. 7180, pp. 809–813, 2008.
[34] S. Jin et al., "Large-area direct laser-shock imprinting of a 3D biomimic hierarchical metal surface for triboelectric nanogenerators," *Advanced Materials*, vol. 30, no. 11, p. 1705840, 2018.
[35] W. Jia, G. Valdés-Ramírez, A. J. Bandodkar, J. R. Windmiller, and J. J. A. C. I. E. Wang, "Epidermal biofuel cells: Energy harvesting from human perspiration," *Angewandte Chemie International Edition*, vol. 52, no. 28, pp. 7233–7236, 2013.
[36] J. Nam et al., "Transfer printed flexible and stretchable thin film solar cells using a water-soluble sacrificial layer," *Advanced Energy Materials*, vol. 6, no. 21, p. 1601269, 2016.
[37] D.-Y. Xie, D. Qian, F. Song, X.-L. Wang, Y.-Z. J. A. S. C. Wang"A fully biobased encapsulant constructed of soy protein and cellulose nanocrystals for flexible electromechanical sensing," *ACS Sustainable Chemistry & Engineering*, vol. 5, no. 8, pp. 7063–7070, 2017.
[38] M. López-Escalante, M. Fernández-Rodríguez, L. Caballero, F. Martín, M. Gabás, and J. J. A. E. Ramos-Barrado, "Novel encapsulant architecture on the road to photovoltaic module power output increase," *Applied Energy*, vol. 228, pp. 1901–1910, 2018.
[39] R. Scott et al., "Encapsulation of fiber optic sensors in 3D printed packages for use in civil engineering applications: A preliminary study," *Sensors*,vol. 19, no. 7, p. 1689, 2019.
[40] X. Zhang, W. Lu, G. Zhou, and Q. J. A. M. Li, "Understanding the mechanical and conductive properties of carbon nanotube fibers for smart electronics," *Advanced Materials*, vol. 32, no. 5, p. 1902028, 2020.
[41] A. C. Pierre and G. M. J. C. R. Pajonk, "Chemistry of aerogels and their applications," *Chemical Reviews*, vol. 102, no. 11, pp. 4243–4266, 2002.
[42] M. Mecklenburg et al., "Aerographite: Ultra lightweight, flexible nanowall, carbon microtube material with outstanding mechanical performance (Adv. Mater. 26/2012)," *Advanced Materials*, vol. 24, no. 26, pp. 3437–3437, 2012.
[43] L. P. Lefebvre, J. Banhart, and D. C. J. A. E. M. Dunand, "Porous metals and metallic foams: Current status and recent developments," *Advanced Engineering Materials*,vol. 10, no. 9, pp. 775–787, 2008.
[44] V. Srivastava and R. J. J. O. M. S. Srivastava, "On the polymeric foams: Modeling and properties," *Journal of Materials Science*, vol. 49, no. 7, pp. 2681–2692, 2014.
[45] Y. Y. Hsu et al., "Effect of polymer foam morphology and density on kinetics of in vitro controlled release of isoniazid from compressed foam matrices," *Journal of Materials Science*, vol. 35, no. 1, pp. 107–116, 1997.
[46] T. H. Mokhothu and M. J. J. C. P. John, "Review on hygroscopic aging of cellulose fibres and their biocomposites," *Carbohydrate Polymers*, vol. 131, pp. 337–354, 2015.

[47] S. Lee, Y. Lee, J. Park, and D. J. N. E. Choi, "Stitchable organic photovoltaic cells with textile electrodes," *Nano Energy*,vol. 9, pp. 88–93, 2014.
[48] S. Park, M. Vosguerichian, and Z. J. N. Bao, "A review of fabrication and applications of carbon nanotube film-based flexible electronics," *Nanoscale*, vol. 5, no. 5, pp. 1727–1752, 2013.
[49] W. Zhang, A. A. Dehghani-Sanij, and R. S. J. J. O. M. S. Blackburn, "Carbon based conductive polymer composites," *Journal of Materials Science*, vol. 42, no. 10, pp. 3408–3418, 2007.
[50] N. Zhang, "A wearable all-solid photovoltaic textile," in *Electrochemical Society Meeting Abstracts 233*, 2018, no. 26: The Electrochemical Society, Inc., pp. 1548–1548.
[51] F. R. Fan, W. Tang, and Z. L. J. A. M. Wang, "Flexible nanogenerators for energy harvesting and self-powered electronics," *Advanced Materials*, vol. 28, no. 22, pp. 4283–4305, 2016.
[52] M. K. Gupta, J.-H. Lee, K. Y. Lee, and S.-W. J. A. N. Kim, "Two-dimensional vanadium-doped ZnO nanosheet-based flexible direct current nanogenerator," *ACS Nano*, vol. 7, no. 10, pp. 8932–8939, 2013.
[53] C. H. Kwon et al., "High-power biofuel cell textiles from woven biscrolled carbon nanotube yarns," *Nature*, vol. 5, no. 1, p. 3928, 2014.
[54] S. Park, J. Kim, C. H. J. J. O. E. F. Park"Superhydrophobic textiles: Review of theoretical definitions, fabrication and functional evaluation," *Fabrics*, vol. 10, no. 4, p. 155892501501000401, 2015.
[55] J. Chen et al., "Micro-cable structured textile for simultaneously harvesting solar and mechanical energy," *Nature Energy*, vol. 1, no. 10, pp. 1–8, 2016.
[56] W. Yan et al., "Thermally drawn advanced functional fibers: New frontier of flexible electronics," *Materials Today*, vol. 35, pp. 168–194, 2020.
[57] W. Yan et al., "Advanced multimaterial electronic and optoelectronic fibers and textiles," *Advanced Materials*, vol. 31, no. 1, p. 1802348, 2019.
[58] M. Y. H. Saty et al., "Recent advances of preparing structure-enhanced conductive yarns to control their performance in potential applications," *Textile Research Journal*, vol. 94, no. 23–24, pp. 2835–2862, 2024.
[59] F. Ahmed, A. C. Mills, C. H. J. J. O. E. F. Kwon, "Impact of weaving parameters on the performance of yarn-based interconnects in woven e-textiles: A comprehensive review," *Fabrics*, vol. 20, p. 15589250251372498, 2025.
[60] G. A. Khan, M. Terano, M. Gafur, and M. S. J. J. O. K. S. U.-E. S. Alam, "Studies on the mechanical properties of woven jute fabric reinforced poly (l-lactic acid) composites," *Journal of King Saud University-Engineering Sciences*, vol. 28, no. 1, pp. 69–74, 2016.
[61] S. Wu et al., "3D printing technology for smart clothing: A topic review," *Materials*, vol. 15, no. 20, p. 7391, 2022.
[62] H. Cui et al., "Three-dimensional printing of piezoelectric materials with designed anisotropy and directional response," *Nature Materials*, vol. 18, no. 3, pp. 234–241, 2019.
[63] J. Qiu et al., "3D Printing of highly textured bulk thermoelectric materials: mechanically robust BiSbTe alloys with superior performance," *Energy & Environmental Science*, vol. 12, no. 10, pp. 3106–3117, 2019.
[64] A. M. L. Marzo, C. C. Mayorga-Martinez, M. J. B. Pumera, and Bioelectronics, "3D-printed graphene direct electron transfer enzyme biosensors," *Biosensors and Bioelectronics*, vol. 151, p. 111980, 2020.
[65] M. Chen et al., "3D nanoprinting of perovskites," *Advanced Materials*, vol. 31, no. 44, p. 1904073, 2019.
[66] A. El-Khateb, W. Issa, T. Mallick, and M. Abusara, "Control strategy for uninterrupted microgrid mode transfer during unintentional islanding scenarios," *University of Exeter. Journal Contribution*, 2017.

[67] M. Alhawari, T. Tekeste, B. Mohammad, H. Saleh, and M. Ismail, "Power management unit for multi-source energy harvesting in wearable electronics," in *2016 IEEE 59th International Midwest Symposium on Circuits and Systems (MWSCAS)*, 2016: IEEE, pp. 1–4.
[68] G. Anastasi, M. Conti, and M. J. I. T. O. I. I. Di Francesco, "Extending the lifetime of wireless sensor networks through adaptive sleep," *University of Exeter: Journal Contribution.*, vol. 5, no. 3, pp. 351–365, 2009.
[69] I. Joseph, H. Louis, T. Unimuke, I. Etim, M. Orosun, and J. J. A. S. E. Odey, "An overview of the operational principles, light harvesting and trapping technologies, and recent advances of the dye sensitized solar cells," *Applied Solar Energy*, vol. 56, no. 5, pp. 334–363, 2020.
[70] Y. Wang, H. Wu, L. Xu, H. Zhang, Y. Yang, and Z. L. J. S. A. Wang, "Hierarchically patterned self-powered sensors for multifunctional tactile sensing," *Science Advances*, vol. 6, no. 34, p. eabb9083, 2020.
[71] M. Kang and W.-H. Yeo, "Advances in energy harvesting technologies for wearable devices," *Micromachines*, vol. 15, no. 7, p. 884. doi: 10.3390/mi15070884
[72] A. Sumboja, J. Liu, W. G. Zheng, Y. Zong, H. Zhang, and Z. J. C. S. R. Liu, "Electrochemical energy storage devices for wearable technology: A rationale for materials selection and cell design," *Chemical Society Reviews*, vol. 47, no. 15, pp. 5919–5945, 2018.
[73] L. Li, Z. Wu, S. Yuan, and X.-B. Zhang, "Advances and challenges for flexible energy storage and conversion devices and systems," *Energy & Environ Sci*, vol. 7, no. 7, pp. 2101–2122, 2014, doi: 10.1039/c4ee00318g
[74] M. Li, J. Lu, Z. Chen, and K. Amine, "30 years of lithium-ion batteries," *Adv Mater*, vol. 30, no. 33, p. e1800561, Jun 14 2018, doi: 10.1002/adma.201800561
[75] H. Li et al., "Nature-inspired materials and designs for flexible lithium-ion batteries," *Carbon Energy*, vol. 4, no. 5, pp. 878–900, 2022.
[76] C.-F. Sun et al., "Weavable high-capacity electrodes," *Nano Energy*, vol. 2, no. 5, pp. 987–994, 2013.
[77] Y. H. Kwon et al., "Cable-type flexible lithium ion battery based on hollow multi-helix electrodes," *Adv Mater*, vol. 24, no. 38, pp. 5192–5197, Oct 2 2012, doi: 10.1002/adma.201202196
[78] H. Dai et al., "Polymer gel electrolytes for flexible supercapacitors: Recent progress, challenges, and perspectives," *Energy Storage Materials*, vol. 34, pp. 320–355, 2021.
[79] T. Khudiyev et al., "Thermally drawn rechargeable battery fiber enables pervasive power," *Materials Today*, vol. 52, pp. 80–89, 2022.
[80] Z. Yang, J. Deng, X. Chen, J. Ren, and H. Peng, "A highly stretchable, fiber-shaped supercapacitor," *Angew Chem Int Ed Engl*, vol. 52, no. 50, pp. 13453–13457, Dec 9 2013, doi: 10.1002/anie.201307619
[81] W. Weng et al., "A gum-like lithium-ion battery based on a novel arched structure," *Adv Mater*, vol. 27, no. 8, pp. 1363–1369, Feb 25 2015, doi: 10.1002/adma.201405127
[82] S. W. Baek et al., "A colloidal-quantum-dot-based self-charging system via the near-infrared band," *Adv Mater*, vol. 30, no. 25, p. e1707224, Jun 2018, doi: 10.1002/adma.201707224
[83] X. Pu et al., "A self-charging power unit by integration of a textile triboelectric nanogenerator and a flexible lithium-ion battery for wearable electronics," *Adv Mater*, vol. 27, no. 15, pp. 2472–2478, Apr 17 2015, doi: 10.1002/adma.201500311
[84] X. Xiao et al., "Rational design of flexible Zn-based batteries for wearable electronic devices," *ACS Nano*, vol. 17, no. 3, pp. 1764–1802, Feb 14 2023, doi: 10.1021/acsnano.2c09509

[85] X. Gao, H. Zhang, X. Liu, and X. J. C. E. Lu, "Flexible Zn-ion batteries based on manganese oxides: Progress and prospect," *Carbon Energy*, vol. 2, no. 3, pp. 387–407, 2020.

[86] D. Wang et al., "Energy density issues of flexible energy storage devices," *Energy Storage Materials*, vol. 28, pp. 264–292, 2020.

[87] Q. Zhang et al., "Overview of fiber-shaped energy storage devices: From fabrication to application," *Nano Energy*, vol. 128, p. 109896, 2024.

[88] M. Liao et al., "A high-capacity aqueous zinc-ion battery fiber with air-recharging capability," *Journal of Materials Chemistry*, vol. 9, no. 11, pp. 6811–6818, 2021.

[89] Y. Sun, Y.-Z. Li, and M. Yuan, "Requirements, challenges, and novel ideas for wearables on power supply and energy harvesting," *Nano Energy*, vol. 115, p. 108715, 2023, doi: 10.1016/j.nanoen.2023.108715

[90] Y. Liu et al., "Engineering integrated structure for high-performance flexible zinc-ion batteries," *Chemical Engineering Journal*, vol. 417, p. 127955, 2021.

[91] Y. Zeng et al., "Recent progress in advanced flexible zinc ion battery design," *Applied Physics Reviews*, vol. 9, no. 2, 2022.

[92] H. Li et al., "Advanced rechargeable zinc-based batteries: Recent progress and future perspectives," *Nano Energy*, vol. 62, pp. 550–587, 2019.

[93] X. Cai et al., "A flexible and safe planar zinc-ion micro-battery with ultrahigh energy density enabled by interfacial engineering for wearable sensing systems," *Advanced Functional Materials*, vol. 33, no. 29, p. 2303009, 2023.

[94] Q. Zhang et al., "A Zinc-Ion Battery-Type Self-Powered pressure sensor with long service life," *Adv Mater*, vol. 34, no. 40, p. e2205369, Oct 2022, doi: 10.1002/adma.202205369

[95] H. Zhang et al., "Advanced fiber-shaped aqueous Zn ion battery integrated with strain sensor," *ACS Appl Mater Interfaces*, vol. 14, no. 36, pp. 41045–41052, Sep 14 2022, doi: 10.1021/acsami.2c11638

[96] H. Liu et al., "Biomimetic honeycomb Zn anode enabled multi-field regulation toward highly stable flexible Zn-ion batteries," *Advanced Functional Materials*, vol. 33, no. 25, p. 2300419, 2023.

[97] Z. Shao et al., "Wearable and fully biocompatible All-in-One Structured "Paper-Like" zinc ion battery," *ACS Appl Mater Interfaces*, vol. 13, no. 29, pp. 34349–34356, Jul 28 2021, doi: 10.1021/acsami.1c08388

[98] J. Zhao et al., "A safe flexible self-powered wristband system by integrating defective MnO_2–x nanosheet-based zinc-ion batteries with perovskite solar cells," *ACS Nano*, vol. 15, no. 6, pp. 10597–10608, 2021.

[99] Z. Zhai et al., "Multi-dimensional materials with layered structures for supercapacitors: Advanced synthesis, supercapacitor performance and functional mechanism," *Nano Energy*, vol. 78, p. 105193, 2020.

[100] Z. Yan, S. Luo, Q. Li, Z. S. Wu, and S. F. Liu, "Recent advances in flexible wearable supercapacitors: Properties, fabrication, and applications," *Adv Sci (Weinh)*, vol. 11, no. 8, p. e2302172, Feb 2024, doi: 10.1002/advs.202302172

[101] M. R. Benzigar, V. D. Dasireddy, X. Guan, T. Wu, and G. J. A. F. M. Liu, "Advances on emerging materials for flexible supercapacitors: Current trends and beyond," *Advanced Functional Materials*, vol. 30, no. 40, p. 2002993, 2020.

[102] X.-Y. Xu et al., "In-situ temperature regulation of flexible supercapacitors by designing intelligent electrode with microencapsulated phase change materials," *Electrochim Acta*, vol. 334, p. 135551, 2020, doi: 10.1016/j.electacta.2019.135551

[103] L. Zhou, Y. Zhang, H. Qiu, J. Xiao, S. Chen, and Y. Liu, "Phase inversion-based microfluidic-fiber-spinning assembly of self-supported rGO/PEDOT FiberFabrics towards wearable supercapacitors," *Adv Fiber Mater*, vol. 6, no. 3, pp. 798–809, 2024, doi: 10.1007/s42765-024-00373-0

[104] L. Wang et al., "Spatial-interleaving graphene supercapacitor with high area energy density and mechanical flexibility," *ACS Nano*, vol. 16, no. 8, pp. 12813–12821, Aug 23 2022, doi: 10.1021/acsnano.2c04989
[105] P. Zhang, F. Wang, S. Yang, G. Wang, M. Yu, and X. J. E. S. M. Feng, "Flexible in-plane micro-supercapacitors: Progresses and challenges in fabrication and applications," *Energy Storage Materials*, vol. 28, pp. 160–187, 2020.
[106] Z. Wu, L. Li, J. M. Yan, and X. B. Zhang, "Materials design and system construction for conventional and new-concept supercapacitors," *Adv Sci (Weinh)*, vol. 4, no. 6, p. 1600382, Jun 2017, doi: 10.1002/advs.201600382
[107] F. Wang et al., "Latest advances in supercapacitors: From new electrode materials to novel device designs," *Chem Soc Rev*, vol. 46, no. 22, pp. 6816–6854, 2017, doi: 10.1039/c7cs00205j
[108] D. Li, S. Yang, X. Chen, W. Y. Lai, and W. J. A. F. M. Huang, "3D wearable fabric-based micro-supercapacitors with ultra-high areal capacitance," *Advanced Functional Materials*, vol. 31, no. 50, p. 2107484, 2021.
[109] Y. Peng et al., "All-in-one integration of polyaniline-polyvinyl alcohol electrode/electrolyte interface for tailorable solid-state supercapacitors," *Journal of Energy Storage*, vol. 61, p. 106701, 2023.
[110] A. Inman et al., "Wearable energy storage with MXene textile supercapacitors for real world use," *Journal of Materials Chemistry*, vol. 11, no. 7, pp. 3514–3523, 2023.
[111] C. Gao et al., "Ultraconformable integrated wireless charging micro-supercapacitor skin," *Nanomicro Lett*, vol. 16, no. 1, p. 123, Feb 19 2024, doi: 10.1007/s40820-024-01352-1
[112] Y. Lin et al., "Printable fabrication of a fully integrated and self-powered sensor system on plastic substrates," *Adv Mater*, vol. 31, no. 5, p. e1804285, Feb 2019, doi: 10.1002/adma.201804285
[113] C. Zhang et al., "High-energy all-in-one stretchable micro-supercapacitor arrays based on 3D laser-induced graphene foams decorated with mesoporous ZnP nanosheets for self-powered stretchable systems," *Nano Energy*, vol. 81, p. 105609, 2021.
[114] Y. Zheng, Y. Yang, S. Chen, and Q. Yuan, "Smart, stretchable and wearable supercapacitors: prospects and challenges," *CrystEngComm*, vol. 18, no. 23, pp. 4218–4235, 2016. doi: 10.1039/c5ce02510a
[115] Z. Niu et al., "Highly stretchable, integrated supercapacitors based on single-walled carbon nanotube films with continuous reticulate architecture," *Adv Mater*, vol. 25, no. 7, pp. 1058–1064, Feb 20 2013, doi: 10.1002/adma.201204003
[116] Z. Niu, L. Zhang, L. Liu, B. Zhu, H. Dong, and X. Chen, "All-solid-state flexible ultrathin micro-supercapacitors based on graphene," *Adv Mater*, vol. 25, no. 29, pp. 4035–4042, Aug 7 2013, doi: 10.1002/adma.201301332
[117] Y. Yang, H. Fei, G. Ruan, C. Xiang, and J. M. Tour, "Edge-oriented MoS2 nanoporous films as flexible electrodes for hydrogen evolution reactions and supercapacitor devices," *Adv Mater*, vol. 26, no. 48, pp. 8163–8168, Dec 23 2014, doi: 10.1002/adma.201402847
[118] X. Sun, T. Chen, S. Huang, L. Li, and H. Peng, "Chromatic polydiacetylene with novel sensitivity," *Chem Soc Rev*, vol. 39, no. 11, pp. 4244–4257, 2010, doi: 10.1039/c001151g
[119] M. Yu et al., "Water surface assisted synthesis of large-scale carbon nanotube film for high-performance and stretchable supercapacitors," *Adv Mater*, vol. 26, no. 27, pp. 4724–4729, Jul 16 2014, doi: 10.1002/adma.201401196
[120] E. H. Kil et al., "Imprintable, bendable, and shape-conformable polymer electrolytes for versatile-shaped lithium-ion batteries," *Adv Mater*, vol. 25, no. 10, pp. 1395–1400, Mar 13 2013, doi: 10.1002/adma.201204182
[121] E. H. Kil, H. J. Ha, S. Y. J. M. C. Lee, "A facile approach to fabricate self-standing gel-polymer electrolytes for flexible lithium-ion batteries by exploitation of UV-cured trivalent/monovalent acrylate polymer matrices," *Physics*, vol. 212, no. 20, pp. 2217–2223, 2011.

[122] N. Guo, S. Chen, J. Tao, Y. Liu, J. Wan, and X. J. J. Li, "Semi-supervised learning for explainable few-shot battery lifetime prediction," *Joule*, vol. 8, no. 6, pp. 1820–1836, 2024.

[123] T. Lu et al., "Robust battery lifetime prediction with noisy measurements via total-least-squares regression," *Integration*, vol. 96, p. 102136, 2024.

[124] J. Li et al., "Machine learning-assisted property prediction of solid-state electrolyte," *Advanced Energy Materials*, vol. 14, no. 20, p. 2304480, 2024.

2 Emergence and Functionality of Artificial Intelligence (AI) and Machine Learning (ML) in Smart Wearables

Kuriyedath Parambil Khadeeja Thanha, Palasseri Ayisha Sana, Vimala Mathew, and Kannissery Pramod

INTRODUCTION

The 21st century has seen a major change in healthcare driven by artificial intelligence (AI) and wearable sensors. Healthcare is moving from rare hospital visits to continuous monitoring. It is a shift toward preventive care where people manage health on their own. At the center are AI-powered wearables like smartwatches, fitness trackers, and medical tools. They track temperature, blood pressure (BP), electrocardiogram (ECG), and heart rate (HR) in real time. The steady flow of data makes healthcare proactive and supports early detection and treatment [1]. This change is possible because of the Internet of Things (IoT). Special systems like the Wearable Internet of Things (w-IoT) and Internet of Medical Things (IoMT) connect devices to give early warnings and fast responses. Advances in wearables and edge computing allow instant action with little delay. This is critical for early signs of cardiac arrest or other urgent problems. IoT and machine learning (ML) add context-aware features for large health platforms. Smartwatches and other wearables can support home care, elderly care, and public health. The goal is to save lives through a quick and smart response [2–4].

The key strength of this technology is turning large amounts of data into useful results. Traditional health checks capture only limited data. AI systems can handle continuous streams, find patterns, and predict outcomes. This is vital for reducing health risks and giving fast care. Stress detection is a key application of AI-powered smart devices. Applying suitable preprocessing methods and selecting the best ML algorithm to suit the data improve model performance. A main challenge is building models that learn general human patterns but also adjust to personal differences.

DOI: 10.1201/9781003507208-2

New deep-learning (DL) methods now combine contrastive learning with metadata. These models improve stress detection and give better health insights across populations [1, 5, 6]. This pipeline connecting data to diagnosis already supports many applications. In noninvasive care, wearables with photoplethysmography (PPG) and bioimpedance (BioZ) sensors often use random forest regression (RFR) models. They give accurate cuffless BP readings for long-term tracking. This is useful for heart care outside clinics. Another system links PPG sensors with a deep model named GlucoNet. It can estimate blood glucose in real time without calibration or limits based on age or background. This is a big step for diabetes care as it avoids invasive checks. These cases show how AI-enabled smart sensors together perform noninvasive monitoring of chronic diseases [2, 7].

Trust is critical in medical AI. Explainable AI (XAI) helps make systems reliable. Accuracy is not enough when decisions affect patients. XAI methods like SHAP (SHapley Additive exPlanations) show how models reach their results. They make decisions easier to understand and apply. In veterinary science, XAI has been used to explain dog behavior from sensor data. It not only classifies behavior but also shows which signals matter most. This moves beyond black box outputs and gives useful insights. Such clarity is important for doctors and researchers to trust AI systems [8].

This chapter reviews the rise and future of AI and ML in wearables. It starts with the roots of this change, looking at many devices and sensors. Along with smartwatches, new tools like fibrous mats, triboelectric sensors, and bioimpedance patches show new ways to measure health. Always, the objective is to develop devices with improved accuracy, stability, and cost. The chapter then turns to AI and ML methods that power wearables. It explains the path from raw signals to predictions, covering preprocessing, feature extraction, and models from support vector machines (SVMs) and random forests (RFs) to deep networks like convolutional neural networks (CNNs) and recurrent neural networks (RNNs). Next, it surveys applications in chronic disease care, mental health, rehabilitation, sports, and human–machine interaction (HMI). It ends with a look at challenges such as model generalization, data privacy, and user commitment. It also points to trends like XAI and sensor fusion that will guide the future of wearable health systems.

THE TECHNOLOGICAL FOUNDATION: SMART WEARABLES AND SENSOR MODALITIES

The Spectrum of Smart Wearables

The foundation of smart wearables includes many types of devices and sensor systems. These are designed for continuous and noninvasive health monitoring. Wrist-worn devices such as the Fitbit are now widely used and provide long-term data for predicting health outcomes. They use common optical sensors like PPG and inertial sensors. One study combined Fitbit data with electronic health records (EHRs) from a cohort of 14,157 people in the All of Us program. It aimed to predict hospital admissions and cardiovascular diseases (CVD). The diseases included heart failure (HF), coronary artery disease and myocardial infarction (CAD-MI), cardiomyopathy (CMP), and atrial fibrillation (AF). The input to ML models included age, sex, average HR, and daily step count. An RF model gave strong performance for hospitalizations with an area under the receiver operating characteristic curve (AUROC) of 0.95, an accuracy of 0.99, and an F1 score of 0.92. Gradient boosting gave fair performance for predicting CVD with an AUROC of 0.80. This shows that simple data like HR and step count from commercial trackers can support predictive analytics. It provides a promising way for early risk assessment and patient care [9].

A flexible epidermal electronic system (EES) was designed as a wireless and smartphone-linked patch. It is thin, light, and can stick closely to the chest for monitoring hypertension. The patch records ECGs using two metal electrodes. It also records seismocardiograms (SCGs) using a vibration sensor made from piezoelectric polymer. The SCG captures chest vibrations linked with heart contractions and blood movement. The EES does not use a battery and is powered by near field communication (NFC) from a smartphone. ML models, especially CNN, were used to detect ECG and SCG peaks. These signals are used to calculate hemodynamic measures like the systolic time interval (STI). It also measures Weissler's index, which is the ratio of pre-ejection period (PEP) to left ventricular ejection time (LVET). All the signal analysis and classification run inside an Android application. This system shows strong promise for prescreening hypertension and identifying risk groups for heart and lung diseases [10].

Smart garments are also being developed for health and communication support. Flexible smart gloves are one such device and help with sign language translation for the hearing impaired. One reported system used a stretchable triboelectric sensor (FS-TS) array combined with ML. The sensors rely on triboelectric nanogenerator (TENG) technology built from low-cost materials such as silicone, fluorinated ethylene propylene (FEP) powder, and hydrogel. These sensors convert hand movements into electrical signals. An artificial neural network (ANN) model was applied for gesture classification. The FS-TS signals were processed by a small printed circuit board (PCB). They were then sent wirelessly through Bluetooth to a computer for ML analysis, as shown in Figure 2.1. The system tested 10 American Sign Language gestures. These included numbers "1" to "7" and words like 'OK," "GOOD," and "LOVE." It reached high recognition accuracy of up to 98.5%. The system is affordable and flexible. It also solves the cost and rigidity problems seen in traditional sign language devices [11].

Ingestible electronic pills are another special type of wearable. They can create 3D maps and measure gas profiles inside the gastrointestinal (GI) tract. This supports diagnosis and monitoring of diseases such as irritable bowel syndrome (IBS). One design used a pill equipped with optoelectronic gas sensors and a magnetic sensor for tracking. External wearable coils were placed on the abdomen to support this process. The pill measured gases using optical absorption spectroscopy. It detected oxygen (O_2) in the 0%–20% range and ammonia (NH_3) up to 100 ppm. NH_3 levels act as a marker for Helicobacter pylori infection. Localization was achieved with a magnetic-based 3D tracking system. This gave millimeter accuracy using lookup tables or neural network (NN) algorithms. The pill communicated with external systems through Bluetooth low energy (BLE). However, BLE signals become weak inside human tissue. This makes reliable communication inside the deep body a major challenge [12].

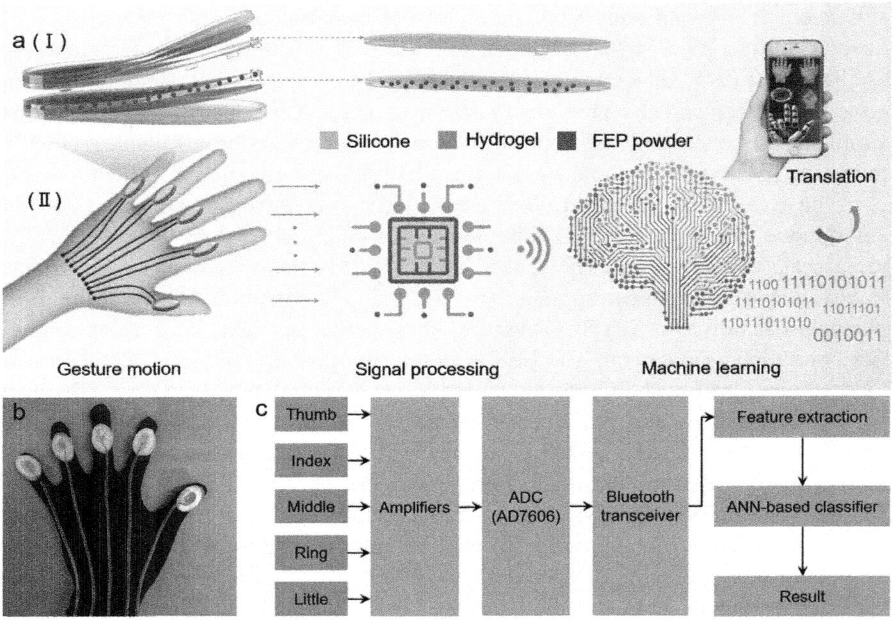

FIGURE 2.1 Wearable system for sign language translation. (a) (I) Diagram of silicone-based FS-TS unit. (II) FS-TS array worn on the hand for translation use. (b) Hand image with FS-TSs. (c) Flow diagram showing amplification, ADC conversion (AD7606 module), wireless data transfer, and ML-based gesture recognition.

(Reprinted with permission from [11]. Copyright (2024) Elsevier Ltd.)

Key Sensor Technologies and Their Signals

Sensors are the basic building blocks that make smart wearables effective health monitoring tools. They convert physiological, biochemical, and mechanical changes into measurable signals. These signals provide continuous and reliable data on the wearer's condition. Different sensor types, such as optical, mechanical, electrical, and biochemical, offer unique insights. They make it possible to track a wide range of health parameters. This section describes the main sensor technologies in smart wearables and the signals they generate for interpretation.

PPG sensors measure changes in blood volume without any invasive procedure. They use LEDs as light sources and photodiodes as detectors. In sleep monitoring, PPG signals are important for measuring blood oxygen saturation (SpO_2) and heart rate variability (HRV). These parameters help identify sleep disorders such as sleep apnea or hypopnea syndrome (SAHS). One DL framework, SleepWatcher, addressed problems like class imbalance and reliance on handcrafted features. It trained 2D CNNs on SpO_2 and HRV signals. These signals were converted into special images called recurrence plots. The first stage classifier distinguished normal from abnormal SAHS events. Using only SpO_2 data, the system achieved 86% accuracy and 89% sensitivity.

Combining SpO_2 with HRV improved performance further. This is important for early detection of SAHS using wearable devices [13].

Inertial measurement units (IMUs) combine accelerometers (ACM) and gyroscopes to capture high-frequency motion signals. They are widely used in wearable systems for activity recognition and event detection. One study developed a two-step DL method to detect severe nocturnal epileptic seizures. The system used a wearable armband that recorded ACM and PPG signals. ACM data from tri-axial accelerometers fed into a prescreening module. This module acted as a lightweight filter. It reduced data volume by 81% while still capturing over 94% of seizures. Threshold-based algorithms on ACM and HR quickly flagged potential events. The second stage applied a CNN–LSTM model to classify these flagged events. Combining ACM and PPG improved accuracy over single-sensor methods. This shows the feasibility of continuous home-based seizure monitoring [14].

ECG monitoring with wearables allows remote cardiac assessment. One study explored predicting serum potassium levels, a risk factor for sudden cardiac death in end-stage renal disease (ESRD). They used single-lead smartwatch ECGs for prediction. The DL model, Kardio-Net, was trained on large datasets of paired 12-lead ECGs and potassium levels. It predicted severe hyperkalemia from Apple Watch ECGs with an AUC of 0.831. Single-lead ECG data mimicked lead I from a 12-lead ECG. The waveform was split into consecutive 5-second intervals for analysis. A noise classifier removed excessively noisy segments. This approach enabled accurate predictions using wearable ECGs in real-world conditions [15].

Electromyography (EMG) sensors, also called surface EMG (sEMG), capture muscle activity signals. These are critical for gesture recognition and control applications. One study optimized electrode placement for a wrist-wearable sEMG hand gesture system. Data was collected from six electrodes placed around the wrist of 43 participants performing 17 gestures. Feature selection methods, such as minimum redundancy maximum relevance (MRMR) and ReliefF, were applied. They found that four electrodes (positions 1, 2, 4, 6) balanced accuracy and speed. An SVM classifier achieved the best results. It reached an accuracy of 85.9% and an F1 score of 86.09%. The optimized setup reduces complexity while maintaining reliable gesture recognition [16].

Wearable electroencephalography (EEG) devices, such as headbands, measure brain activity noninvasively. They capture cognitive signals that reflect mental states like attention. One study used a single-channel Bluetooth EEG headband from Neurosky. It predicted student attentiveness during e-learning sessions. EEG signals were analyzed across five frequency bands: Delta, Theta, Alpha, Beta, and Gamma. Power spectral density (PSD) was used to extract features. ML classifiers, especially SVM, predicted attentiveness with 91.68% accuracy. Beta waves (12–30 Hz) associated with alertness and alpha waves (8–12 Hz) associated with passive attention were most important. This study demonstrates that wearable EEG can assess attention in real time [17].

Bioimpedance analysis measures body resistance by sending a small, safe electrical current through tissue. This indicates fluid variations in blood vessels or tissues. One study developed an IoT-enabled wearable using transthoracic BioZ to classify smoking habits. Measurements were collected from 341 smokers. Data were analyzed using the Cole-Cole plot to extract parameters such as cell membrane capacitance and tissue heterogeneity. Combined with age, these features identified lung damage caused by smoking. ML models like k-nearest neighbor (kNN), RF, and SVM were tested. The SVM achieved the highest accuracy of 97% for classifying smoking dependence according to the Fagerstrom Test for Cigarette Dependence (FTCD). This shows that BioZ-based wearables can effectively monitor health risks related to smoking [18].

Wearable hydrogel-based patches allow noninvasive sensing of biomarkers in sweat. One patch featured a bilayer hydrogel design for continuous monitoring of sweat pH and glucose levels. The first layer was a polyvinyl alcohol (PVA) hydrogel functionalized with colorimetric reagents and the glucose oxidase (GOx) enzyme. The second layer combined PVA with sucrose to improve adhesion and protect the sensor from air moisture. The system worked with AI-assisted smartphones for visual sensing. Figure 2.2 illustrates the design of an ML-assisted wearable biosensor

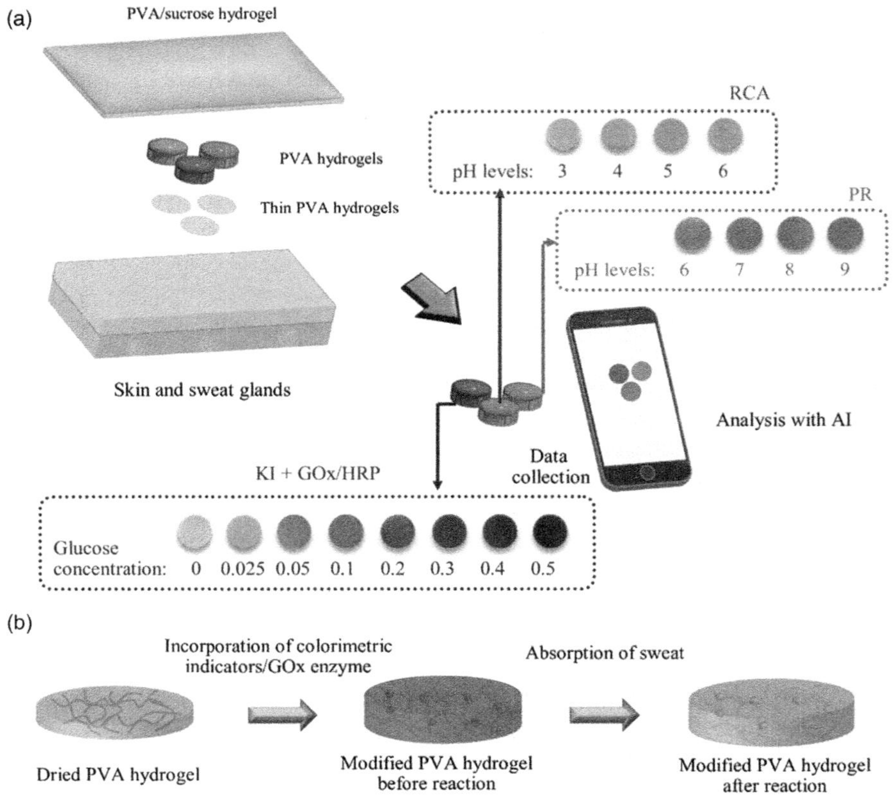

FIGURE 2.2 (a) Wearable ML-based colorimetric biosensor for glucose and pH monitoring in sweat. (b) Schematic of hydrogel fabrication for colorimetric sensing.

that measures glucose and pH in sweat, along with the schematic of its colorimetric hydrogel fabrication process. Colorimetric images captured by the phone were processed using ML models such as RF and CNN. This setup allowed real-time and continuous monitoring of sweat biomarkers [19].

Breath analysis is an important tool for monitoring systemic diseases. A novel wearable porous eutectogel was developed to measure ammonia levels in exhaled breath in real time. High ammonia concentrations can indicate Chronic Kidney Disease (CKD). The eutectogel was created using 3D printing and included components like bromocresol green (BCG). It binds strongly to ammonia, causing a rapid color change from yellow to blue when exposed. The system had a fast response time of 3 seconds and a low detection limit of 0.5 ppm. DL algorithms, specifically CNNs, were applied for classification in clinical tests. The approach achieved an overall accuracy of 96.5%, demonstrating its potential for noninvasive disease monitoring [20].

Thermal sensors in wearables can measure skin temperature, an important indicator of the body's thermophysiological state and comfort. A wristband-type sensor was developed for three-point wrist measurement, including the upper wrist, radial artery, and ulnar artery. It was integrated into an AI-enabled Human-in-the-Loop (HIL) HVAC control system. The system uses an RF classifier trained on skin temperature, environmental factors (air temperature and humidity), and personal information (age, sex, and clothing). It predicts the occupant's thermal comfort in real time. This multi-sensing strategy achieved 93.9% accuracy in thermal comfort prediction. The HIL system demonstration showed that it could automatically enhance comfort and potentially reduce building energy consumption by 20% compared to traditional fixed-set control [21].

Wearable strain sensors using triboelectric and electromagnetic principles are advancing for motion recognition and HMI. One example is the wavelength-thickness dual-gradient microwrinkled structure (WTGM) flexible strain sensor. It uses MXene/reduced graphene oxide (rGO) composite films to achieve high sensitivity and wide stretchability up to 220%. The sensor has robust mechanics and a fast response time of 42 ms. It can detect small signals like pulse and respiration, as well as large movements like joint bending. This sensor was paired with an ANN model for sign language recognition. It achieved 98.29% accuracy across 13 gestures, demonstrating strong potential for wearable HMI applications [22]. Figure 2.3 presents the dual-gradient microwrinkled strain sensor design, its fabrication, and potential uses.

Wearable stethoscopes overcome the limitations of traditional digital stethoscopes by providing wireless and continuous auscultation. One device, the lung sound monitoring patch (LSMP), uses a flexible patch embedded with micro-electro-mechanical systems (MEMS) microphones. It monitors lung sounds continuously for long periods. This allows detection of Adventitious Breathing Sounds, such as wheezing, which are critical for diagnosing and managing respiratory diseases like asthma and chronic obstructive pulmonary disease (COPD). The LSMP integrates an AI-based event-counting algorithm developed using a 2D CNN deep-learning model. It quantifies wheezing events over time, greatly reducing the labor required for long-term lung function assessments. The system demonstrated high classification accuracy, exceeding 80% in detecting abnormal breathing in elderly COPD patients during long-term monitoring sessions [23].

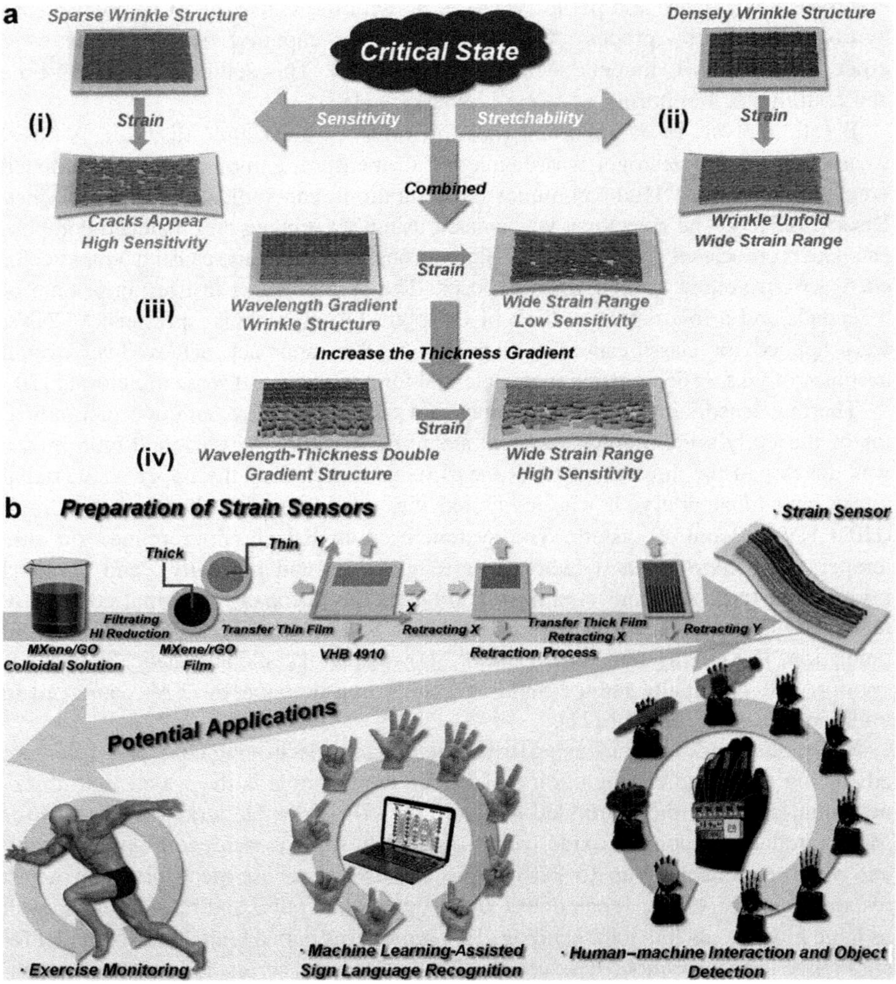

FIGURE 2.3 Schematic of the WTGM strain sensor design, process, and uses.

(Reprinted with permission from [22]. Copyright (2025) Elsevier B.V.)

FUNCTIONALITY IN PRACTICE: A SURVEY OF APPLICATIONS

Comprehensive Health and Wellness Monitoring

A study focused on cuffless BP prediction developed a multi-modal wrist-worn wearable device. This device used an RFR algorithm. High BP (hypertension) is a major risk factor for CVD such as stroke, myocardial infarction, and HF. Traditional cuff-based monitoring is inconvenient and cannot provide continuous tracking. The developed device employed two PPG sensors and two BioZ sensors. Redundancy in sensors helped enhance prediction accuracy. Multiple pulsatile waveforms were input into the ML model. The model performed better than single-sensor models. Results from training and validation datasets achieved mean absolute errors (MAE) below 3.3 mmHg for both systolic and diastolic BP. This demonstrates the potential for a universal cuffless BP estimation model for long-term cardiovascular health management [7].

Another approach involved a thin, lightweight, AI-enabled EES for continuous, noninvasive monitoring of hypertension. The EES attaches tightly to the chest and simultaneously records first-lead ECGs and SCGs. ECG signals are collected through metal electrodes, and SCG signals are captured via a thin piezoelectric polymer vibration sensor. The system is battery-less and powered by smartphone NFC functionality. The smartphone also hosts the AI algorithm for automatic signal analysis and classification. Key hemodynamic parameters, particularly Weissler's index (ratio of PEP to LVET), are extracted to assess hypertension risk. Clinical tests in patients with cardiopulmonary hypertension demonstrated promising prognostic performance. This indicates that EES can be used for the prescreening of hypertension [10].

A systematic review evaluated ECG-AI models for predicting fatal coronary heart disease (FCHD) risk using wearable devices such as Apple Watch ECGs. The goal was to develop a single-lead ECG-based DL model for FCHD prediction. The study also assessed concordance between clinical ECGs and Apple Watch ECGs. The ECG-AI model used single-lead data corresponding to lead I of a 12-lead ECG. On holdout data, the model achieved an AUC of 0.76. A Cox proportional hazards model combining ECG-AI predictions with age, sex, and race achieved an AUC of 0.79. Paired clinical and Apple Watch ECGs showed strong correlations: Pearson $R = 0.74$, Spearman $r = 0.67$, and Cohen's kappa $\kappa = 0.58$. These results confirm that single-lead wearable ECGs can predict FCHD risk with strong concordance to clinical ECGs. This enhances the potential for remote monitoring and early intervention [24].

Noninvasive blood glucose estimation is being explored via an IoMT platform. The system integrates a wearable PPG device, a smart analytics cloud for estimation models, and a mobile/web application for diabetes monitoring. GlucoNet, a lightweight one-dimensional input-reinforced deep neural network, performs blood glucose computation. It captures long and short temporal and spatial features from PPG signals. On a dataset of 283 participants, GlucoNet achieved a mean absolute percentage error (MAPE) of 17.8% $\pm$ 12.8%. All predictions fell within clinically acceptable zones A and B of the Clarke-error grid. The model is non-calibrated, noninvasive, demography-agnostic, and can run on edge devices for real-time and offline monitoring [2].

Another study evaluated AI models for estimating blood glucose levels (BGL) in diabetic patients using noninvasive wearable data. The dataset included BGL readings, HR, SpO_2, BP, body temperature, sweating, and shivering from 13 participants. Data were collected using a continuous glucose monitoring kit (Freestyle LibrePro) and a smart band (Riversong Wave O2). Traditional ML models and DL models were employed for BGL estimation. Results showed strong correlations between glycemic metrics and wearable-derived features. Root mean square error (RMSE) values ranged from 0.099 to 0.197. These findings confirm the feasibility of using commercially available wearables for BGL estimation in diabetic patients [25].

Noninvasive sweat monitoring is possible using stretchable, adhesive bilayer hydrogel patches combined with AI-assisted smartphone applications. The bottom layer contains PVA hydrogels with colorimetric reagents and glucose oxidase for sensing. The top layer is a PVA–sucrose layer that enhances adhesion and protects the sensor from moisture. The hydrogel offers high mechanical robustness with a tensile strength of 2.22 MPa and a fracture strain of 440%. Colorimetric changes captured via smartphone are processed by RF and CNN models. This approach achieved $R^2 \approx 0.99$ for pH (3–9) using RF and $R^2 = 0.988$ for glucose concentrations (up to 0.5 mM) using CNN. The integration of AI enhances detection accuracy and robustness for continuous biomarker monitoring [19].

A wearable patch-type stethoscope (LSMP) enables wireless, skin-attachable, continuous monitoring of lung sounds. The system integrates an AI algorithm for automatic wheezing detection. MEMS microphones acquire the signals. Audio data is converted into Log-Mel spectrograms and analyzed using a 2D CNN deep-learning model. A counting algorithm identifies wheezing events in real time, independent of the respiratory cycle. Clinical tests on elderly COPD patients showed the system detected over 80% of wheezing events. This allows continuous evaluation of disease severity and progression over time [23]. SleepWatcher, a DL method, detects SAHS using wearable biomedical signals. The method addresses class imbalance and dependence on handcrafted features in traditional SAHS detection. Single-lead ECG-derived HRV and SpO_2 signals are transformed into 2D recurrence plots for CNN analysis. Stage 1 classifies normal vs abnormal events, and Stage 2 classifies apnea vs hypopnea. The ensemble classifier achieved 87% accuracy and 91% sensitivity. Combining HRV and SpO_2 signals improves performance over using either signal alone [13].

A neck-wearable piezoelectric sensor (NPS) enhances obstructive sleep apnea detection. The NPS captures snoring vibrations (70–250 Hz) and carotid artery pulsations (0.01–1.5 Hz). A hybrid DL model combining a 1D CNN and gated recurrent unit (GRU) processes spatial and temporal features. The model achieved 0.90 classification accuracy for severe SAS, with a recall of 1.00 and an F1 score of 0.91. Continuous monitoring captures vibrations rather than sounds, addressing privacy concerns from microphone-based methods [26]. A wearable armband detects nocturnal major epileptic seizures using tri-axial accelerometry (ACM) and PPG signals. A prescreening step uses threshold-based algorithms to flag suspected events. ACM captures motion signals, while PPG captures HR changes, which often precede motor activity. A CNN–LSTM hybrid with attention processes flagged events. Validation on 6,304 hours of data from 68 patients achieved a sensitivity of 0.716 and 0.165 false alarms per hour. This supports long-term out-of-hospital seizure monitoring and reduces SUDEP risk [14].

AiCareGaitRehabilitation monitors Parkinson's disease (PD) patients for freezing of gait (FoG). It uses multi-modal sensors: strain sensors, 18 plantar piezoelectric sensors, and IMUs (accelerometer and gyroscope). The system computes a "Freezing Index" using CNN with SVM output to classify FoG in real time. Functional electrical stimulation (FES) improves gait control, reducing FoG episodes from 0.63 to 0.23 per minute. This demonstrates potential for enhancing mobility and fall prevention in PD patients [27]. The Multi-Level Meta Learner (MLML) algorithm automates post-stroke assessment using wearable sensors. It classifies impaired vs non-impaired body parts, focusing on hands. MLML combines multiple base learners in a multi-level decision process. Inputs are time-series linear acceleration and angular velocity data from Xsens IMUs on the upper limb. Experimental results show that MLML improved the classification accuracy to up to 95.64% for impaired hands. This validates its use for automated post-stroke diagnosis [28].

The iPREDICT framework uses AI and crowd-sensed wearable data to predict pandemics. Individual biomarker anomalies are monitored and correlated with other nearby individuals. Time-series anomaly detection is performed using autoregressive integrated moving average (ARIMA), seasonal autoregressive integrated moving average with exogenous regressors (SARIMAX), LSTM, or autoencoders. Graph neural networks (GNNs) model pathogen and population dynamics. The framework adapts epidemic alarm thresholds using demographic, social, and geographic data. Smartphones and smartwatches act as a personalized biosensing mesh. They monitor skin temperature, SpO_2, HRV, and cough sounds [29]. The WE SENSE study predicts systemic inflammation after live attenuated influenza vaccine (LAIV) exposure. Healthy adults were monitored for 12 days using a smart ring, smartwatch, and smart shirt. Physiological and activity data included HR, breathing rate, and acceleration. ML models, specifically extreme gradient boosting (XGBoost), predicted surges in systemic inflammation defined by 48 blood biomarkers. Night-time data from the Oura ring achieved an ROC–AUC of 0.89 for 24-hour prediction. This method outperformed symptom-based detection. It enables early detection of viral respiratory tract infections, including influenza and SARS-CoV-2, even in asymptomatic individuals [30].

Mental and Cognitive State Assessment

A novel DL framework called StressCon was proposed for automated stress detection. It fully integrates contrastive learning with user metadata fusion. This allows the model to capture both general stress patterns and individual physiological variations. Stress is prevalent and linked to major health risks like depression and CVD. Continuous monitoring is therefore important. StressCon uses two supervised contrastive loss functions derived from sensor data. The sensors include electrocardiograms (ECG), electrodermal activity (EDA), and electromyograms (EMG). These functions extract robust, subject-invariant features. The model also incorporates individual metadata such as age, weight, gender, and height into the classification head. This accounts for person-specific physiological responses. It improves personalized stress detection accuracy. Evaluation on public datasets (WESAD and UTD) included binary (stress vs. non-stress) and three-class (normal, stress, and amusement) tasks. StressCon outperformed baseline models. It achieved up to a 3.49% improvement in accuracy and a 3.48% improvement in F1-score on the three-class WESAD task [5].

Another study applied ML to detect stress in free-living environments. It goes beyond traditional laboratory studies. The SWEET dataset includes data from 240 subjects collected via wearable sensors. The sensors measured ECG, skin temperature (ST), and skin conductance (SC). Stress detection prevents chronic health problems such as vascular disease, headaches, and depression. Several ML models were tested, including RF, KNN, and XGBoost. Tests included binary and multi-class classification with two or three stress levels. Some scenarios used the Synthetic Minority Over-sampling Technique (SMOTE) to handle imbalance. RF performed best in unbalanced datasets. It achieved 98.29% accuracy and a 97.89% F1-score for binary classification without SMOTE. For three-level classification without SMOTE, accuracy reached 97.98%. ML models detect subtle patterns in physiological parameters. These include HR changes reflecting the body's "fight-or-flight" stress response [6].

A separate study focused on using instantaneous HR from wearable fitness trackers. This simplifies stress detection compared to ECG or HRV. Stress is widespread and linked to CVD. Wearable adoption is high, making this approach scalable. Raw ECG signals from a spider-fearful participant dataset were converted into HR data. A sliding window of 40 consecutive HRs (~1 minute) augmented the dataset. This captured dynamic temporal changes indicative of stress. To improve the model's accuracy, feature engineering was used to generate new variables by calculating the ratios and differences between local (chunked) and global (entire session) HR statistics. Simple statistics alone poorly distinguished stress from rest. A Histogram-based Gradient Boosting Classifier achieved 73% accuracy. This shows the potential for using wearable HR data for stress prediction. It simplifies monitoring for general use [31].

Mental fatigue in construction equipment operators causes severe attention failure. This endangers site safety. Traditional methods like NASA-TLX questionnaires are slow, biased, and less accurate. This study collected raw EEG signals from 15 operators using flexible Muse headband sensors (four channels). Data were recorded for one hour during excavation tasks. Over 18 million data points allowed DL models to be applied effectively. NASA-TLX scores classified mental states into alert, mild fatigue, and fatigue. Three DL models LSTM, bidirectional long short-term memory (Bi-LSTM), and 1D-CNN were compared. The experimental procedure is shown in Figure 2.4. Bi-LSTM achieved the highest accuracy at 99.941%. F1-scores ranged from 99.917% to 99.993%. Bi-LSTM effectively processes time-series data and retains past and future information [32].

A Decision Support System (DSS) was proposed to monitor learner attention during online sessions. The system uses a wearable EEG to determine if students are attentive or inattentive. EEG data were collected from a single-channel, Bluetooth-enabled Neurosky device at FP1. Five frequency bands were recorded: Delta, Theta, Alpha, Beta, and Gamma. Alpha (8–12 Hz) indicates passive attention. Beta (12–30 Hz) signals active, external attention. Data from 100 participants were analyzed using ML models. SVM achieved the highest classification accuracy of 91.68%. It outperformed logistic and ridge regression. Feature importance analysis confirmed that Beta and Alpha waves were key predictors. K-means clustering of these bands identified clear clusters representing attentive and inattentive states. This provides a robust tool for decision support in e-learning [17].

Wearable sensor data combined with AI was used to generate digital phenotypes for classifying psychiatric disorders. Data from Fitbit smartwatches in the ABCD cohort were collected. HR, activity intensity (METs), sleep, and accelerometry were recorded. Data were processed into static (summary) and dynamic (time-series) features. An interpretable AI framework combined XGBoost for static features and Xception CNN for dynamic features. It classified externalizing (ADHD) and internalizing (anxiety) phenotypes versus healthy controls. Models using dynamic features

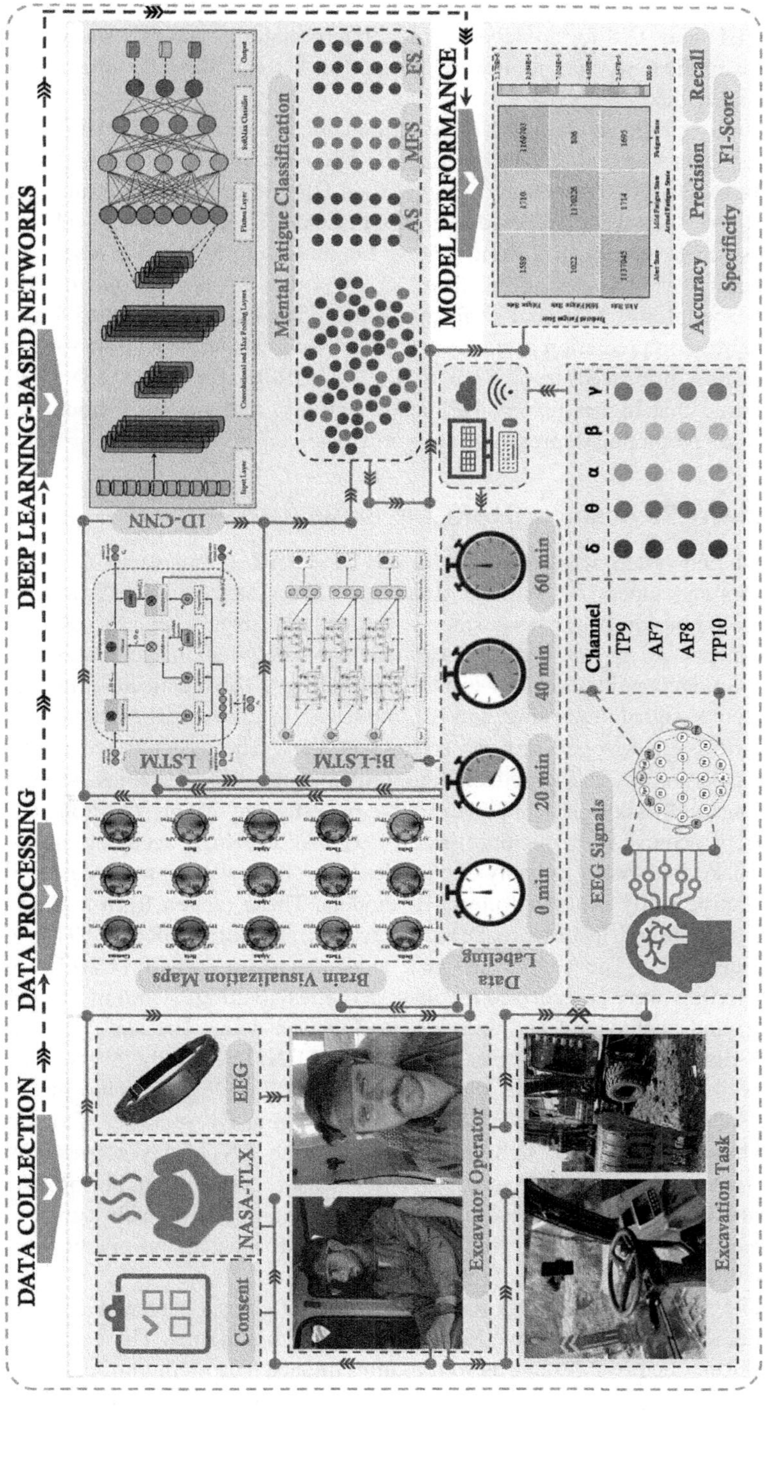

FIGURE 2.4 Research methodology to detect mental fatigue.

(Reprinted with permission from [33]. Copyright (2023) Elsevier Ltd.)

outperformed baseline models relying on clinical covariates. AUROC values were 0.89 for ADHD and 0.78 for anxiety disorder. Interpretability analysis showed that HR dynamics in the early afternoon were critical for ADHD. Sleep features at night predicted anxiety disorder [33].

Another study predicted early depressive and manic symptoms in bipolar disorder (BD) patients. Data were collected using Garmin Vivosmart 4 smartwatches. Metrics included motor activity, sleep duration/stages, and HR (minimum, maximum, average, and resting). Data were collected over months and mapped to weekly self-rated Beck Depression Inventory (BDI) and Young Mania Rating Scale (YMRS) scores. The hypothesis was that deviations from an individual's personal mean would predict mood symptoms. XGBoost performed best. Including individualized features improved depression prediction accuracy from 79% to 83% (AUROC 0.85 → 0.89). Manic prediction improved from 83% to 91% (AUROC 0.84 → 0.88). SHAP analysis showed that higher resting HR, lower activity, and sleep <6 hours were key features for predicting depressive symptoms [34].

Human Activity Recognition (HAR) and Biomechanics

A novel IoT-based system was developed for real-time HAR. It uses wearable accelerometer sensors and supervised learning algorithms. The system provides a low-cost, responsive solution for health assessments, particularly for the elderly or newly recovered patients. A major challenge was the high computational complexity for accurate real-time recognition on low-cost microprocessors. The architecture integrates electronic circuits, signal processing, and ML models. This ensures efficiency for real-time requirements on low-performance microcontrollers. The system used a single three-axis accelerometer. It focused on four fundamental daily activities: standing, sitting, running, and walking. By applying simple but effective feature selection and an RF classifier, it achieved a recognition rate of 99.2%. The high accuracy and low complexity make this system broadly applicable due to its low equipment cost [35].

A comparative study evaluated nine ML models. These ranged from traditional SVMs with handcrafted features to DL architectures, including GRU, LSTM, and transformer networks. The study used two HAR datasets: OPPORTUNITY and CogAge. Activities were categorized as "state activities" (sitting, standing, walking, and running) and "behavioral activities" (like closing a door). For state activities, without the distracting "NULL" class in OPPORTUNITY, nearly every model performed well. Average F1-scores ranged from 94.10% to 96.48%. This suggests that simple state activities can be modeled even with non-DL methods. For more complex behavioral activities, hybrid DL models with LSTM layers performed better. The research shows that multiple body sensors yield high performance. However, single, well-placed devices, like a smartphone or smartwatch, can sometimes suffice. This highlights the trade-off between performance and practicality in everyday healthcare HAR systems [36].

A wearable sensor-based system for fall detection in elderly care was proposed. It uses inertial sensors and ensemble ML to provide prompt alarms. The goal was to detect falls during the pre-fall stage as well as after impact. The methodology used an ensemble classifier combining RF for feature extraction and SVM for classification. Each composite model identifies a specific event type: Non-Fall, Pre-Fall, or Fall.

This maximizes overall accuracy. Using the KFall database, the system achieved high precision: 95% for Non-Falls, 96% for Pre-Falls, and 98% for detectable Falls. The ensemble algorithm outperformed single RF and SVM models for Pre-Fall and Fall detection [37].

Another fall risk monitoring system targeted construction workers. It integrates wearable data (HR and brainwaves) with Building Information Modeling (BIM). The SOS–NN–GNN hybrid model was used. The NN processes time-independent variables, and the GNN handles time-dependent sequential data. Optimization was guided by the Symbiotic Organisms Search (SOS) algorithm. This system classified fall risk levels (High, Moderate, and Low) based on HR, heat load, location hazards, and mental fatigue. SOS–NN–GNN achieved 90.98% accuracy in assessing fall risk. It outperformed other DL models in robustness [38]. A separate IoT-based elderly fall prediction system uses low-power wearables. It couples with an LSTM classifier running on a smart gateway. The system focuses on indoor fall detection using a three-axis accelerometer. Local edge analysis reduces reliance on cloud processing, improving efficiency and lowering cost. Testing on the MobiAct dataset achieved 95.87% real-time accuracy. Sensor placement on both the wrist and the waist improved performance to 97.8% [39].

A graph-based hybrid DL model, CGRU–dGCN, was introduced for abnormal gait detection. Multi-sensor data were placed on the lower limb joints. The model represents joint interactions as a graph, capturing irregular nonlinear features in non-Euclidean space. The cumulative GRU module extracts temporal dependency features. The deeper GCN module captures local spatial features using residual connections and preactivation techniques. Tested on simulated abnormal gaits, it achieved 99.02% accuracy for nine gait patterns. This includes normal walking, walking with shoe lifts, and carrying sandbags. The model can precisely measure subtle gait differences, aiding early diagnosis, rehabilitation, and injury detection [40]. Research assisting post-stroke assessment proposed an MLML ensemble algorithm. It uses heterogeneous ML classifiers and Xsens wearable sensors. The focus was on classifying impaired versus non-impaired body parts, particularly hands. The MLML uses a multi-level decision process to select optimal classifier combinations. It minimizes complexity and maximizes accuracy. Linear acceleration and angular velocity data from IMUs were used. MLML achieved up to 95.64% accuracy in identifying impaired hands with boosting learning [28].

A self-powered triboelectric nanogenerator sensor (PN-Sensor) was developed for soccer training. It was fabricated using recycled nylon 66. The sensor monitored hip, knee, and ankle movements. Four soccer maneuvers were analyzed: directional changes, dribbling (La Croqueta), aerial duels, and shooting. A CNN algorithm enabled real-time motion analysis, as illustrated in Figure 2.5. The posture recognition system achieved 97.5% accuracy. It provides biomechanical data to optimize training and reduce injury risk [41]. A self-powered wearable device (MLTS) combining triboelectric and electromagnetic sensors was developed. It recognizes joint motion features to prevent exercise injuries and assist rehabilitation. ML (decision tree model) analyzes dynamic limb movement signals. It achieved 99.58% accuracy across five motion patterns, supporting real-time personalized training [42].

Research predicted functional recovery after musculoskeletal injury using wearable activity data. Step counts before and after injury were analyzed using logistic regression.

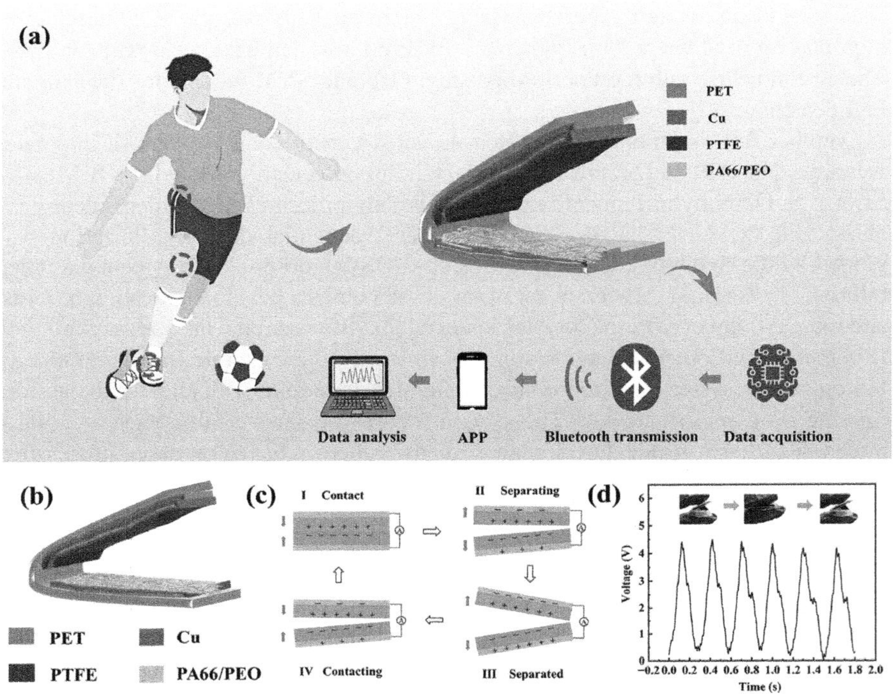

FIGURE 2.5 Illustration of PN-Sensor: training application, structure, principle, and cyclic response.

(Reprinted with permission from [42]. Copyright (2025) Elsevier Ltd.)

The model predicted the likelihood of achieving at least 50% pre-injury activity by six weeks. Accuracy exceeded 80% with activity data from three weeks post-injury. AUC was 0.86 at week 3. This enables early identification of patients needing extra care or rehabilitation [43]. A hybrid DL framework estimated vertical ground reaction force (vGRF) during ACL rehabilitation. Wearable IMUs captured motion during walking, running, and stair descent. vGRF estimates provide a practical alternative to lab-based force plates. The CNN-BiGRU-Attention model achieved R^2 values of 0.953 (walking), 0.971 (running), and 0.979 (descending stairs). This allows continuous, real-world monitoring and personalized rehabilitation planning [44].

An intelligent sensing system (GSIS) uses a degradable hydrogel-based triboelectric nanogenerator (GE-TENG) for HAR and HMI applications. High electrical and mechanical performance enabled 100% signal recognition. Covariance matrix feature analysis and CNN classification were integrated. The system supports sports health monitoring, intelligent robotic control, and interactive HMI applications [45]. A multi-channel flexible optical fiber wearable system with ML was developed for motion monitoring. It is suitable for rehabilitation and sports biomechanics. A 1D-CNN analyzes strain and pressure signals from TPU optical fiber arrays. Two prototypes were demonstrated: a smart glove for gesture recognition and a smart carpet for full-body monitoring. The glove achieved 98.12% accuracy for eight gestures. The carpet achieved 96.82% for six body movements. Multi-signal fusion and DL provide quantitative motion assessment beyond conventional single-sensor platforms [46].

Advanced HMI

A wearable sign language translation system was developed. It uses flexible and stretchable triboelectric sensor (FS-TS) arrays combined with ML algorithms. The system accurately transforms sign language movements into spoken language. Cost-effective materials like silicone, hydrogel, and FEP powders provide rapid response and high sensitivity. FS-TS arrays convert finger movements into analog signals. A PCB processes these signals and transmits them via Bluetooth to a computer terminal. An ANN model classifies the gestures. It achieved an average recognition accuracy of 98.5% for 10 American Sign Language gestures using a dataset of 1000 instances. The system has a user-friendly interface that translates gestures into text and speech. This provides an affordable and scalable solution for overcoming communication barriers [11].

A study optimized electrode configuration for a wrist-wearable EMG-based hand gesture recognition system. The system classifies 17 hand gestures using sEMG signals from six bipolar electrodes around the wrist. Several ML models were tested, including SVM, KNN, ANN, and RF. The optimal configuration used four electrodes (positions 1, 2, 4, 6). This setup achieved 85.9% accuracy and an F1-score of 86.09% with the SVM classifier. The design balances high accuracy with user comfort and reduced complexity. It covers key flexor and extensor muscles involved in wrist and finger movement, including the flexor digitorium superficialis and extensor digitorium superficialis [16].

A driver gesture command digital twin system (DC-SS) was developed. It uses a friction electric ring as a self-powered sensor to recognize gestures in real time. The system converts train driver gestures into digital signals. These signals create a digital twin in a virtual environment. The triboelectric ring sensor (TENG) has a PLA shell and an Ecoflex inner pad. It captures electrical signals generated by finger movements. Gesture recognition employed GRU, LSTM, and 1D-CNN models. The Conv1D model achieved 98.9% accuracy for eight standard train driver commands. This system overcomes the limitations of camera-based methods and enables accurate HMI in smart transportation and VR [47].

A highly sensitive ultra-thin ZnO–PDMS dental patch was developed for periodontitis screening. It detects volatile sulfur compounds (VSCs) like H_2S produced by bacteria beneath the gingiva. Blue-emitting ZnO quantum dots (QDs) enhance sensitivity via surface defect modulation. A U-Net CNN model identifies fluorescence quenching areas corresponding to hidden lesions. The system detects and maps lesions in vivo within 15 minutes. This provides a fast, accurate method for routine screening and disease monitoring [48]. A triboelectric sensor array (TSA) mimicking neural synapses was developed for motion sensing and user recognition. The system integrates DL CNN algorithms, using an optimized AlexNet model. TSA identifies complex behaviors like foot type and gait (99.23%) and even individuals (99.75%). Applications include smart homes and rehabilitation for limb-injured patients. TSA offers high accuracy and privacy for personalized interaction systems [49]. Table 2.1 illustrates how diverse sensor technologies are paired with different ML methods to achieve high accuracy in real-world tasks.

TABLE 2.1
Functional Applications of Wearable Sensors Integrated with AI/ML Methods, Highlighting Sensor Technologies, Computational Models, and Key Outcomes across Various Fields

Application/Goal	Technology	AI/ML Method	Outcome/Result	References
HMI Morse code → speech translation	Conductive electrospun fibrous mats Strain sensors (graphene + TPU)	Deep learning stacked Bi-LSTM	Recognition accuracy >98.5%	[50]
Early detection of mood changes Bipolar disorder symptoms	Smartwatch Tracks activity, sleep, HR	XGBoost	83% accuracy (AUROC 0.89, depression) 91% accuracy (AUROC 0.88, mania)	[34]
Real-time HR monitoring Adaptive thresholding for cardiac issues	IoT smartwatch (Fitbit Ionic) PPG sensors	KNN DT XGBoost Logistic regression (LR)	KNN F-score = 91%	[3]
Personalized seizure detection Focal epileptic seizures	Wearable ECG device HRV analysis	Patient-adaptive LRML classifier	False alarm rate ↓ 31% (0.62/24h) Sensitivity 78.2%	[51]
Soccer biomechanics Lower limb training	Self-powered TENG PN-Sensor Recycled Nylon 66	CNN (deep-learning)	Accuracy 97.5% (4 movements)	[41]
HAR	Wearable multi-sensors: accelerometer, magnetometer, gyroscope	UC fusion (CNNs for features)	96.84% (UCI HAR) 98.85% (WISDM)	[52]
HMI Dual-language recognition Motion detection	All-nanofiber self-powered TENG sensor	1D-CNN	100% recognition accuracy by the 11th epoch	[53]
Noninvasive pulmonary capillary wedge pressure (PCWP) estimation	CardioTag wearable ECG, SCG, PPG	CNN + large language model (LLM) encoder–decoder	Error = 1.04 ± 5.57 mmHg	[54]
Gait disorder recognition Fall risk in aging and Parkinson's	Wearable IMU Three-axis accelerometer (L4 vertebra)	Decision tree (DT) RF	DT achieved AUC = 100%	[55]

Motion recognition, handwriting and movements HMI applications	3D-printed flexible triboelectric sensor AgNWs in TPU	1D CNN (AI, deep learning)	Detection reliability 97%	[56]
Epileptic seizure forecasting in pediatric patients	Wrist-worn Empatica E4 EDA, TEMP, BVP, ACC	Deep LSTM model 24-hour cycle integration	IoC improvement in 68% Overall IoC = 27.3%	[57]
Diabetes screening Noninvasive sweat glucose sensing	Wearable strap SPCE electrode + CNTs CNFs Prussian blue	XGBoost regressor	$R^2 = 0.86$ RMSE = 0.177	[58]
Fault detection in WSN outlier/faulty sensor estimation	DHT22 (humidity/temp) ESP32 microcontroller	OCC models Iterative ARIMA	MAE = 0.4274 (humidity)	[59]
Tyrosine detection Early diagnosis of markers	Flexible wearable biosensor	ANN SVM Ridge regression	$R^2 = 0.9828$ Recovery 99.24%	[60]
Remote fetal heart rate monitoring	Fetal phonocardiography device Wearable belt	BiGRU MBCapsNet VAE-GAN	Accuracy 81.34% Sensitivity 72% F1 = 83%	[61]
Friction and wear prediction	Tribometer (epoxy + MWCNTs)	ANN RNN (LSTM) CNN	$R^2 = 0.9999$ (RNN-Bi-LSTM)	[62]
Stress and fitness pressure monitoring	Wearable BSNs RTLS	LSTM	F1 score = 92% Accuracy = 87%	[63]
Predicting harmful behaviors (suicide, self-harm, aggression)	URBAN HR wearable HR, acceleration, location data	Temporal fusion transformer (TFT)	Accuracy = 95.1% Recall = 74.9% F1 = 78.1%	[64]
Hand gesture recognition HMI	sEMG armband (16 channels)	SVM CNN	Accuracy = 87% (10 gestures)	[65]
Epilepsy detection	EEG sensor headset (Neurosky MindWave 2.0)	RF classifier Multi-layer LSTM (deep learning)	RF accuracy = 90.74%	[66]
Pedestrian emotion detection Emotional valence (pleasant/unpleasant)	Wearable sensors EDA, HRV, ACC data	ML algorithms (sensor + anomaly detection features)	Accuracy = 0.7092 F1 = 0.7011	[67]

(Continued)

TABLE 2.1 (Continued)
Functional Applications of Wearable Sensors Integrated with AI/ML Methods, Highlighting Sensor Technologies, Computational Models, and Key Outcomes across Various Fields

Application/Goal	Technology	AI/ML Method	Outcome/Result	References
Sign language recognition Real-time translation	Smart glove MF@AC foam + pressure sensors	CNN + BiGRU + attention (CBGA)	Accuracy = 99.4% (26 letters)	[68]
Energy optimization, rehabilitation robot gait recognition	Lower limb rehab robot, motion sensors	Deep-learning gait model + energy optimization	Gait recognition = 96%	[69]
Urine biomarker monitoring (DMA, creatinine, glucose, H^+)	Diaper-integrated flexible potentiometric sensors	MLP calibration system	High accuracy (pH error correction)	[70]
Physiological monitoring Sweat biomarkers (tyrosine, tryptophan, pH)	Sweat sensor (LIG-based electrodes)	KNN (classification) BPNN (regression)	Accuracy = 99.3% (KNN with pH factor)	[71]

CRITICAL ANALYSIS: CHALLENGES, LIMITATIONS, AND ETHICAL CONSIDERATIONS

The analysis of surface-enhanced Raman spectroscopy (SERS) spectra in complex human sweat presents significant analytical challenges. Target molecules often exist at very low concentrations. Individual variability further complicates detection. The intricate nature of sweat composition also adds complexity. The biological sample matrix can interfere with sensor signals. This interference affects the overall accuracy and reliability of results. Current wearable sweat sensors frequently show poor sensitivity. Temporal variability is another issue. These factors make continuous monitoring difficult. Achieving high selectivity, sensitivity, and environmental stability is challenging [72, 73].

A primary obstacle in developing reliable ML models for stress detection is the lack of large, diverse public datasets. Many datasets include fewer than fifty subjects. Most datasets come from single-study protocols. Models trained on these small datasets often fail to generalize. They perform poorly on new, unseen data. Limited statistical power is another concern. Physiological biomarkers vary widely between individuals. Capturing heterogeneous responses is critical for robust generalization. This limitation is particularly relevant outside controlled laboratory settings. Dataset scarcity also hampers reproducibility. It reduces the practical utility of stress detection models in real-world applications [74]. The practical success of wearable devices hinges on user comfort and adherence, which directly impact the quality and continuity of data collection. Device design often needs refinement. Reducing the device size can improve comfort. Comfortable devices encourage long-term usage. Data acquisition may suffer from inconsistencies caused by physiological factors. Skin tone can affect sensor performance. Conventional PPG sensors may perform poorly on darker skin tones. Developing technologies that enhance sensor consistency across all skin tones is vital. Optimal placement of devices also improves data quality and comfort [7].

Advanced on-chip AI wearable systems require energy-efficient circuits. These circuits must provide high-performance processing. Extended battery life is also necessary. Conventional computing technologies, like CMOS, are reaching their limits. Miniaturization increases power consumption due to leakage currents. Thermal management becomes more complicated. Ternary logic, using three logic states, is one proposed solution. Graphene Nanoribbon Field-Effect Transistors (GNRFETs) are another. These approaches reduce power dissipation. They also improve computational efficiency and sustainability [75]. Device heterogeneity presents additional challenges. BYOD (Bring Your Own Device) introduces differences in hardware and software. Patients using different wearables may not be comparable. Accelerometric sensor technology is improving. Consumer-grade wearable measurements are becoming more precise. However, variations in device accuracy and capability remain significant [43].

Gaining user adoption is another challenge, especially among older adults. Many older adults are reluctant to use wearables. Anxiety and perceived discomfort contribute to low adoption. Lack of motivation and technical knowledge is another factor. High costs also limit acceptance. Alternative monitoring strategies are needed.

These strategies should measure key metrics, like activity levels, without relying solely on wearables [76]. System on Chip (SoC) platforms in wearable devices face severe computational limitations. Processing power, memory, and energy supply are constrained. Running classical ML algorithms on these platforms is often impractical. Execution times or latencies may be prohibitively long. Safety-critical applications, like fall detection, require real-time responses. For example, RF may take over 2,200 seconds for a single prediction on a medium dataset. Compressed ML or DL algorithms optimized for resource-constrained execution (TinyML) are necessary [77].

Traditional cloud-based monitoring systems introduce latency. Performance may degrade due to connectivity failures. This is a major barrier for essential healthcare applications. Immediate decision-making is often required. Life-threatening scenarios, like stroke or cardiac arrest, demand quick responses. Fall detection typically requires response delays under one second. Decentralized processing on edge nodes addresses this issue. Efficient AI models on the edge enable instantaneous responses. They allow autonomous decision-making without constant cloud reliance [78]. Privacy and security are critical in automated worker safety systems. Continuous collection of sensitive data raises ethical concerns. Measures such as Bluetooth limitations can reduce privacy risks. For instance, monitoring can automatically disable when workers leave the site. High adoption depends on addressing these privacy concerns. Gaining sufficient worker compliance is essential [38].

Algorithmic bias is another challenge. Predictive models rely on representative training data. Medical conditions are inherently variable and complex. Consumer-grade fitness tracker data may introduce bias. User adherence, device usage, and demographics affect predictions. Physiological features may vary depending on the condition. AF may cause irregular, high HRs. CAD may cause lower HRs and reduced activity due to sedentary behavior [9]. The transition from polysomnography (PSG) to AI-powered wearable sleep devices requires rigorous validation. Devices like the Belun Ring must be compared with in-lab PSG. FDA clearance as a home sleep testing device requires strict evaluation. Translating AI innovations into clinical practice demands ongoing assessment. Continuous investigations ensure optimal clinical use of these systems [79].

CONCLUSION

The core conclusion of this chapter is that advanced AI and DL models like CNNs and Bi-LSTMs, combined with wearable sensors, have advanced healthcare. These sensors range from consumer PPG and IMU trackers for CVD risk to specialized systems like EES for hypertension and BioZ sensors for smoking detection. This fusion has shifted healthcare toward continuous monitoring and personalized prevention. The utility of these systems is shown by high accuracy in tasks such as cuffless BP estimation with sensor fusion, mental fatigue detection with EEG, and Human–Machine Interaction (HMI) using triboelectric arrays. These advances are enabled by AI's ability to process continuous and complex data streams. Scaling these systems requires solving challenges like weak generalization from small datasets, variations in sensor performance due to factors such as skin tone, and strict limits of power and computation in wearable SoC devices. The way forward involves edge computing to reduce delays in emergencies, XAI methods like SHAP to build clinical trust, and energy-efficient hardware such as GNRFETs and ternary logic to enable powerful AI on miniaturized devices.

REFERENCES

[1] M. Fatima, A. Sahebkar, P. Kesharwani, From data to diagnosis: How AI-enabled wearable sensors are leading the charge, *Microchem. J.* 205 (2024) 111397. https://doi.org/10.1016/j.microc.2024.111397

[2] D. Padmavilochanan, R.K. Pathinarupothi, K.A.U. Menon, H. Kumar, R. Guntha, M.V. Ramesh, P.V. Rangan, Personalized diabetes monitoring platform leveraging IoMT and AI for non-invasive estimation, *Smart Health* 30 (2023) 100428. https://doi.org/10.1016/j.smhl.2023.100428

[3] G. Singh, Wearable IoT (w-IoT) artificial intelligence (AI) solution for sustainable smart-healthcare, *Int. J. Inf. Manag. Data Insights* 5 (2025) 100291. https://doi.org/10.1016/j.jjimei.2024.100291

[4] A. Hannan, S.M. Cheema, I.M. Pires, Machine learning-based smart wearable system for cardiac arrest monitoring using hybrid computing, *Biomed. Signal Process. Control* 87 (2024) 105519. https://doi.org/10.1016/j.bspc.2023.105519

[5] K. Oh, H. Park, G.H. Lee, J.K. Choi, Supervised contrastive learning-based stress detection for wearable sensor-based healthcare applications, *Future Gener. Comput. Syst.* 175 (2026) 108058. https://doi.org/10.1016/j.future.2025.108058

[6] M. Abd Al-Alim, R. Mubarak, N. M. Salem, I. Sadek, A machine-learning approach for stress detection using wearable sensors in free-living environments, *Comput. Biol. Med.* 179 (2024) 108918. https://doi.org/10.1016/j.compbiomed.2024.108918

[7] C.-T. Hsiao, S. Hong, K.L. Branan, J. McMurray, G.L. Coté, Predicting blood pressure without a cuff using a unique multi-modal wearable device and machine learning algorithm, *Comput. Biol. Med.* 192 (2025) 110357. https://doi.org/10.1016/j.compbiomed.2025.110357

[8] B. Cetintav, A. Yalcin, Decoding canine behaviors through explainable ai and wearable sensor data, *J. Vet. Behav.* 81 (2025) 1–9. https://doi.org/10.1016/j.jveb.2025.07.001

[9] J. Kundrick, A. Naniwadekar, V. Singla, K. Kancharla, A. Bhonsale, A. Voigt, A. Shalaby, N.A.M. Estes, S.K. Jain, S. Saba, Machine learning applied to wearable fitness tracker data and the risk of hospitalizations and cardiovascular events, *Am. J. Prev. Cardiol.* 22 (2025) 101006. https://doi.org/10.1016/j.ajpc.2025.101006

[10] M.E. Hesar, N.S. Seyedsadrkhani, D. Khan, A. Naghashian, M. Piekarski, H. Gall, R. Schermuly, H.A. Ghofrani, S. Ingebrandt, AI-enabled epidermal electronic system to automatically monitor a prognostic parameter for hypertension with a smartphone, *Biosens. Bioelectron.* 241 (2023) 115693. https://doi.org/10.1016/j.bios.2023.115693

[11] C. Yan, S. Jiang, Y. Wang, J. Deng, X. Wang, Z. Chen, T. Chen, H. Huang, H. Wu, A wearable sign language translation device utilizing silicone-hydrogel hybrid triboelectric sensor arrays and machine learning, *Nano Energy* 133 (2025) 110425. https://doi.org/10.1016/j.nanoen.2024.110425

[12] A. Abdigazy, M. Arfan, J. Shao, M.S. Islam, M.F. Hassan, Y. Khan, 3D gas mapping in the gut with AI-enabled ingestible and wearable electronics, *Cell Rep. Phys. Sci.* 5 (2024) 101990. https://doi.org/10.1016/j.xcrp.2024.101990

[13] H. Kim, H. Lee, M. Kim, Y.D. Chung, SleepWatcher: Detecting sleep apnea/hypopnea syndrome from wearable devices using deep learning, *Biomed. Signal Process. Control* 103 (2025) 107460. https://doi.org/10.1016/j.bspc.2024.107460

[14] C. Dong, J.P. Van Dijk, R.M. Aarts, Y. Wang, X. Long, Detection of nocturnal epileptic seizures using a wearable armband: A deep learning approach combining accelerometry and photoplethysmography signals, *Comput. Methods Programs Biomed.* (2025) 109087. https://doi.org/10.1016/j.cmpb.2025.109087

[15] I.-M. Chiu, P.-J. Wu, H. Zhang, J.W. Hughes, A.J. Rogers, L. Jalilian, M. Perez, C.-H.R. Lin, C.-T. Lee, J. Zou, D. Ouyang, Serum potassium monitoring using AI-enabled smartwatch electrocardiograms, *JACC Clin. Electrophysiol.* 10 (2024) 2644–2654. https://doi.org/10.1016/j.jacep.2024.07.023

[16] K.S. Prakash, N. Kunju, An optimized electrode configuration for wrist wearable EMG-based hand gesture recognition using machine learning, *Expert Syst. Appl.* 274 (2025) 127040. https://doi.org/10.1016/j.eswa.2025.127040

[17] S. Gupta, P. Kumar, R. Tekchandani, A machine learning-based decision support system for temporal human cognitive state estimation during online education using wearable physiological monitoring devices, *Decis. Anal. J.* 8 (2023) 100280. https://doi.org/10.1016/j.dajour.2023.100280

[18] A. Mondal, D. Dutta, S. Sen, N. Chanda, S. Mandal, AI-IoT assisted wearable bio-impedance sensor for classification of smoking habits on Fagerstrom scale, *Measurement* 256 (2025) 118171. https://doi.org/10.1016/j.measurement.2025.118171

[19] H. Chenani, Z. Razaghi, M. Saeidi, A.H. Aghaii, M.A. Rastkhiz, M. Orouji, A. Ekrami, A. Simchi, A stretchable, adhesive, and wearable hydrogel-based patches based on a bilayer PVA composite for online monitoring of sweat by artificial intelligence-assisted smartphones, *Talanta* 287 (2025) 127640. https://doi.org/10.1016/j.talanta.2025.127640

[20] X. Zhang, Y. Hu, H. Li, J. Chen, C. Lv, X. Yang, F. Liu, X. Chen, H. Dong, Artificial intelligence-assisted wearable porous eutectogel with high-performance NH3 enrichment and visual sensing enables non-invasive monitoring of chronic kidney disease, *Chem. Eng. J.* 507 (2025) 160678. https://doi.org/10.1016/j.cej.2025.160678

[21] S. Cho, H.J. Nam, C. Shi, C.Y. Kim, S.-H. Byun, K.-C. Agno, B.C. Lee, J. Xiao, J.Y. Sim, J.-W. Jeong, Wireless, AI-enabled wearable thermal comfort sensor for energy-efficient, human-in-the-loop control of indoor temperature, *Biosens. Bioelectron.* 223 (2023) 115018. https://doi.org/10.1016/j.bios.2022.115018

[22] Z. Chu, L. Fu, D. Huang, Y. Song, J. Liu, X. Fan, J. Wang, W. Zhang, Rigid-flexible synergy dual-gradient microwrinkled strain sensor for machine learning-assisted wearable technology, *Chem. Eng. J.* 515 (2025) 163818. https://doi.org/10.1016/j.cej.2025.163818

[23] K.-R. Lee, T. Kim, S. Im, Y.J. Lee, S. Jeong, H. Shin, H. Cho, S.-H. Park, M. Kim, J.G. Lee, D. Kim, G.-S. Choi, D. Kang, S. Seo, S.H. Lee, A wearable stethoscope for accurate real-time lung sound monitoring and automatic wheezing detection based on an AI algorithm, *Engineering* (2025) S2095809925000086. https://doi.org/10.1016/j.eng.2024.12.031

[24] L. Butler, A. Ivanov, T. Celik, I. Karabayir, L. Chinthala, M.M. Hudson, K.K. Ness, D.A. Mulrooney, S.B. Dixon, M.S. Tootooni, A.J. Doerr, B.C. Jaeger, R.L. Davis, D.D. McManus, D. Herrington, O. Akbilgic, Feasibility of remote monitoring for fatal coronary heart disease using Apple Watch ECGs, *Cardiovasc. Digit. Health J.* 5 (2024) 115–121. https://doi.org/10.1016/j.cvdhj.2024.03.007

[25] A. Ahmed, S. Aziz, U. Qidwai, A. Abd-Alrazaq, J. Sheikh, Performance of artificial intelligence models in estimating blood glucose level among diabetic patients using non-invasive wearable device data, *Comput. Methods Programs Biomed. Update* 3 (2023) 100094. https://doi.org/10.1016/j.cmpbup.2023.100094

[26] Y.-P. Chao, H.-H. Chuang, Z.-H. Lee, S.-Y. Huang, W.-T. Zhan, L.-Y. Shyu, Y.-L. Lo, G.-S. Lee, H.-Y. Li, L.-A. Lee, Distinguishing severe sleep apnea from habitual snoring using a neck-wearable piezoelectric sensor and deep learning: A pilot study, *Comput. Biol. Med.* 190 (2025) 110070. https://doi.org/10.1016/j.compbiomed.2025.110070

[27] H. Ghayvat, M. Awais, R. Geddam, M.T. Quasim, S.A. Khowaja, K. Dev, AiCareGaitRehabilitation: Multi-modalities sensor data fusion for AI-IoT enabled realtime electrical stimulation device for pre-Fog and post-Fog to person with Parkinson's Disease, *Inf. Fusion* 122 (2025) 103155. https://doi.org/10.1016/j.inffus.2025.103155
[28] N. Razfar, R. Kashef, F. Mohammadi, An Artificial Intelligence model for smart post-stroke assessment using wearable sensors, *Decis. Anal. J.* 7 (2023) 100218. https://doi.org/10.1016/j.dajour.2023.100218
[29] M.S. Riaz, M. Shaukat, T. Saeed, A. Ijaz, H.N. Qureshi, I. Posokhova, I. Sadiq, A. Rizwan, A. Imran, iPREDICT: AI enabled proactive pandemic prediction using biosensing wearable devices, *Inform. Med. Unlocked* 46 (2024) 101478. https://doi.org/10.1016/j.imu.2024.101478
[30] A. Hadid, E.G. McDonald, Q. Ding, C. Phillipp, A. Trottier, P.C. Dixon, O. Jlassi, M.P. Cheng, J. Papenburg, M. Libman, D. Jensen, Development of machine learning prediction models for systemic inflammatory response following controlled exposure to a live attenuated influenza vaccine in healthy adults using multimodal wearable biosensors in Canada: A single-centre, prospective controlled trial, *Lancet Digit. Health* 7 (2025) 100886. https://doi.org/10.1016/j.landig.2025.100886
[31] A. Mandal, M. Paradkar, P. Panindre, S. Kumar, AI-based detection of stress using heart rate data obtained from wearable devices, *Procedia Comput. Sci.* 230 (2023) 749–759. https://doi.org/10.1016/j.procs.2023.12.129
[32] I. Mehmood, H. Li, Y. Qarout, W. Umer, S. Anwer, H. Wu, M. Hussain, M. Fordjour Antwi-Afari, Deep learning-based construction equipment operators' mental fatigue classification using wearable EEG sensor data, *Adv. Eng. Inform.* 56 (2023) 101978. https://doi.org/10.1016/j.aei.2023.101978
[33] J.J. Liu, B. Borsari, Y. Li, S.X. Liu, Y. Gao, X. Xin, S. Lou, M. Jensen, D. Garrido-Martín, T.L. Verplaetse, G. Ash, J. Zhang, M.J. Girgenti, W. Roberts, M. Gerstein, Digital phenotyping from wearables using AI characterizes psychiatric disorders and identifies genetic associations, *Cell* 188 (2025) 515–529. https://doi.org/10.1016/j.cell.2024.11.012
[34] C.-T. Wu, M.H. Hsieh, I.-M. Chen, L.-Y. Jhao, D.-S. Liu, S.-M. Wang, C.-T. Wu, Y.-L. Chien, Using wearable device and machine learning to predict mood symptoms in bipolar disorder: Development and usability study, *JMIR Med. Inform.* 13 (2025) e66277. https://doi.org/10.2196/66277
[35] N. Tran Thi Hong, G.L. Nguyen, N. Quang Huy, D. Viet Manh, D.-N. Tran, D.-T. Tran, A low-cost real-time IoT human activity recognition system based on wearable sensor and the supervised learning algorithms, *Measurement* 218 (2023) 113231. https://doi.org/10.1016/j.measurement.2023.113231
[36] G. Ciortuz, H. Hozhabr Pour, M.T. Irshad, M.A. Nisar, X. Huang, S. Fudickar, Machine learning models for wearable-based human activity recognition: A comparative study, *Neurocomputing* 650 (2025) 130911. https://doi.org/10.1016/j.neucom.2025.130911
[37] C. Gangadhar, P.P. Roy, R. Dinesh Kumar, J.V. Naga Ramesh, S. Ravikanth, N. Akhila, Wearable sensor-based fall detection for elderly care using ensemble machine learning techniques, *Meas. Sens.* 39 (2025) 101870. https://doi.org/10.1016/j.measen.2025.101870
[38] M.-Y. Cheng, D.V.N. Soegiono, A.F.K. Khitam, Automated fall risk classification for construction workers using wearable devices, BIM, and optimized hybrid deep learning, *Autom. Constr.* 172 (2025) 106072. https://doi.org/10.1016/j.autcon.2025.106072
[39] P. Kulurkar, C.K. Dixit, V.C. Bharathi, A. Monikavishnuvarthini, A. Dhakne, P. Preethi, AI based elderly fall prediction system using wearable sensors: A smart home-care technology with IOT, *Meas. Sens.* 25 (2023) 100614. https://doi.org/10.1016/j.measen.2022.100614

[40] J. Wu, J. Huang, X. Wu, H. Dai, A novel graph-based hybrid deep learning of cumulative GRU and deeper GCN for recognition of abnormal gait patterns using wearable sensors, *Expert Syst. Appl.* 233 (2023) 120968. https://doi.org/10.1016/j.eswa.2023.120968
[41] J. Zhang, Q. Li, J. Li, Y. Zhang, Y. Shen, L. Zeng, G. Sun, C. Xiao, High-sensitivity flexible triboelectric nanogenerator sensor based on recycled PA66 for the monitoring of soccer player lower limb training, *Mater. Res. Bull.* 194 (2026) 113774. https://doi.org/10.1016/j.materresbull.2025.113774
[42] L. Zhao, S. Jia, C. Fang, B. Qin, Y. Hu, X. Yang, Machine learning-assisted wearable triboelectric-electromagnetic sensor for monitoring human motion feature, *Chem. Eng. J.* 503 (2025) 158637. https://doi.org/10.1016/j.cej.2024.158637
[43] B.J. Braun, T. Histing, M.M. Menger, S.C. Herath, G.A. Mueller-Franzes, B. Grimm, M.T. Marmor, D. Truhn, Wearable activity data can predict functional recovery after musculoskeletal injury: Feasibility of a machine learning approach, *Injury* 55 (2024) 111254. https://doi.org/10.1016/j.injury.2023.111254
[44] T. Chen, D. Xu, M. Wang, Z. Zhou, T. Jie, H. Zhou, Y. Yuan, J.S. Baker, Z. Gao, Y. Gu, Wearable monitoring for rehabilitation: Deep learning-driven vertical ground reaction force estimation for anterior cruciate ligament reconstruction, *Clin. Biomech.* 130 (2025) 106663. https://doi.org/10.1016/j.clinbiomech.2025.106663
[45] Z. Xie, Y. Dai, Y. Wen, M. Zhang, M. Tu, F. Sun, Z. An, T. Zhao, B. Liu, Y. Mao, Hydrogel-based flexible degradable triboelectric nanogenerators for human activity recognition, *Sustain. Mater. Technol.* 40 (2024) e00967. https://doi.org/10.1016/j.susmat.2024.e00967
[46] F. Ding, W. Zhan, B. Liu, X. Zhou, S. Li, W. Fu, M. Wang, B. Yu, Z. Hu, A machine learning-integrated multi-channel flexible optical fiber wearable system for high-accuracy human motion monitoring in dynamic environments, *Opt. Laser Technol.* 192 (2025) 113497. https://doi.org/10.1016/j.optlastec.2025.113497
[47] K. Xiong, C. Fan, D. He, H. Chen, X. Zhao, X. Feng, Z. Zhang, X. Wu, A wearable driver gesture recognition system enabled AI application in virtual reality interaction for smart traffic, *Energy* 327 (2025) 136434. https://doi.org/10.1016/j.energy.2025.136434
[48] Y. Liu, S. Wang, X. Qu, Y. Li, C. Zhou, D. Xie, J. Dong, L. Ji, J. Xu, J. Zhou, Highly-sensitive ultra-thin dental patches assisted with artificial-intelligence recognition for mapping hidden periodontitis lesions, *Sens. Actuators B Chem.* 435 (2025) 137648. https://doi.org/10.1016/j.snb.2025.137648
[49] D. Zhang, Z. Xu, Z. Wang, H. Cai, J. Wang, K. Li, Machine-learning-assisted wearable PVA/Acrylic fluorescent layer-based triboelectric sensor for motion, gait and individual recognition, *Chem. Eng. J.* 478 (2023) 147075. https://doi.org/10.1016/j.cej.2023.147075
[50] H. Chen, J. Zhou, X. Long, F. Zhuo, Y. Liu, Y. Zhao, J. Xie, H. Duan, Y. Fu, Fibrous mats based skin sensor with ultra-sensitivity and anti-overshooting dynamic stability enhanced with artificial intelligence, *Chem. Eng. J.* 473 (2023) 145054. https://doi.org/10.1016/j.cej.2023.145054
[51] J. Jeppesen, J. Christensen, P. Johansen, S. Beniczky, Personalized seizure detection using logistic regression machine learning based on wearable ECG-monitoring device., Seizure Eur. *J. Epilepsy* 107 (2023) 155–161. https://doi.org/10.1016/j.seizure.2023.04.012
[52] K. Liu, C. Gao, B. Li, W. Liu, Human activity recognition through deep learning: Leveraging unique and common feature fusion in wearable multi-sensor systems, *Appl. Soft Comput.* 151 (2024) 111146. https://doi.org/10.1016/j.asoc.2023.111146
[53] L. Wu, J. Huang, Y. Chen, T. Wang, J. Chen, X. Chang, Z. Liu, Z. Liu, Y. Zhu, Breathable and highly sensitive self-powered pressure sensors for wearable electronics and human-machine interaction, *Compos. Sci. Technol.* 262 (2025) 111078. https://doi.org/10.1016/j.compscitech.2025.111078

[54] L. Klein, M. Fudim, M. Etemadi, R. Gordon, A. Tibrewala, J. Hernandez-Montfort, P. McCann, L. Zier, K. Shah, A. Rollins, D. Karia, A. Quyyumi, S.R. Motiwala, N. Diakos, J. Rommel, A.P. Ambrosy, V.G. Ganti, P. Soni, K. Larimer, A.M. Carek, O.T. Inan, Noninvasive pulmonary capillary wedge pressure estimation in heart failure patients with the use of wearable sensing and AI, *JACC Heart Fail.* 13 (2025) 102513. https://doi.org/10.1016/j.jchf.2025.102513

[55] M. Chen, L. Yu, Q. Zhao, X. Wang, E.H.K. Yeung, K.L. Tsui, H. Wang, Y. Zhao, A data-knowledge driven approach with wearable sensors for intelligent gait disorders identification, *Biomed. Signal Process. Control* 111 (2026) 108289. https://doi.org/10.1016/j.bspc.2025.108289

[56] M. Zhou, J. Li, P. Reyes, M. Erkoç, G. Wang, M. Edeleva, N. Zhu, M. Deng, L. Cardon, D.R. D'Hooge, Self-powered flexible ultralong electrode sensor made by material-extrusion for artificial intelligence driven accurate motion recognition, *Nano Energy* 135 (2025) 110629. https://doi.org/10.1016/j.nanoen.2024.110629

[57] M. Halimeh, M. Jackson, T. Loddenkemper, C. Meisel, Training size predictably improves machine learning-based epileptic seizure forecasting from wearables, *Neurosci. Inform.* 5 (2025) 100184. https://doi.org/10.1016/j.neuri.2024.100184

[58] N. Promphet, C. Thanawattano, C. Buekban, T. Laochai, P. Lormaneenopparat, W. Sukmas, P. Rattanawaleedirojn, P. Puthongkham, P. Potiyaraj, W. Leewattanakit, N. Rodthongkum, Smartphone based wearable sweat glucose sensing device correlated with machine learning for real-time diabetes screening, *Anal. Chim. Acta* 1312 (2024) 342761. https://doi.org/10.1016/j.aca.2024.342761

[59] P.S. Shakunt, S.K. Udgata, A novel unsupervised machine learning-inspired framework for detection and estimation of faulty wireless sensor data, *Measurement* 257 (2026) 118877. https://doi.org/10.1016/j.measurement.2025.118877

[60] Q. Bao, G. Li, W. Cheng, Z. Yang, Z. Qu, J. Wei, L. Lin, Machine learning-assisted flexible wearable device for tyrosine detection, *RSC Adv.* 13 (2023) 23788–23795. https://doi.org/10.1039/D3RA02900J

[61] R. Abburi, I. Hatai, R. Jaros, R. Martinek, T.A. Babu, S.A. Babu, S. Samanta, Adopting artificial intelligence algorithms for remote fetal heart rate monitoring and classification using wearable fetal phonocardiography, *Appl. Soft Comput.* 165 (2024) 112049. https://doi.org/10.1016/j.asoc.2024.112049

[62] R. Jayasinghe, M. Ramezani, AI-powered prediction of friction and wear in functionalized Epoxy-MWCNT composites, *Wear* 582–583 (2025) 206333. https://doi.org/10.1016/j.wear.2025.206333

[63] R. Liu, W. Shen, Data acquisition of exercise and fitness pressure measurement based on artificial intelligence technology, *SLAS Technol.* 33 (2025) 100328. https://doi.org/10.1016/j.slast.2025.100328

[64] R.-R. Kang, Y. Kim, M. Hong, Y. Min Ahn, K. Lee, AI-based personalized real-time risk prediction for behavioral management in psychiatric wards using multimodal data, *Int. J. Med. Inf.* 198 (2025) 105870. https://doi.org/10.1016/j.ijmedinf.2025.105870

[65] R. Zhang, Y. Hong, H. Zhang, H. Dong, L. Dang, A wearable electromyography armband for gesture recognition, *Biomed. Signal Process. Control* 112 (2026) 108378. https://doi.org/10.1016/j.bspc.2025.108378

[66] S. Dash, D.K. Dash, R.K. Tripathy, R.B. Pachori, Time–frequency domain machine learning for detection of epilepsy using wearable EEG sensor signals recorded during physical activities, *Biomed. Signal Process. Control* 100 (2025) 107041. https://doi.org/10.1016/j.bspc.2024.107041

[67] T. Kim, S. Kim, M. Lee, Y. Kang, S. Hwang, Assessing human emotional experience in pedestrian environments using wearable sensing and machine learning with anomaly detection, *Transp. Res. Part F Traffic Psychol. Behav.* 109 (2025) 540–555. https://doi.org/10.1016/j.trf.2024.12.031

[68] W. He, F. Wan, Y. Liu, G. Wu, P. Zhang, Y. Xiong, X. Zhang, T. Hang, W. Chen, K. Chen, B. Xue, R. Li, G. Hu, Z. Li, Y. Wu, J. Zhu, T. Xiang, J. Zheng, Y. Zhang, Deep learning-powered composite foam smart glove with cross-dimensional conductive networks for sign language recognition, *Compos. Part B Eng.* 308 (2026) 112985. https://doi.org/10.1016/j.compositesb.2025.112985
[69] W. Ling, Energy optimization of a wearable lower limb rehabilitation robot based on deep learning, *Sustain. Energy Technol. Assess.* 56 (2023) 103123. https://doi.org/10.1016/j.seta.2023.103123
[70] Y. Dong, W. An, Y. Zhang, Z. Kang, B. Gao, J. Lv, Y. Jiang, C. Niu, Y. Mao, D. Zhang, An artificial intelligence-assisted, kilometer-scale wireless and wearable biochemical sensing platform for monitoring of key biomarkers in urine, *Biosens. Bioelectron.* 288 (2025) 117844. https://doi.org/10.1016/j.bios.2025.117844
[71] Z. Zhou, X. He, J. Xiao, J. Pan, M. Li, T. Xu, X. Zhang, Machine learning-powered wearable interface for distinguishable and predictable sweat sensing, *Biosens. Bioelectron.* 265 (2024) 116712. https://doi.org/10.1016/j.bios.2024.116712
[72] Z. Chen, S. Liu, W. Yu, L. Wang, F. Lv, L. Yang, H. Yu, H. Shi, Y. Huang, Hydrogel based flexible wearable sweat sensor for SERS-AI monitoring treatment effect of lung cancer, *Sens. Actuators B Chem.* 427 (2025) 137155. https://doi.org/10.1016/j.snb.2024.137155
[73] Y. Yang, L. Dingqi, S. Guilin, G. Yan, W. Dong, X. Chongcheng, F. Quansheng, Research on the precise design of lung cancer-specific receptors and intelligent optimization strategies for sensing interfaces based on the fusion technology of chemical sensors and generative AI, *Chin. J. Anal. Chem.* (2025) 100599. https://doi.org/10.1016/j.cjac.2025.100599
[74] G. Vos, K. Trinh, Z. Sarnyai, M. Rahimi Azghadi, Ensemble machine learning model trained on a new synthesized dataset generalizes well for stress prediction using wearable devices, *J. Biomed. Inform.* 148 (2023) 104556. https://doi.org/10.1016/j.jbi.2023.104556
[75] J. Li, Z. Tang, H.A. Mahmoud, An efficient ternary multiplier for enhanced on-chip AI wearable systems using graphene nanoribbon field-effect transistors, *AEU - Int. J. Electron. Commun.* 191 (2025) 155658. https://doi.org/10.1016/j.aeue.2024.155658
[76] J. Lee, S. Seo, S. Han, C. Koo, A simplified machine learning model to forecast individual thermal comfort in older adults' residential spaces without relying on wearable devices, *Sustain. Cities Soc.* 119 (2025) 106085. https://doi.org/10.1016/j.scs.2024.106085
[77] P. Nandi, K.R. Anupama, A. Bajaj, S. Shukla, T. Musale, S. Kachadiya, Performance evaluation of Machine Learning algorithms on System on Chips in Wearables for Healthcare Monitoring, *Procedia Comput. Sci.* 218 (2023) 2755–2766. https://doi.org/10.1016/j.procs.2023.01.247
[78] A. Alshuhail, A. Alshahrani, H. Mahgoub, M. Ghaleb, A.A. Darem, N.O. Aljehane, M. Alotaibi, F. Alzahrani, Machine edge-aware IoT framework for real-time health monitoring: Sensor fusion and AI-driven emergency response in decentralized networks, *Alex. Eng. J.* 129 (2025) 1349–1361. https://doi.org/10.1016/j.aej.2025.08.030
[79] Z. Strumpf, W. Gu, C.-W. Tsai, P.-L. Chen, E. Yeh, L. Leung, C. Cheung, I.-C. Wu, K.P. Strohl, T. Tsai, R.J. Folz, A.A. Chiang, Belun Ring (Belun Sleep System BLS-100): Deep learning-facilitated wearable enables obstructive sleep apnea detection, apnea severity categorization, and sleep stage classification in patients suspected of obstructive sleep apnea, Sleep *Health* 9 (2023) 430–440. https://doi.org/10.1016/j.sleh.2023.05.001

3 From Eco-Design to Intelligent Functionality
A Chapter on Modern Smart Wearables

Kuriyedath Parambil Khadeeja Thanha, Palasseri Ayisha Sana, Vimala Mathew, and Kannissery Pramod

INTRODUCTION

Modern society faces a serious environmental problem caused by technological growth. Two major concerns are electronic waste (e-waste) and pollution from the textile industry. The textile sector is the world's second most polluting industry, producing nearly 92 million tons of waste each year. Around 85% of textiles go to landfills, although 95% can be recycled. At the same time, e-waste contains toxic materials like heavy metals and halogenated compounds, which harm both people and the environment. When electronics are added to textiles, recycling becomes harder, since separating them is difficult. This shows the urgent need for eco-friendly design and sustainable materials [1]. Artificial intelligence (AI) combined with wearable sensors is changing healthcare. Instead of single clinical check-ups, these devices give real-time health data. They are widely used in healthcare, sports training, and rehabilitation. Wearables track body signals continuously, helping to detect diseases early, predict risks, and give personal health advice. For better use, these devices must be comfortable, light, and self-powered. This is especially important for rehabilitation and smart healthcare systems [2, 3]. Power supply of devices is another big challenge for Internet of Things (IoT). Triboelectric nanogenerators (TENGs) offer a green solution to replace normal batteries. TENGs can collect energy from vibrations, movements, and daily activities, then turn it into electricity. Unlike normal batteries, they do not cause pollution or need frequent replacement. If TENGs are made from recycled or natural materials, they become low-cost, light, and sustainable. They can then provide power to IoT devices without harming the environment [4].

Researchers are now focusing on natural polymers for wearable devices. Cellulose, the most common biopolymer, is cheap, biodegradable, and rich in hydrophilic groups. It works well for humidity sensors and comes from renewable sources like wood. Cellulosic paper is breathable, safe for the body, and already used in wound

DOI: 10.1201/9781003507208-3

dressings. Using such natural materials is key to building flexible and eco-friendly sensors [5, 6]. To turn simple biopolymers into functional devices, scientists often engineer them into hydrogels. Conductive hydrogels are highly useful for flexible electronics because they are strong, safe, and adjustable. For example, combining polyvinyl alcohol (PVA) with zwitterionic polymers, cross-linked by phytic acid, creates self-adhesive ionic hydrogels. These hydrogels have high conductivity, flexibility, and durability. They are ideal as soft electrodes for tracking body signals such as the electrocardiograms (ECG) [7–9].

Alongside TENGs, piezoelectric nanogenerators (PENGs) are another option for energy harvesting. PENGs can convert mechanical energy into electricity and power small sensors. When subjected to mechanical stress like vertical compression, a significant piezoelectric potential develops between the electrodes due to polarization shifts, generating an output voltage signal. They are suitable for remote areas or wearable self-powered devices. Flexible PENGs made from ceramic composites can be attached to the skin for noninvasive sensing. They can even be used in advanced applications like eye tracking, security systems, or helping disabled people [10, 11].

Green manufacturing methods are essential for future electronics. Digital inkjet printing (DIP) is a precise and sustainable technique for fabricating wearable electronics, especially e-textiles. It enables rapid and accurate deposition of conductive materials exactly where needed, reducing material waste significantly. DIP is more efficient than traditional methods like screen printing and coating, which consume more energy and water and generate large amounts of waste. Its key advantages include direct, mask-free patterning, high resolution, and compatibility with various active materials and substrates. Water-based inks, such as those containing graphene or poly(3,4-ethylenedioxythiophene):poly(styrene sulfonate) (PEDOT:PSS), are often used, and these inks are generally nontoxic and safer for both humans and the environment than solvent-based inks. Centrifugal spinning can produce porous polylactic acid fibers on a large scale, which can then be used for multifunctional sensors. These processes make wearable devices practical while staying environmentally responsible [1, 12].

The aim is to merge sustainable materials and green manufacturing into smart, self-powered systems. For example, an AI-based lower-limb motion detection system can use TENG sensors to monitor muscles for rehabilitation in the metaverse. Another example is a smart mask powered by a TENG textile, with an NH_3-sensing unit for toxic gas monitoring. It uses energy from breathing to power itself. These devices show how energy harvesting, sensing, and AI can work together for next-generation healthcare wearables [3, 13]. The move from simple biopolymers to smart fibers is clear in new research. One example is a self-healing thermoelectric aerogel fiber (STDF) made with coaxial wet spinning. It has a core–shell design, with a sensing core and a self-healing protective layer. This fiber can sense temperature, warn of damage, and repair itself. Another example is a pH-responsive hydrogel from starch and sodium alginate. It can track pH changes in sweat, making it useful for personal health checks [14, 15]. The final stage is developing green electronics that fully degrade after use. For example, flexible pressure sensors made from cornstarch and lignocellulose can break down within 24 hours. Wearable electronic skin made from β-glycine and gelatin can disappear in tap water within a week. These new devices solve the e-waste problem by safely returning to nature after their lifetime. This marks a step toward smart, sustainable, and biodegradable electronics [9, 16]. The following sections discuss the development of sustainable materials, fabrication processes, and their integration into intelligent, eco-friendly wearable systems.

ADVANCED ECO-FRIENDLY MATERIALS FOR WEARABLE SENSORS

To realize sustainable smart wearables, eco-friendly and biodegradable materials form the foundation of device design. A multifunctional conductive hydrogel was developed for green and recyclable electronics. It is made by adding starch into a PVA/glycerol (Gly)/LiCl system. Starch with hydroxyl groups forms a double network with PVA. The double network structure refers to the architecture formed within the hydrogel by integrating two distinct polymer networks. This results in a tensile stress of 0.92 MPa and an elongation of 294%. Gly and LiCl provide anti-freezing down to −40°C. The hydrogel also resists drying, keeping 82.6% weight after 30 days. Density functional theory simulations (DFT) studies show that Gly makes strong hydrogen bonds with water. These simulations were utilized to reveal the mechanism of Gly and LiCl in inhibiting the freezing and dehydration of the fabricated hydrogels. Li^+ and Cl^- ions give stable hydration. The hydrogel exhibits a conductivity of 40.65 mS cm^{-1}. It is recyclable and retains 80% conductivity after remolding. It works well in harsh conditions. As a strain sensor, it exhibits a gauge factor (GF) of 1.68 at 0%–100% strain. Response time is 0.2 s. It detects human motion after freezing or drying. Linked to a microcontroller, it controls a robotic arm and makes calls, as shown in Figure 3.1. As a supercapacitor electrolyte, it shows 120.0 mF cm^{-2} capacitance. Energy density is 6.0 μWh cm^{-2}. Power density is 50 μW cm^{-2}. It keeps 98.17% capacitance after 10,000 cycles. These starch-based devices are useful for flexible electronics, energy storage, and soft robotics [7].

Another composite hydrogel fiber based on bovine serum albumin (BSA) was developed to solve the low biocompatibility of synthetic polymers and the weak strength of pure proteins. The design uses both ionic and covalent cross-linking. Ionic bonds form between sodium alginate and Ca^{2+}, giving the material continuous spinnability. Covalent bonds form through a thiol–ene click reaction between BSA and 4-arm PEG-VS, giving high strength. This dual network makes strong fibers by wet spinning. The fibers reach 846 kPa tensile strength and over 700% fracture strain. A solvent exchange with glycerol/water and salts (NaCl, $AlCl_3$, and $CaCl_2$) adds freezing resistance. The fibers remain flexible at −80°C. They can be used in smart textiles as strain and pressure sensors. These sensors support human–machine interaction (HMI). The technology was applied to assistive systems for disabled people. One example is a "position-voice" HMI for deaf-mute users. This shows strong potential for electronic skin and next-generation wearables [17].

A flexible electronic skin (e-skin) was made for green and high-performance wearables. It is built from regenerated chitin (RCH), a common natural polymer, mixed with sulfonated carbon nanotubes (SCNTs). The process is simple, cheap, and eco-friendly. The film is very strong, with a mechanical strength of up to 50 MPa. It also exhibits a high sensing ability. Sensitivity is 1.75 kPa^{-1} within the 0–5 kPa range. Response and recovery times are below 10 ms. The lowest detection limit is 1.39 Pa,

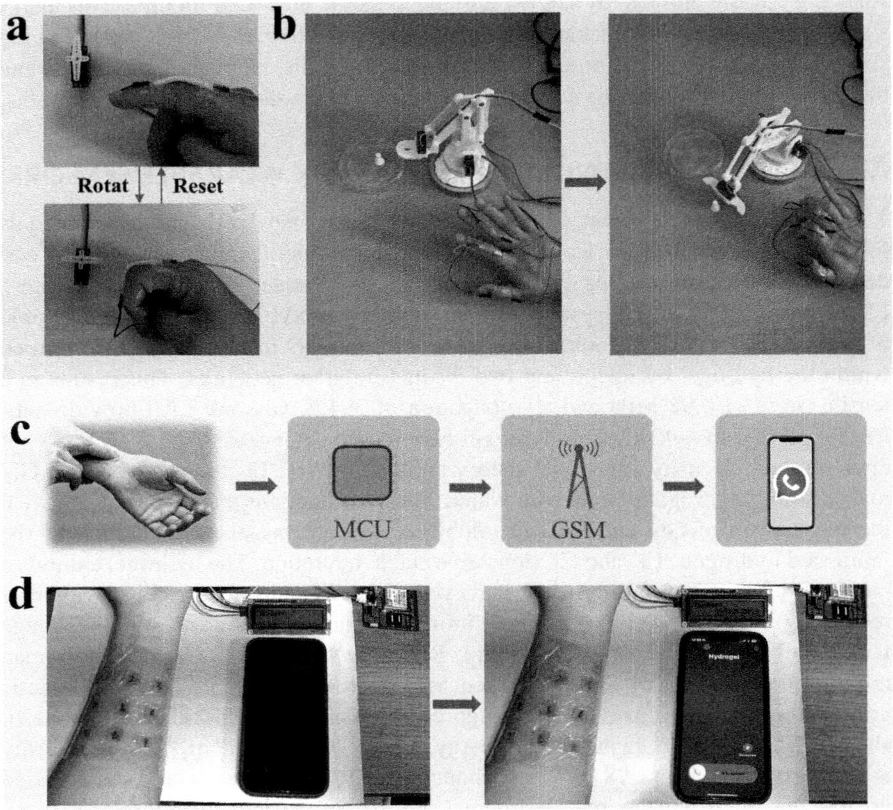

FIGURE 3.1 (a) Finger bending controls the servo rotation. (b) Finger bending directs the robotic arm to grip an object. (c) Concept of dialing through PSG2L2 hydrogel sensors. (d) Dialing a phone number by pressing the PSG2L2 hydrogel sensors.

(Reprinted with permission from [7]. Copyright (2025) Elsevier B.V.)

corresponding to 5 mg. The e-skin is stable for 11,000 loading cycles. It can sense small motions and heavy activity. This makes it useful for wearable devices, health checks, sports tracking, smart sensing, and AI systems [18].

A new multifunctional lignocellulose bio-dressing was made using a green and scalable method. It combines ammonia–oxygen pretreatment with papermaking. Tea polyphenols (TP) were added to the lignocellulose base. Leftover lignin was aminated during the process. This improves dielectric features and triboelectric polarity. The lignocellulose/TP TENG reaches 210.43 mW/m^2. This is 890.72% higher than normal lignocellulose TENGs. The bio-dressing works as a self-powered sensor. It detects pressure, humidity, and material type. Machine learning (ML) helps analyze the signals. It also speeds wound healing in 12 days. Figure 3.2 depicts

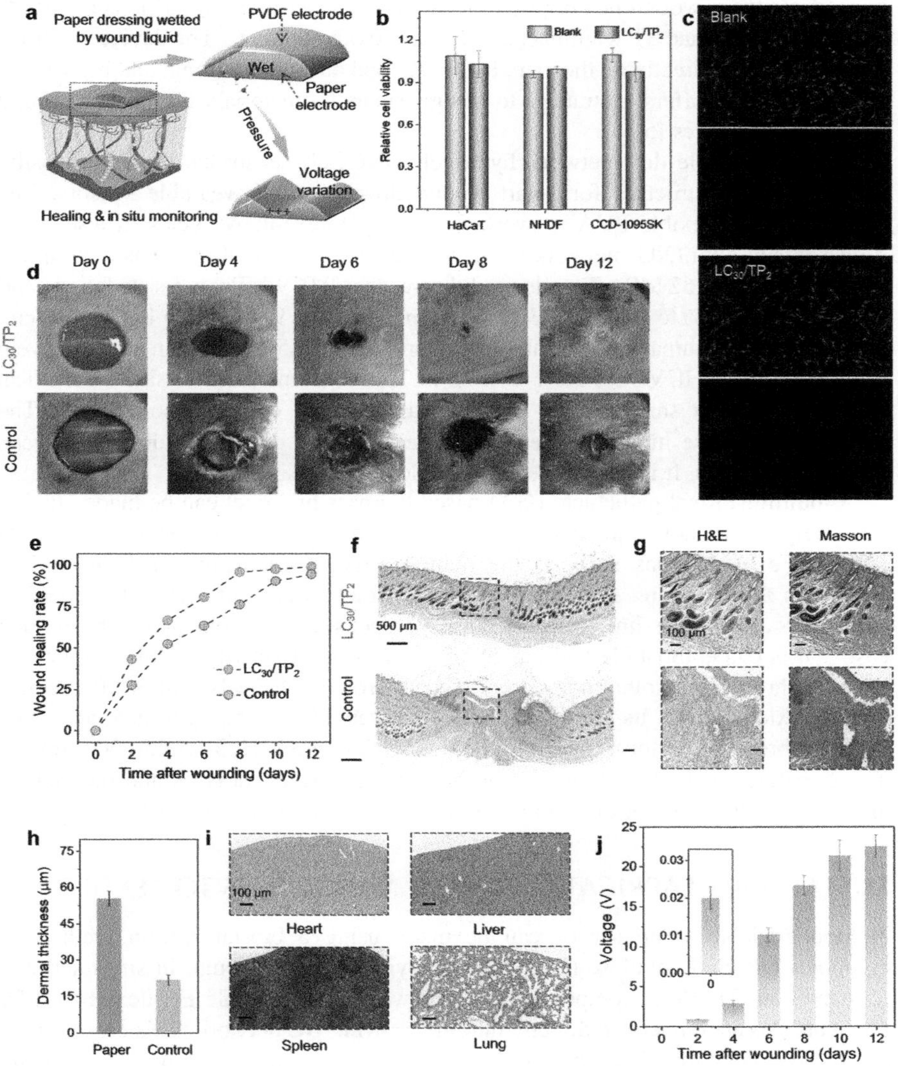

FIGURE 3.2 Lignocellulose paper dressing enhances wound closure. (a) Schematic of dressing for promoting and monitoring healing. (b) Cell viability across groups after incubation. (c) Live/dead fluorescence of cells with LC30/TP2 dressing versus control. (d) Photos showing wound healing in the excisional model. (e) Change in wound size over time. (f) Cross-sectional histology of skin with or without dressing after 12 days. (g) H&E and Masson staining of wounds at day 12. (h) Dermal thickness comparison in treated and untreated groups. (i) Histology of heart, liver, spleen, and lung after 12 days. (j) LC30/TP2 TENG voltage output at different healing stages.

(Reprinted with permission from [5]. Copyright (2025) Elsevier Ltd.)

improvements in wound healing by lignocellulose paper dressing. This is due to its 3D fiber network and TP's antibacterial and antioxidant effects. The device can track wound recovery directly on the skin. It allows good airflow, is strong, and biodegradable. This work offers a strategy to design smart biomaterials for healthcare and wearable electronics [5].

A biocompatible dual-network hydrogel (DNCGel) sensor can be made easily. It has a unique structure for smart wound dressings and wearable sensors. The hydrogel has a flexible PVA network formed by freeze-thaw cycles. It also has a rigid xanthan gum (XG) network chelated with Fe^{3+}. The hydrogel has a compressive modulus of 0.62 MPa. Tensile modulus reaches 0.079 MPa. It resists fatigue and swells from 24.14% to 104.56%. Electrical conductivity is 9×10^{-4} to 1×10^{-3} S/cm. This is similar to human skin. It inhibits bacteria up to 98.56%. Human umbilical vein endothelial cells (HUVECs) grow well on it. This confirms that it is safe for medical use. In vivo tests show 97.8% wound healing under electrical stimulation. The hydrogel can sense human movements in real time. It detects bending of fingers, wrists, and elbows. It is ideal for sensory biological dressings [19].

A multifunctional gallic acid (GA)-based biomass hydrogel can be made quickly and safely. It uses nontoxic materials and a green method. The hydrogel, called PVA/B/KGM/Fe^{3+}/GA3, has PVA as the main matrix. Konjac glucomannan (KGM) reinforces the structure. GA and Fe^{3+} give photothermal antibacterial properties. Borax acts as the cross-linking agent. The hydrogel stretches well and can be molded easily. It is cytocompatible and has 58.08% antioxidant activity. It kills 99.94% of bacteria through photothermal action. The hydrogel self-heals within 60 seconds without external help. Its 3D network allows reversible borate, hydrogen, and ionic bonds to break and reform. It is fully biodegradable in 3% H_2O_2 within 60 minutes. These features make it suitable for strain sensing in smart devices and biomedical applications. It overcomes the limits of single-function hydrogels [20].

SUSTAINABLE FABRICATION: FROM NANOFIBERS TO TEXTILES

A biodegradable strain sensor can be made using a bio-fabrication method. It combines bacterial cellulose (BC) and polypyrrole (PPy) through in situ microbial fermentation. The final composite is coated with biodegradable Ecoflex (EF). PPy nanoparticles are embedded directly into the growing BC network. This creates a 3D conductive network. The structure supports a crack-based sensing mechanism. Sensitivity increases as fibers straighten and cracks form during stretching. The BC/PPy@EF sensor exhibits a GF of 3.21 to 4.86. It can stretch up to 90% and detect as low as 0.05%. The sensor is durable and stable. It keeps its sensitivity after 1,000 cycles at 90% strain. It can track large motions like walking and squatting. It can also sense subtle signals like pulse, breathing, and voice. The sensor can be integrated into textiles. It can be used in smart gloves for human–computer interaction. This method offers a green way to make advanced wearable devices [21].

A bionic STDF can be made using coaxial wet spinning. The design is inspired by the protective and regenerative properties of skin. It has a core–shell structure with a temperature-sensing core and a self-healing sheath. The core uses flexible TPU with rigid $Ti_3C_2T_x$ MXene. This keeps conductive pathways even during repeated bending. The sheath is made from oxidized alginate (OSA), sericin (SS), and tannic acid (TA). It forms reversible bonds, including Schiff-base and hydrogen bonds. This gives the fiber self-healing ability. Water triggers healing with 97.51% stress recovery. Mechanical healing efficiency is 86.96% after five cycles. The fiber has a visual damage warning from the Fe^{2+} and 1,10-phenanthroline reaction. It senses temperatures from 100°C to 500°C. It can respond to fire in under two seconds. This fiber combines autonomous healing, damage detection, and sensitive sensing. It is useful for wearable applications like firefighting clothing [14].

A biodegradable, reusable, and multi-responsive shape memory foam can be made from poly(butylene adipate-co-terephthalate) (PBAT) and polycaprolactone (PCL). It is prepared using one-step supercritical CO_2 foaming and sinter molding. PBAT acts as the permanent phase, and PCL as the reversible phase. Carbon nanotubes (CNTs) form a conductive network. This gives the foam shape memory and electrical conductivity. The foam exhibits a shape fixing ratio of 91.6% and a recovery ratio of 95.8%. These values stay above 85% after 10 cycles, showing reusability. The foam responds to heat, near-infrared light, and microwaves. It can act as a self-sensing actuator and lift 40 times its weight. It works as a contact-separation nanogenerator, producing up to 1.3 V. As a piezoresistive sensor, it exhibits a GF of 25.54 and a response time of 175 ms. It remains stable over 1,000 cycles. Its sensitivity can be tuned by the shape memory effect. With ML, it reaches 100% accuracy in multimodal recognition. This makes it suitable for smart healthcare, HMI, and early fire warning systems [22].

A breathable pressure-sensitive fully textile tribovoltaic nanogenerator can be made using low-cost, scalable screen printing. This smart fabric uses cheap, eco-friendly carbon black (CB) as a new p-type semiconductor. CB forms a Schottky junction when it rubs against a screen-printed silver (Ag) layer and produces direct current (DC). Adding thermoplastic polyurethane (TPU) as an interface layer helps CB spread evenly on the fabric. This boosts the output four times and keeps it stable for over 12 months. When used in everyday clothes, it can detect human movements like bending knees or elbows. It can also monitor body signals such as breathing with high sensitivity (29–37.8 dB). This method offers a practical, scalable way to make next-generation wearable devices. It improves air flow comfort and performance for low-force wearable and medical uses [23].

A sustainable method was developed to make fully inkjet-printed Smart, Wearable, and Eco-friendly Electronic Textiles (SWEET). They can continuously monitor human health. The process uses precise eco-friendly inkjet printing to deposit water-based graphene and PEDOT:PSS inks on biodegradable Tencel fabric. The e-textiles work as sensors for skin temperature and heart rate via ECG. The temperature sensor has a TCR of about $-4.4\%°C^{-1}$. The ECG sensor measures heart rate at around 74 beats per minute. In soil tests, the electrodes lost about 48% of their weight and 98% of their strength in four months. A life cycle assessment shows the graphene electrodes have a climate impact of 0.037 kg CO_2 eq. This is nearly 40 times lower than metal-based electrodes. This method offers a path for eco-friendly health monitoring devices [1].

FUNCTIONALITY AND PERFORMANCE OF SMART WEARABLE DEVICES

The fabricated wearable platforms demonstrate diverse functionalities ranging from energy harvesting and biosensing to environmental monitoring. Eco-friendly compact and low-cost TENGs have been developed for renewable energy harvesting. They use waste and recycled materials like recycled PET sheets. Three types of nopal powder were tested as triboelectric layers. The TENGs work in contact-separation mode, converting environmental vibrations into electricity. The design is small and portable and can be placed on uneven surfaces. The best-performing TENG used dehydrated nopal powder and generated 2.145 mWm^{-2} at 39.97 MΩ, 3g acceleration, and 25 Hz. These devices are durable and stable over 22,500 cycles. They can power commercial devices, such as 116 blue LEDs and a digital calculator. TENGs can also sense motion; a fresh nopal powder device powered a chess clock linked to a phone app. Using waste materials, these nanogenerators can convert ambient kinetic energy into electricity for smart devices [4].

The field of analytical chemistry has advanced with paper-based electrochemical biosensors. They provide pump-free fluid management. They can preconcentrate analytes. They allow task-free sample treatment. They maintain high sensitivity and accuracy. A recent investigation highlights paper as an eco-designed and smart material. It can be used for sample preparation integrated with electrochemical biosensors. Paper has unique properties like porosity, capillarity, and foldability. These properties make devices cost-effective, environmentally friendly, and easy to use. The platforms can perform many sample preparation steps. These include pump-free fluid handling, filtration, separation, centrifugation, chemical treatments like pH adjustment, and preloading reagents. By integrating these preanalytical processes, the devices pave the way for fully integrated autonomous analytical platforms. They are suitable for biomedical, environmental, and agrifood applications [24].

To tackle e-waste, a biodegradable wearable electronic skin (e-skin) with precise tactile sensing has been developed. It is made using a simple, cost-effective solution-casting method. This creates β-glycine structures within a gelatin matrix. The design boosts the piezoelectric properties of the layer, giving a high coefficient of 22.5 pm/V. The self-powered e-skin performs well. It generates 2.1 ± 0.1 V output voltage. It has a dynamic pressure sensitivity of 41.3 ± 1.3 mV/kPa. Its response time is ultralow at 1.0 ± 0.1 ms. It can detect static pressure as low as 0.35 Pa, equal to a single rice grain. Its peak sensitivity in this range is 1.74 ± 0.07 Pa^{-1}. Practical uses include a 3×3 sensor array for 2D object projection. It also enables real-time wireless HMI. The e-skin degrades completely in 7 days in tap water. This makes it a green candidate for next-generation biodegradable flexible electronics for healthcare and human–machine interfaces [9].

A flexible, environmentally friendly, and highly sensitive capacitive humidity sensor has been developed. It is made from a composite of cellulose, MXene, and sodium polyacrylate (PAAS). The sensor is fabricated using vacuum filtration and impregnation. This composite film sensor, called CMPP, has a porous structure formed by vacuum filtration. The porous structure allows loading of sodium polyacrylate and adsorption of water molecules. This increases the dielectric constant. The sensor's high sensitivity in high humidity comes from proton transfer, ionization, and ionic polarization. Sodium polyacrylate enhances these effects by forming multiple water layers and releasing sodium ions. The sensor works over a wide range from 10% RH to 90% RH. It exhibits a low hysteresis of 1.42% and excellent cyclic stability. Its sensitivity is 2.11 pF/%RH in the low humidity range (10%–41% RH). In the high humidity range (41%–84% RH), sensitivity reaches 60.97 pF/%RH. The sensor exhibits a maximum linearity of 0.96. It is suitable for health monitoring. It can detect breathing patterns and rates. It can sense the distance to a humidity source. It can distinguish between nose and mouth breathing. It can detect dry and moist skin. This work provides new insights for making flexible, sensitive, and eco-friendly capacitive humidity sensors for many applications [6].

A highly sensitive, flexible, and intelligent microfluidic glucose biosensor has been developed. It integrates ML-optimized laser-induced graphene (LIG) electrodes with an electrodeposited layer of polyamine saccharide-functionalized glucose oxidase (CS/GluOx) crosslinked with polyethene glycol (PEG). The LIG fabrication parameters were optimized using a supervised ML model. This reduced fabrication time by 60% and improved reproducibility. The biosensor is housed in a flexible sodium alginate hydrogel microfluidic system. It shows excellent electrochemical performance and robustness. The CS/GluOx-PEG layer enhances catalytic efficiency and mechanical stability. It maintains over 90% of its signal after 250 bending cycles at 45° and remains stable for months. Electrochemical tests show a linear glucose detection range from 0.4 to 10 mM. Sensitivity is 18.0 $\mu A\ (mM\ cm^2)^{-1}$ when flat and 17.6 $\mu A\ (mM\ cm^2)^{-1}$ when bent. A key innovation is ML-based signal deconvolution. This allows simultaneous real-time measurement of glucose and lactate in mixed samples. The combination of high sensitivity, flexibility, ML-guided fabrication, and accurate analyte discrimination makes this platform promising for next-generation wearable continuous glucose monitoring systems to improve diabetes management [25].

A dual-network (DN) ion-conductive hydrogel called phytic acid-poly(sulfobetaine methacrylate-co-hydroxyethyl methacrylate) (PHS-PA) has been developed. It combines a PVA network with a zwitterionic poly(sulfobetaine methacrylate (SBMA) -co-2-hydroxyethyl methacrylate(HEMA)) matrix. Phytic acid (PA) acts as a physical cross-linker to improve structural integrity and ion transport. The dual-network design gives excellent mechanical properties and flexibility. The hydrogel maintains stable performance under various frequencies and strains. It also has strong environmental stability, including anti-freezing and dehydration resistance. It works across a temperature range from −20°C to 50°C and even in a dehydrated state. Ionic conductivity is high, ranging from 10.6 to 15.0 S/m. The response time is fast at 310 ms. PHS-PA strongly adheres to various surfaces, including skin. It also has inherent antibacterial activity, useful for prolonged skin contact. These features make it suitable for wearable bioelectronics. It has been used as a strain sensor in a wearable insole for real-time gait monitoring. It has also served as a soft electrode in a three-lead device for reliable ECG signals. The hydrogel performs well even under extreme conditions. This work provides a robust, versatile hydrogel platform for durable and environmentally stable sensors with broad potential in wearable

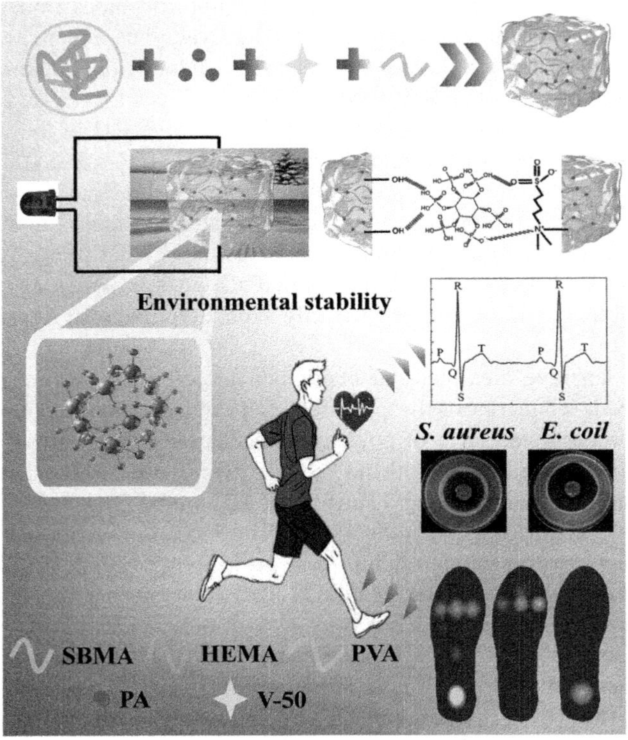

FIGURE 3.3 Design and synthesis of the PHS-PA composite hydrogel, created by embedding a PVA network into poly(SBMA-co-HEMA) with PA as a physical cross-linker. The hydrogel shows stability under different environments, good mechanics, ionic conductivity, and antibacterial activity, suitable for monitoring motion and physiological signals.

(Reprinted with permission from [8]. Copyright (2025) Elsevier Inc.)

electronics and health monitoring [8]. The design, synthesis, stability, mechanics, conductivity, antibacterial function, and monitoring applications of PHS-PA hydrogel are shown in Figure 3.3.

INTELLIGENT SYSTEMS: INTEGRATING AI AND HUMAN–MACHINE INTERFACES

To translate raw sensor data into meaningful insights, AI and ML are integrated into wearable systems for advanced HMI. An AI-enabled self-sustaining lower-limb motion detection system (SS-LMD) has been developed. It meets the need for accurate and portable HMI systems. It is useful for healthcare, virtual fitness, and rehabilitation in the metaverse. The system uses wearable TENG-based sensors to monitor lower-limb muscle movements. These sensors convert human motion into electrical signals to power the device. The self-sustaining feature allows it to work without external power. This improves portability and convenience. A real-time data processing method with a Double Long Short-Term Memory (LSTM) deep learning model is integrated. The system achieves 99.8% accuracy in motion recognition. Usability tests were conducted with users of different ages. It has been applied to fitness and rehabilitation in metaverse scenarios, providing accurate motion feedback. A digital twin-based monitoring platform was developed for real-time observation of user status. This innovation is a major advance in wearable technology. It offers an accurate, low-power, and intelligent monitoring solution for smart healthcare and elderly rehabilitation [3].

A renewable, low-cost, all-textile, biodegradable AI-enabled pressure sensor has been demonstrated. It uses natural plant materials, specifically cornstarch and lignocellulose. The sensor avoids nonbiodegradable components like metallic electrodes and polymer films. Cotton with tunable conductivity serves as both the electrode and the sensitive layer. The device shows high performance. It exhibits a sensitivity of 2.427 kPa^{-1} at 0–2 kPa. Response and recovery times are fast at 20/24 ms. The detection limit is very low at around 10 Pa. It remains stable over more than 6,000 fatigue tests. When combined with AI algorithms and wireless technology, it forms a dual-functional intelligent speech recognition system. It can recognize audible speech by detecting vocal cord vibrations. It can detect silent speech by capturing facial muscle activity. Recognition rates exceed 95%. The Lego-like all-textile design allows it to be cut and reassembled for various wearable uses. Applications include real-time health monitoring and human–computer interaction. The sensor is highly biocompatible. It biodegrades completely in a cellulase solution within 24 hours, turning into sugars without toxic by-products. This work offers a strategy for sustainable high-performance textile sensors. It has great potential for next-generation smart green electronics in medical and HCI applications [16]. Figure 3.4 illustrates the SLCPS design, highlighting both the preparation process and its role in speech recognition.

A fully self-powered smart mask has been developed. It integrates a TENG textile and a $Ti_3C_2T_x$/polyaniline (PANI)-based NH_3-sensing module onto the mask. The TENG textile uses the mechanical energy from breathing to power the sensor. No external power source is needed. This setup allows precise sensing and warning for ammonia (NH_3) in breath and the environment. The NH_3 sensor shows high performance. It exhibits a 550% response at 100 ppm and a low detection limit of 50 ppb. It also exhibits long-term stability and high selectivity for NH_3. This wearable device uses green energy. It has potential for noninvasive disease diagnosis. It can detect and warn about food spoilage. It can alert to high NH_3 levels in the surroundings [13].

A multifunctional biocompatible wearable electronic skin (e-skin) has been fabricated. It uses a TENG made from a flexible chitosan (CTS) film and an edible gold leaf electrode. A plasticizer was added to the CTS film to improve mechanical strength and flexibility. The device is fully biocompatible. Both the CTS layer and the gold electrode can dissolve completely in acetic acid or H_2O_2. The e-skin works

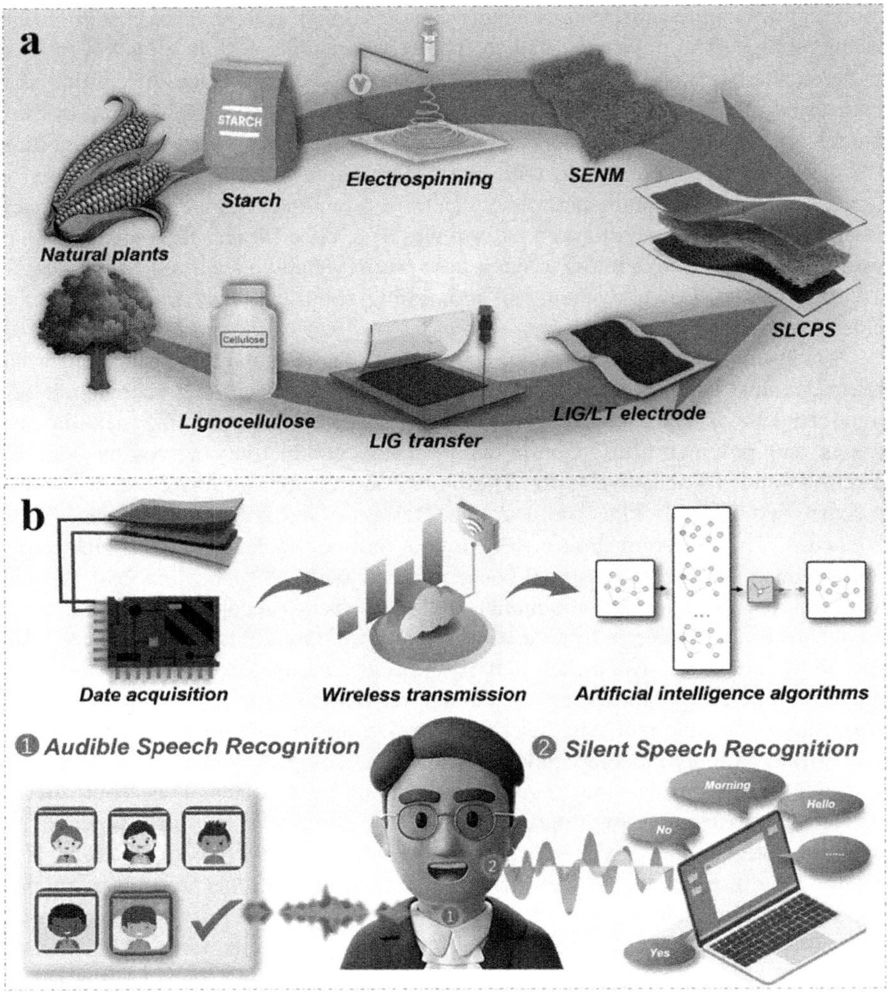

FIGURE 3.4 Design of SLCPS: (a) preparation process and (b) use in an intelligent speech recognition system.

(Reprinted with permission from [16]. Copyright (2025) Elsevier Ltd.)

as a TENG in vertical contact-separation and single-electrode modes. It converts mechanical energy into electrical output. The device has real-time applications. It can serve as a wearable Morse code transmitter for sending emergency signals and short messages by finger tapping. It also functions as a self-powered sleep monitoring system. It can be attached to a bed to track movements of patients and elderly people without discomfort. The TENG-based e-skin's performance and biocompatibility make it promising for healthcare, haptics, and HMI. It could be particularly useful for individuals with disabilities [26].

The integration of advanced materials with AI is demonstrated in a high-performance antibacterial textile pressure sensor for intelligent systems and human–machine interfaces. The sensor is made from a CTS/MXene fiber composite. It shows a sensitivity of 1.16 kPa^{-1} over an ultrawide linear range of 1.5 MPa. It also exhibits excellent mechanical stability and low fatigue after 1,000 cycles. The sensor enables real-time detection of human actions. A deep neural network based on a convolutional neural network (CNN) processes its output. The system achieves 98.61% classification accuracy for six distinct human actions. This combination of a biodegradable sensor and deep learning shows great potential. It can support intelligent interfaces that interpret human movement with high accuracy [27].

In the pursuit of advanced wearable platforms, a multifunctional smart leather collagen fiber (S-LCF) has been developed. It combines high-output signaling, electromagnetic shielding, and temperature regulation for intelligent clothing. The material is made from natural leather collagen fibers (LCF) cross-linked with polyionic liquids and metal-organic framework-layered double hydroxide (MOF-LDH) composites. A PPy conducting network is also added. When used as the positive triboelectric layer of a TENG, the S-LCF harvests mechanical energy from human movement. It delivers a power density of 795 mW m^{-2} and a sensitivity of 0.25 $V{\cdot}kPa^{-1}$. The S-LCF also provides electromagnetic shielding of 58 dB. It offers photothermal performance for thermal management, raising the temperature to 81°C in 600 s. It further shows antibacterial activity with more than 99% inhibition. The textile is strong and resists humid-thermal stress. It can also capture human motion data with high precision. These features make it suitable for human–machine interfaces and AI systems. By combining energy harvesting, shielding, thermal control, and antibacterial functions, the S-LCF creates a path for high-performance flexible devices for smart health monitoring [28].

A safe and eco-friendly humidity sensor has been developed. It is flexible, antibacterial, and degradable. The sensor is designed for health tracking and intelligent systems. It is called the boot-fruited sterculia polysaccharides/silk nanofibers/graphene/silver nanoparticles (BSGA) film. The film is made from sterculia seed polysaccharides (BSP) and silk nanofibers (SNF). Graphene (GN) and silver nanoparticles (AgNPs) are also added. The fabrication uses in situ reduction of silver ions. These ions bind with carboxyl and amino groups in the biopolymers. This creates a cross-linked network. The network improves stability and water resistance. The sensor exhibits a very high sensitivity of 416.4%/% RH. It responds in 0.3 s and recovers in 0.4 s. It detects humidity from 22% to 94%. Its hysteresis is only 0.96%. The film also shows strong antibacterial activity. It is fully biodegradable. The BSGA film is suitable for intelligent devices. It works for noncontact human–computer interaction. It is very effective for respiratory monitoring. This technology supports intelligent healthcare. It also helps sustainable environmental monitoring [29].

A new multifunctional conductive hydrogel has been created. It is designed for flexible wearable sensors in AI, soft robotics, and electronic skin. Conventional hydrogels have poor strength, weak adhesion, and limited antibacterial activity. This hydrogel solves those problems. It is called LGS@PANI-Ag-PVA hydrogel. It is made of lignosulfonate-doped polyaniline (LGS@PANI), silver nitrate, and PVA. Lignosulfonate improves the electrical properties of polyaniline. It also increases hydrophobicity and conductivity. In addition, it gives the hydrogel strong mechanical strength (0.71 MPa), good self-adhesion (81.27 J/m^2), and nearly 100% UV resistance. The hydrogel shows many useful features for human–machine interfaces. It exhibits high sensitivity with a GF of 4.18. Its response time is fast at 100 ms. It resists large tensile strain, up to 397%. It also shows strong antimicrobial activity, with up to 100% inhibition of *Escherichia* coli and *Staphylococcus aureus*. Its ionic conductivity is 0.43 S/m. The sensor can track subtle and complex body movements. These features make it very suitable for wearable devices. It can be used for bio-signal monitoring, human–computer interaction, and AI applications [30].

A new functional sensor has been developed. It uses natural polymers, sodium alginate, and gelatin, to form a sponge electrode. The sensor is designed for flexible wearable electronics and HMI. It can detect multiple external stimuli. These include mechanical forces such as stretching, bending, and compression. It can also sense magnetic stimuli. The sensor shows excellent sensitivity and stability. A major advancement is its ability to separate different signals in real time. It produces opposite electrical responses to magnetic and compression stimuli. This helps the sensor recognize and decouple signals without overlap. It solves a key challenge in multimodal sensing. The device shows high magnetic sensitivity of 0.45 T^{-1}. It also demonstrates strong durability, staying stable over 1000 cycles of mechanical stress. The sensor can be used as electronic skin and in smart sensor arrays. It can monitor human physiological signals. With its multimodal response and signal discrimination, it has strong potential for future intelligent multifunctional systems and AI [31].

A new hydrogel-based triboelectric nanogenerator (GE-TENG) has been developed. It is stretchable, durable, and degradable. The device addresses the power supply and performance limitations of traditional wearable electronics. The GE-TENG is made from a gelatin–NaCl organohydrogel and an EF elastomer. It functions as a self-powered sensor. It can accurately detect human movements and transmit signals wirelessly. A key innovation is its integration into a self-powered intelligent sensing system (GSIS). The GSIS combines the sensor with AI. It enables advanced HMI, such as controlling a tank model or manipulator through gesture monitoring. The system uses covariance matrix feature analysis with CNN algorithms. This combination achieves 100% signal recognition accuracy. The GE-TENG also shows strong electrical performance. It produces an open-circuit voltage of 142 V and a current of 3.8 μA. Its response time is only 85 ms. This intelligent platform demonstrates great potential for wearable hydrogel-based devices. It can harvest biomechanical energy and support interactive human–machine interfaces. It also paves the way for intelligent robots, digital twins, and virtual environments [32]. An overview of various devices, their features, AI/HMI integration, and biocompatibility considerations is provided in Table 3.1.

TABLE 3.1
Summary of Device Description, Key Features, AI/HMI Integration with Applications, and Sustainability/Biocompatibility Aspects

Device Description	Key Features	AI/HMI Integration and Application	Sustainability/Biocompatibility Aspect
Eco-friendly TENG made from nopal (cactus) powder and recycled materials for energy harvesting and smart sensing.	The device uses a contact-separation mechanism and achieves a peak power density of 2.145 mWm^{-2} with dehydrated nopal powder. It demonstrates stable output voltage over 22,500 cycles.	The TENG functions as a self-powered smart motion sensor, demonstrated as a chess clock connected to a mobile app. It can also power small electronics like 116 LEDs and a calculator.	Fabricated from eco-friendly and sustainable materials, including waste nopal powder and recycled PET, supporting a circular economy [4].
A "smart" alginate hydrogels that serve as a matrix for responsive components like enzymes and nanoparticles for various biomedical applications.	The hydrogels are biocompatible, biodegradable, and stable in biological fluids. Their functions can be triggered by endogenous (e.g., pH, biomolecules) or exogenous (e.g., magnetic fields) signals.	Applications include biosensors for glucose and lactate, bioactuators for "smart" drug delivery, and biocomputing systems that use Boolean logic to process multiple signals and control drug release.	Alginate is a natural, biocompatible, and biodegradable biopolymer derived from brown algae. It is highly suitable for biomedical devices [33].
A multifunctional, wearable TENG created from a composite of 1D collagen fibers, 2D MXene nanosheets, and 2D zirconium phosphate nanoplatelets.	The TENG exhibit excellent mechanical strength, flame retardancy, and water vapor permeability due to a polyphenol-mediated interface. It achieves a power density of 234.63 $mW{\cdot}m^{-2}$ and a fast response time (170 ms).	The device serves as a self-powered sensor for human motion monitoring, handwriting recognition, and can function as a distress signaling system (e.g., using Morse code) in emergencies.	The sensor is biodegradable, using natural collagen fibers as the substrate. It can be decomposed in protease or NaOH solutions, offering a green end-of-life disposal method [34].
A biodegradable medical implant made from a Mg–5.5Zn–1.5Si alloy with a plasma electrolytic oxidation (PEO) coating for enhanced performance.	The PEO coating improves corrosion resistance by 99%, increases surface hardness six-fold, and enhances bioactivity by promoting apatite formation.	AI is used in the material design process. A Gaussian Process Regression (GPR) ML model accurately predicts and optimizes the implant's surface topography (e.g., roughness, waviness) to enhance its function.	The Mg–Zn–Si alloy is a biocompatible and biodegradable material that naturally degrades in the body. The PEO coating is also biocompatible and supports cell adhesion [35].

(Continued)

TABLE 3.1 (Continued)
Summary of Device Description, Key Features, AI/HMI Integration with Applications, and Sustainability/Biocompatibility Aspects

Device Description	Key Features	AI/HMI Integration and Application	Sustainability/Biocompatibility Aspect
A Smart Wearable Sweat Patch (SWSP) made from cellulose-based materials for noninvasive, fluorescent multi-sensing of five sweat biomarkers (e.g., glucose, lactate, pH).	The sensor is ultralow-cost (~$0.03 per patch), flexible, and disposable. It uses cotton threads as microfluidic channels and paper discs with fluorescent probes for detection.	A smartphone app with a detection algorithm processes images from a 3D-printed module to quantify biomarkers. An Internet of Things (IoT) model is proposed for remote health monitoring via cloud data transfer.	The patch is made entirely from low-cost, biodegradable, and biocompatible materials such as paper and cotton, making it environmentally friendly [36].
A flexible and stretchable self-powered TENG smart sensor based on a graphene oxide–polyacrylamide (GO–PAM) hydrogel.	The GO–PAM hydrogel provides high stretchability (>768%), self-healing properties, and enhances the TENG's output power to 26 mW, which can power 353 LEDs.	The sensor is integrated into a wearable in-shoe system for gait recognition. An artificial neural network (ANN) model processes the data, achieving 99.5% accuracy for daily gaits and 98.2% for pathological gaits (e.g., Parkinson's), enabling HMI for medical diagnostics.	The hydrogel base offers potential for biodegradability, a desirable feature for flexible and wearable electronics [37].
A multifunctional flexible sensor based on a PVA/silk fibroin/graphene nanocomposite hydrogel with self-healing and ultra-stretchable properties.	The sensor has ultrahigh stretchability (>6000%), rapid self-healing, fast response times (12.5 ms), and a wide sensing range (>400%). It is also functional at low temperatures.	Used for human motion monitoring and HMI, including a smart combination lock and a system to assist in basketball training. Its large strain-sensing capability is suitable for applications like parachute descent monitoring.	The hydrogel is eco-friendly and biodegradable, degrading completely in a bio-enzyme solution in 12 hours. Its natural components, such as silk fibroin, ensure biocompatibility [38].

A rapidly moldable thermogalvanic organohydrogel (TGH) sensor made from a PVA/PEG polymer network and an $Fe^{3+/2+}$ redox pair for self-powered temperature sensing.	Fabricated via a rapid (<6 hours) and green synthesis method, it has a high thermoelectric potential (0.86 mV/K), excellent anti-freezing properties, and good mechanical stability.	Demonstrated as an e-skin device ("smart finger cuffs") that uses human body heat to monitor the temperature and volume of water in a cup. Multichannel sensing provides spatial information on the water level.	The sensor is fabricated using a green and safe method with nontoxic, biocompatible materials (PVA, PEG), avoiding harmful chemical cross-linkers [39].
A high-performance and biodegradable TENG made from glutathione-functionalized cellulose (CF-Gx) composite films.	Glutathione reinforces the cellulose matrix via hydrogen bonding, boosting mechanical and triboelectric performance. The TENG achieves a peak-to-peak voltage of up to 798.8 V and is durable against artificial sweat.	The TENG is used in a gesture-recognition system for wireless computer control (e.g., copy–paste commands) via wrist movements, demonstrating its potential for smart HMI and drone control.	Fabricated from eco-friendly and biodegradable materials (cellulose and glutathione). Glutathione accelerates degradation in soil and enzymatic solutions, and the by-products support healthy plant growth [40].
A scalable, 3D imitation skin-based TENG (PCR-TENG) constructed from polypyrrole-modified cotton yarns and glycerol-plasticized silk fibroin film.	Inspired by human skin structure, the sensor has high stability (8000 compression cycles) and an output voltage of >350 mV from a 2 cm^2 area. The 3D yarn structure allows for editable morphology.	The self-powered sensor identifies body motions and gestures, with applications in tiptoe posture monitoring, somatosensory feedback for prosthetics, and rehabilitation monitoring for hand spasticity.	The device is made from inexpensive, biodegradable, and biocompatible materials like silk fibroin and cotton, making it environmentally friendly and suitable for wearable use [41].
A biodegradable strain sensor composed of bacterial cellulose (BC) and polypyrrole (PPy), fabricated via in situ fermentation and encapsulated with Ecoflex.	Features a 3D conductive network with a crack-based sensing mechanism, offering high sensitivity (GF up to 4.86), a large strain range (90%), a low detection limit (0.05%), and high durability (>10,000 cycles).	The sensor detects a full range of motions from vital signs (pulse, respiration) to large movements. A smart glove integrated with the sensors controls a robot wirelessly, achieving 99.2% gesture-recognition accuracy using a k-nearest neighbor model.	Made from biodegradable materials, including bacterial cellulose and Ecoflex biopolymer. The sensor degrades in natural soil over approximately 90 days [21].

BIODEGRADABLE SMART WEARABLES FOR WOUND-HEALING APPLICATIONS

Among the various biomedical applications, wound healing represents a key area where AI-integrated biodegradable wearables show great potential. The effective management of complex and chronic wounds is a major and growing challenge in modern healthcare. It places a heavy burden on both patients and medical systems worldwide. Wound healing is a dynamic process that progresses through inflammation, proliferation, and remodeling. Each stage carries distinct biochemical and biophysical signatures. Traditional wound dressings act as passive barriers and offer only basic protection. They cannot provide real-time diagnostic feedback or adaptive therapy. This often delays detection of complications such as infection. It also slows healing and requires frequent, painful dressing changes that may damage tissue further. To overcome these limits, research is moving toward intelligent or "smart" wound care. These solutions integrate advanced materials with bioelectronic systems. A major development in this field is the use of AI and ML. These tools can interpret complex biological signals and automate therapeutic responses. When combined with sustainable design (SD) and biodegradable, biocompatible materials, they create a new class of smart wearables. Such systems can improve healing, ensure safety, and reduce environmental impact.

The success of AI in biomaterials depends on robust data and predictive modeling. Conventional empirical methods are limited in handling the complexity of functional biomaterials such as hydrogels. ML, combined with high-throughput experiments and simulations, enables more efficient development. Yet, data scarcity remains a major constraint. Creating and optimizing high-value functional systems is both costly and time-intensive. To address this, researchers are adopting transfer learning. A novel approach called Partial Layer Freezing and Re-initialization (PLFRi) has been developed for small datasets. It predicts the viscosity of hybrid hydrogels, a critical factor for their performance and printability. Using a multilayer perceptron architecture, this model adapts knowledge from a simple precursor system, such as hyaluronic acid/gelatin, to a complex target like hyaluronic acid methacryloyl/gelatin methacryloyl. This method greatly improves prediction accuracy even with limited data. It shows how AI can make biodegradable wound care material development faster, cheaper, and more resource-efficient [42].

The foundation of any smart wearable is the sensor technology that detects physiological changes. In wound care, pH monitoring is crucial because changes indicate healing status and infection risk. Healthy skin is acidic with a pH of 4–6, which prevents bacterial growth. Infected or chronic wounds often become alkaline with a pH of 7–9. Thus, biocompatible and flexible pH sensors are essential for smart wound care. PANI, a conductive polymer, is widely used for pH sensing. It can be integrated into bandages or tattoos. For example, a bandage-based potentiometric sensor was made by screen printing a PANI pH sensor and a polyvinyl butyral (PVB) reference electrode directly on the bandage. This device showed a Nernstian response across the relevant range. It provided a simple, low-cost, and robust method for real-time monitoring without complex equipment. The continuous pH data can then be analyzed by AI and ML models to detect infection early and trigger

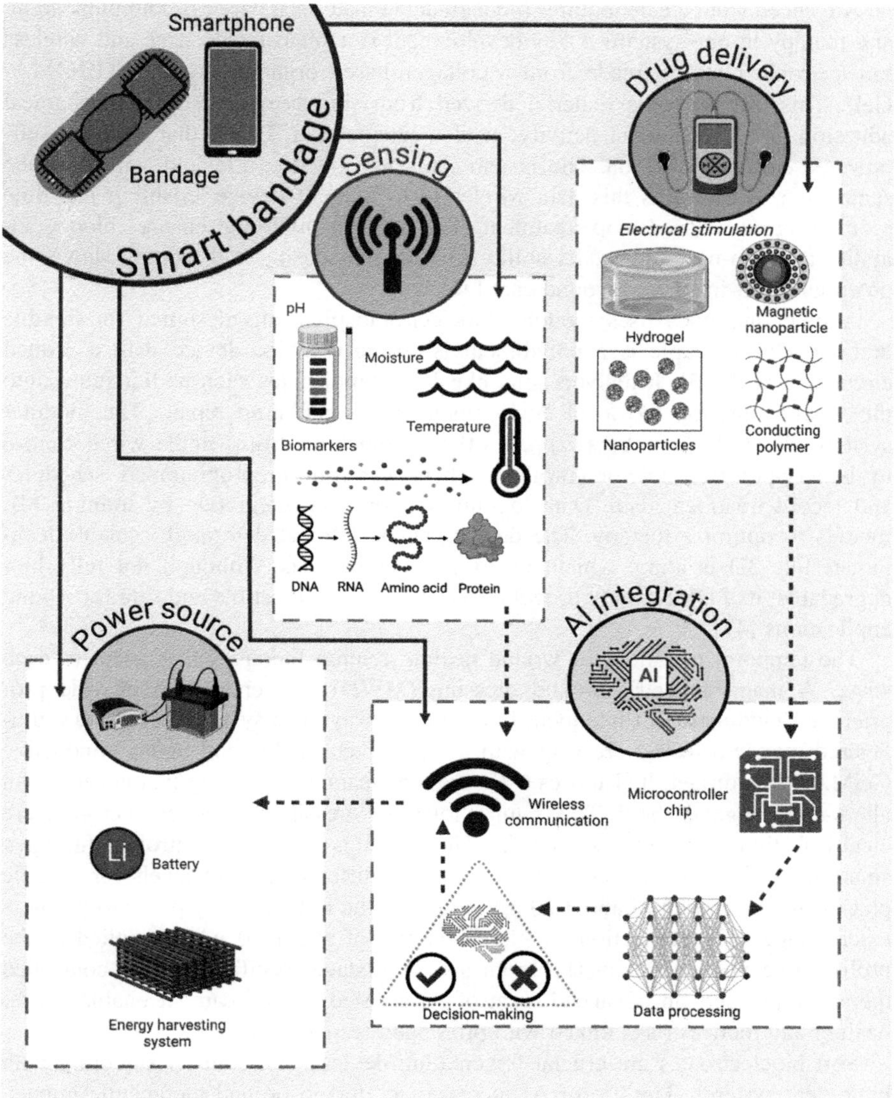

FIGURE 3.5 Schematic of a multifunctional smart bandage integrating sensors, drug delivery systems, power sources, and AI for wireless monitoring and automated therapy.

(Reprinted with permission from [45]. Copyright (2025) Elsevier B.V.)

automatic therapy [43–45]. Figure 3.5 shows a multifunctional smart bandage system with various sensors (pH, temperature, moisture, and biomarkers), drug delivery methods (electrical stimulation, nanoparticles, and hydrogels), a power source (battery and energy harvesting), and AI-enabled processing for wireless communication and automated treatment.

Advanced wound care requires biodegradable materials, wireless communication, and therapy in one system. A key development is a smart battery-free and wireless bioelectronic platform made from a collagen-based organohydrogel (SGC@MA-Gel). This bioresorbable material derived from skin provides strong mechanical adhesion and antibacterial activity. It also integrates a TENG that supplies self-powered electrostimulation. The system can sense glucose, uric acid, and pH at the same time. It transmits this data wirelessly to a smart device, enabling real-time monitoring and closed-loop treatment. The collagen hydrogel ensures biodegradability and eco-friendliness. Its ability to combine sensing and therapy shows the potential of AI-integrated wound care [46].

Another approach uses modular bioelectronic platforms designed for flexible therapy. One example is a polydimethylsiloxane (PDMS) device with a printed circuit board (PCB). It delivers ions like H+ or molecules such as fluoxetine into the wound, altering the local environment and stimulating repair. The modular system allows PCBs to be swapped, offering functions from simple wired control to battery-powered microcontrollers. These can run preprogrammed schedules and record treatment data. Data logging supports AI integration by training ML models to optimize therapy. The device can be fabricated through scalable techniques like 3D printing, which lowers production costs. Although not fully biodegradable, its PDMS and hydrogel materials are biocompatible and safe for wound applications [47].

The temporal dynamics of wound healing require therapies that adapt to each stage. A magnetoelectric wound dressing (MWD) was created to provide programmed modulation of treatment in a wireless way. The system includes a multilayered magnetoelectric dressing with a liquid metal (LM) coil and a conductive GelMA-BP hydrogel. It also uses a 3D-printed wearable accessory that generates an alternating magnetic field. This setup allows the dressing to deliver micro-ampere electrical stimulation at the wound site. The current is wirelessly controlled through a smartphone application. The system can apply different current levels for specific phases. A low current of about 20 μA is used in the inflammatory phase to promote macrophage M2 polarization. A higher current of about 50 μA is applied in the proliferative phase to activate fibroblasts. This stage-specific and user-controlled therapy represents an advanced form of AI-assisted wound care. It enables faster healing and reduces risks linked with prolonged recovery [48].

Soft bioelectronics are crucial for creating devices that connect smoothly with biological systems. The use of AI increases its diagnostic and therapeutic impact. Soft bioelectronics made from conductive polymers and hydrogels can record and control neurophysiological signals with high resolution and low invasiveness. In wound care, these devices must adapt to the fragile wound environment. They also generate large and complex datasets. Manual analysis of this data is difficult, so ML becomes essential. ML models can interpret signals from multimodal sensors, enabling automatic decision-making and closed-loop care. In wounds, this means systems could detect infection early and trigger the release of antimicrobials. This approach personalizes therapy and improves healing. The development of biodegradable implants that dissolve in the body removes the need for extra surgery and increases patient comfort [49].

A photothermal-enhanced thermoelectric nanosheet has been added to a zwitterionic hydrogel to create a multifunctional wound dressing. The hydrogel is made of a poly(sulfobetaine methacrylate)/silk fibroin (PSBMA/SF) network and poly(3,4-ethylenedioxythiophene)-modified polydopamine-functionalized bismuth telluride (PEDOT@PBT) nanosheets. It shows strong photothermal and thermoelectric effects. These features give it broad antibacterial activity against *E. coli*, *S. aureus*, and methicillin-resistant Staphylococcus aureus (MRSA). The thermoelectric materials allow the dressing to produce a voltage when exposed to a temperature gradient. Near-infrared light further boosts this effect. The generated field can reshape the wound's electric field, promoting cell migration and faster healing. Its conductivity (0.13 S/m) also enables bioelectronic sensing. It can monitor ECG, electromyography, and wound motion in real time. Motion monitoring is critical, as it prevents secondary injury and guides regeneration. While no AI is used yet, the complex data produced makes it ideal for future ML integration in wound care [50].

Smart bandages that give real-time wound monitoring mark a key step toward personalized medicine. A flexible wound-healing system has been developed with pro-regeneration features, temperature sensing, and early infection alerts. It has a double-layer Band-Aid shape. The upper layer is a flexible temperature circuit. The lower layer is a collagen–chitosan dermal equivalent that helps skin regrow. Data is sent to a smartphone app via Bluetooth. The app tracks wound temperature in real time. Rising, stable, and falling trends match biological stages like inflammation and angiogenesis. An AI algorithm analyzes this data to give a reliable wound status. It can also warn about abnormal changes linked to infection. This creates a window for timely treatment before infection worsens. It shows how AI can convert raw sensor data into clinical insights [51].

More advanced bandages can monitor multiple biomarkers at once. A smart dressing with an MXene-functionalized porous graphene scaffold tracks uric acid, pH, and temperature. Its 3D–2D hybrid scaffold provides a large surface area and high conductivity. The UA sensor works over a wide range (50–1200 μM) with high sensitivity. The pH sensor gives a stable linear response from 4 to 9. The temperature sensor also shows consistent readings. Together, these parameters give a fuller picture of wound health. High UA signals severe inflammation. Shifts in pH and temperature can confirm infection. This creates a rich dataset that AI can use. ML algorithms can detect hidden patterns and predict healing progress more accurately [51–53].

A battery-free smart bandage goes one step further by detecting bacterial virulence factors. It identifies sortase A from Gram-positive bacteria and pyocyanin from P. aeruginosa. The sensors are built on a Ti3C2Tx MXene electrode array for high sensitivity. Sortase A is detected with a peptide probe tagged with ferrocene. The enzyme cuts the probe, producing a measurable signal change. Pyocyanin is redox-active and can be detected directly. The system uses near field communication (NFC) for wireless power and data transfer. This avoids bulky batteries and improves patient comfort. It can distinguish infections by type, enabling targeted treatment. If linked with AI, this data could guide precise antimicrobial therapy. It is a step toward personalized wound care [54].

The value of smart wearables increases when they are sustainable and biodegradable. A flexible triboelectric patch has been created using a cellulose-based TENG

(C-TENG) combined with a SnS_2 nanocomposite for improved output. This eco-friendly patch serves multiple roles. It works as a self-powered biomechanical sensor for motion tracking and as an electrical stimulator for wound therapy. Its intelligence is supported by a one-dimensional geometric fast data density functional transform deep learning model. The model improves accuracy in predicting motion from sensor data. For wound healing, the C-TENG provides stable electrical stimulation. Tests in vitro and in animal models show it accelerates wound closure by boosting cell proliferation and migration. By combining deep learning, therapeutic stimulation, and biodegradability, this patch offers a holistic system for smart, sustainable wound management [45, 55]. As shown in Figure 3.6, the deep learning-assisted C-TENG, made from environmentally friendly materials, can be used for sustainable biomechanical energy harvesting, healthcare sensing, and wound healing applications.

A smart hydrogel bandage has been developed using biocompatible polymers for multisensor monitoring and therapy. It is built on a PVA substrate with zig-zag

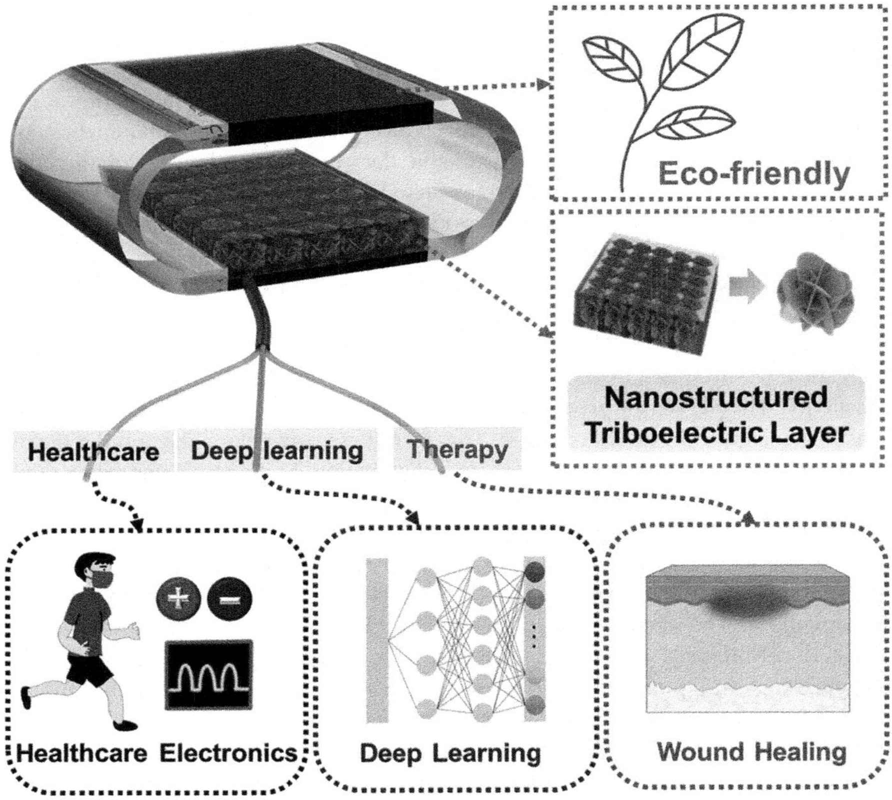

FIGURE 3.6 Deep learning-assisted C-TENG fabricated from eco-friendly materials for sustainable energy harvesting, healthcare sensing, and wound healing.

(Reprinted with permission from [55]. Copyright (2024) Elsevier Ltd.)

elastic hydrogel fibers. Each fiber has a distinct function. Phenol red fibers provide pH sensing. Dual-enzyme fibers with glucose oxidase and horseradish peroxidase allow glucose monitoring. CNT-embedded fibers support photothermal antimicrobial therapy and act as strain sensors to measure wound tension. A smartphone app interprets colorimetric changes from the sensors by converting them to RGB values, giving clear wound information. This integration of chemical and mechanical sensing with therapeutic functions creates a highly versatile bandage. Data from the sensors could be analyzed by AI to build a closed-loop system. Such a system would not only monitor wounds but also trigger treatment in real time [45, 56].

Stage-controlled therapy is crucial for wound healing. A double-layer hydrogel has been designed for sequential drug delivery. The lower layer, made of hyaluronic acid, carries mupirocin. It degrades quickly to release the antibiotic for rapid antibacterial action. The upper gelatin-based hybrid hydrogel contains ibuprofen and MXene nanosheets. MXene acts as a photothermal agent. When exposed to an 808 nm laser, it heats up and triggers the controlled release of ibuprofen. This creates an initial antibacterial burst followed by a sustained anti-inflammatory effect. Although AI is not yet part of this system, the light-triggered release makes it a strong candidate for automation. With sensor data, AI could decide when to activate the laser, enabling precise, stage-specific wound therapy [57].

In conclusion, AI and ML with biodegradable smart wearables are changing wound care. These systems replace passive dressings with active intelligent platforms. They provide personalized, real-time adaptive treatment. Eco-friendly materials like cellulose, chitosan, and hydrogels ensure safety and sustainability. The dressings include sensors for pH, temperature, glucose, and bacterial markers. The data is complex and hard to interpret manually. AI and ML turn it into useful insights. They predict healing, detect infection early, and trigger therapy like drug release or stimulation. Challenges remain in stability, power security, and scale-up. Even so, combining materials science, bioelectronics, and AI is creating closed-loop autonomous wound care. This will improve outcomes, cut costs, and transform the treatment of complex wounds.

SUSTAINABILITY LIFECYCLE: BIODEGRADABILITY AND ECO-DESIGN

The sustainability of smart wearables extends beyond use; it includes their biodegradation, recyclability, and eco-design. A biodegradable TENG has been developed using glutathione-functionalized cellulose (CF-Gx) composite films. It provides a scalable and eco-friendly solution for wearable electronics. Glutathione reinforces the cellulose matrix by forming a stable network through hydrogen bonding via –COOH and $-NH_2$ groups. This improves mechanical strength, charge-trapping, and triboelectric performance. The optimized TENG CF-G2 with 2 wt% glutathione achieved a peak-to-peak voltage of 798.8 V under manual operation. It maintained strong output even after exposure to artificial sweat. Biodegradability was tested via cellulase immersion and soil burial. Glutathione accelerated decomposition. CF-G2 lost 97.6% of its mass after six weeks in soil compared to 89.4% for pure cellulose. The polar groups increase hydrophilicity, which allows better molecular diffusion and enzyme access. This process is eco-friendly as basil plants grow well in soil containing degraded CF-G2 [40].

A novel eco-design approach for wearable e-textiles has been proposed. It addresses challenges in material sustainability, production processes, and end-of-life (EoL) management. The approach focuses on three core constructs: SD, performance measurement (PM), and EoL solutions. These constructs are integrated throughout the product lifecycle. To validate the framework, the study used Partial Least Squares Structural Equation Modeling (PLS-SEM). Data from a survey of experts was analyzed to examine the relationships between the constructs. The results showed a significant positive correlation between them. SD directly influences both PM and EoL solutions. PM also positively affects EoL solutions. A key finding is that PM partially mediates the relationship between SD and EoL solutions, accounting for 18% of the total effect. This framework advances theoretical, managerial, and societal understanding of sustainability in the e-textile industry. It shows how integrating sustainable strategies can drive scientific progress and support global sustainability goals [1].

A cost-effective and environmentally sustainable approach has been developed to make multifunctional wearable electronics. It uses upcycled waste lyocell fabric made from 100% natural wood pulp. The fabric is converted into a flexible, stretchable material through carbonization. This material is suitable for sensing and energy harvesting. The carbonized fabric, inspired by the crack-shaped slit sensilla of a spider, acts as a high-performance strain sensor. It can detect both subtle and large strains. The sensor has a high GF of about 72.17 and a wide working range of up to 600% strain with good linearity. It can also serve as an electrode in a TENG. The TENG shows excellent output performance and can harvest mechanical energy. This approach offers a strategy to make wearable devices from textile waste. It contributes to eco-friendly advanced electronics for health monitoring, soft robotics, and self-powered systems [58].

Paper can act as a functional, eco-designed smart material for sample preparation with electrochemical (bio)sensors. It offers advantages for monitoring in biomedical, environmental, and agri-food fields. Paper overcomes the limits of traditional time-consuming laboratory methods. Its natural capillarity enables pump-free microfluidics. Its porosity allows easy preloading of reagents. This makes reagent-free and user-friendly devices possible. Paper can perform on-site sample treatments. These include physical processes like filtration, separation, and centrifugation. It also allows chemical treatments such as pH control, chemical addition, and precipitation to remove interfering compounds. For example, the devices can monitor food freshness like strawberries. They can also detect analytes in wastewater, aerosols, and soil. The goal is to create fully integrated autonomous analytical platforms. These combine sample collection, treatment, and detection in one cost-effective and eco-friendly device. This approach advances point-of-need diagnostics [24].

CONCLUSION

The convergence of materials science, bioelectronics, and AI addresses e-waste and textile pollution. Smart functional wearables are key to this transformation. They use sustainable biodegradable polymers like cellulose, starch, chitin, and collagen.

Conductive hydrogels are also incorporated for flexibility and performance. These materials ensure comfort and safety during prolonged skin contact. Some devices fully degrade in soil or water within days or weeks. Conventional batteries are being replaced by self-powered systems. Triboelectric and PENGs harvest energy from walking, breathing, or daily movements. This approach improves portability and eliminates polluting power sources. AI and ML processes complex sensor data efficiently. Models detect human activities from large movements to subtle physiological signals. This enables advanced human–machine interfaces for real-time interaction. Intelligent bandages are transforming wound care treatments. Multimodal sensors monitor biomarkers like pH, temperature, glucose, and bacteria. AI detects infection early, predicts healing, and triggers adaptive therapies. Controlled drug release or electrical stimulation personalizes treatment and accelerates recovery. Fully transient high-performance electronics integrate with the body seamlessly. These devices monitor, treat, and return safely to nature. This shift promotes sustainable and intelligent technology.

REFERENCES

[1] M. Dulal, H.R.M. Modha, J. Liu, M.R. Islam, C. Carr, T. Hasan, R.M.S. Thorn, S. Afroj, N. Karim, Sustainable, wearable, and eco-friendly electronic textiles, *Energy Environ. Mater.* 8 (2025) e12854. https://doi.org/10.1002/eem2.12854

[2] M. Fatima, A. Sahebkar, P. Kesharwani, From data to diagnosis: How AI-enabled wearable sensors are leading the charge, *Microchem. J.* 205 (2024) 111397. https://doi.org/10.1016/j.microc.2024.111397

[3] H. Chen, D. He, K. Xiong, X. Zhao, Z. Fang, R. Zou, J. Zhi, Z. Zhang, An AI-enabled self-sustaining sensing lower-limb motion detection system for HMI in the metaverse, *Nano Energy* 136 (2025) 110724. https://doi.org/10.1016/j.nanoen.2025.110724

[4] E. Delgado-Alvarado, J. Martínez-Castillo, E.A. Morales-González, J.A. González-Calderón, E.F. Armendáriz-Alonso, G.M. Rodríguez-Liñán, R. López-Esparza, J. Hernández-Hernández, E.A. Elvira-Hernández, A.L. Herrera-May, Eco-friendly, compact, and cost-efficient triboelectric nanogenerator for renewable energy harvesting and smart motion sensing, *Heliyon* 10 (2024) e28482. https://doi.org/10.1016/j.heliyon.2024.e28482

[5] C. Li, J. Du, L. Zhu, J. Hu, C. Fu, J. Lu, H. Du, H. Wang, D. Lv, Natural lignocellulose fibers-based bio-dressing for accelerated wound healing and machine learning-assisted smart multimodal sensing, *Biomaterials* 325 (2026) 123603. https://doi.org/10.1016/j.biomaterials.2025.123603

[6] C. Huang, J. Chen, Y. Zhang, C. Jiang, Y. Wang, J. Qi, H. Yang, K. Gao, M. Jiang, F. Liu, Wide-detection-range, highly-sensitive, environmental-friendly and flexible cellulose-based capacitive humidity sensor, *Carbohydr. Polym.* 358 (2025) 123507. https://doi.org/10.1016/j.carbpol.2025.123507

[7] J. Ma, J. Zhu, S. Zhou, C. Zhao, C. Liu, Z. Xin, J. Cai, J. He, P. Feng, L. Guo, X. Tao, Multifunctional starch-based conductive hydrogels for smart sensors and flexible supercapacitors, *Int. J. Biol. Macromol.* 302 (2025) 140430. https://doi.org/10.1016/j.ijbiomac.2025.140430

[8] C. Jiang, Y. Zhang, W. Li, X. Li, P. Zhou, H. Zhang, Flexible, self-adhesive and eco-stable bioelectronics with dual-network phytic acid-based ionic hydrogel for biomechanical and physiological signal monitoring, *J. Colloid Interface Sci.* 701 (2026) 138720. https://doi.org/10.1016/j.jcis.2025.138720

[9] S. Maity, R.K. Singh, M. Gadhewal, S.P. Tiwari, Highly biodegradable piezoelectric flexible wearable tactile sensors with amino acid crystals: a paradigm shift towards smart transient electronics, *Chem. Eng. J.* 512 (2025) 162531. https://doi.org/10.1016/j.cej.2025.162531

[10] R. Gowdaman, A. Deepa, Flexible and eco-friendly piezoelectric nanogenerator based on PVDF-HFP: CBT: Ag fibers for self-powered sensing applications, *Results Eng.* 28 (2025) 107184. https://doi.org/10.1016/j.rineng.2025.107184

[11] R. Sasikumar, H. Jung, J. Kim, B. Kim, Real-time eye tracking and eye movement decoding using a flexible and wearable piezoelectric nanogenerator for a smart IoT security system, *Nano Energy* 144 (2025) 111409. https://doi.org/10.1016/j.nanoen.2025.111409

[12] J. Hou, D. Zhang, Q. Ran, G. Yin, M. Wang, W. Yang, H. Wang, Y. Xiong, X. Liu, H. Pan, L. Xia, W. Xu, Multifunctional bending sensor based on PLA hierarchical porosity: smart wearable device with wide range response, self-powered capability and biodegradable, *Chem. Eng. J.* 522 (2025) 166925. https://doi.org/10.1016/j.cej.2025.166925

[13] J. Huang, Y. Huang, T. Li, H. Xu, H. Wu, Z. Su, Respiration-Driven Ammonia Sensing Mask for Multifunctional Self-powered Monitoring Application, *Chem. Eng. J.* 507 (2025) 160598. https://doi.org/10.1016/j.cej.2025.160598

[14] H. He, Q. Jiang, Y. Wan, M.H. Mia, X. Qu, M. Zhou, X. He, X. Li, M. Hong, Z. Yu, S. Huo, Biological skin-inspired damage warning and self-healing thermoelectric aerogel fiber via coaxial wet spinning for wearable temperature sensing, *J. Mater. Sci. Technol.* 250 (2026) 257–271. https://doi.org/10.1016/j.jmst.2025.06.038

[15] J. Lei, K. Chen, H. Qiu, Preparation of pH-responsive starch/sodium alginate hybrid hydrogels for smart monitoring applications, *Sustain. Mater. Technol.* 45 (2025) e01551. https://doi.org/10.1016/j.susmat.2025.e01551

[16] X. Zheng, B. Yi, Q. Zhou, R. Xu, X. Liu, H.-K. Sung, L. Chernogor, M. Cao, Z. Yao, Y. Li, Y. Li, Ecofriendly and high-performance flexible pressure sensor derived from natural plant materials for intelligent audible and silent speech recognition, *Nano Energy* 126 (2024) 109701. https://doi.org/10.1016/j.nanoen.2024.109701

[17] Z. Zhang, L. Wang, T. Wu, J. Jiang, Y. Zhang, Smart textile interfaces based on protein structure-reconstructed bovine serum albumin hydrogel fibers for disability support, *J. Colloid Interface Sci.* 702 (2026) 138864. https://doi.org/10.1016/j.jcis.2025.138864

[18] Z. Yang, X. Yu, Y. Song, Y. Hu, Q. Yang, C. Xiong, Z. Shi, Flexible and ultrasensitive piezoresistive electronic skin based on chitin/sulfonated carbon nanotube films, *Int. J. Biol. Macromol.* 259 (2024) 129103. https://doi.org/10.1016/j.ijbiomac.2023.129103

[19] L. Hu, Y. Wang, Q. Liu, M. Liu, F. Yang, C. Wang, P. Pan, L. Wang, L. Chen, J. Chen, Real-time monitoring flexible hydrogels based on dual physically cross-linked network for promoting wound healing, *Chin. Chem. Lett.* 34 (2023) 108262. https://doi.org/10.1016/j.cclet.2023.108262

[20] W. Zeng, M. Qiu, X. Zeng, X. Shi, R. You, T. Yu, M. Jin, Z. Hong, N. Ma, F.-C. Tsai, A biomass hydrogel with self-healing and photothermal antibacterial activity based on gallic acid for strain sensing, *Chem. Eng. Sci.* 320 (2026) 122435. https://doi.org/10.1016/j.ces.2025.122435

[21] C. Gao, Y. Liu, F. Gu, Z. Chen, Z. Su, H. Du, D. Xu, K. Liu, W. Xu, Biodegradable Ecoflex encapsulated bacterial cellulose/polypyrrole strain sensor detects motion with high sensitivity, flexibility and scalability, *Chem. Eng. J.* 460 (2023) 141769. https://doi.org/10.1016/j.cej.2023.141769

[22] H. Qian, Z. Gong, X. Gao, Y. Chen, L. Zhao, D. Hu, Self-sensing actuators and piezoresistive sensors based on biodegradable and multi-responsive shape memory PBAT/PCL foams for machine learning-enabled multimodal intelligent monitoring, *Compos. Part Appl. Sci. Manuf.* 199 (2025) 109267. https://doi.org/10.1016/j.compositesa.2025.109267

[23] Y. Wu, Y. Wu, C. Menon, Screen-printing tribovoltaic textile with enhanced interface layer for motion tracking and respiration monitoring, *Nano Energy* 141 (2025) 111095. https://doi.org/10.1016/j.nanoen.2025.111095

[24] V. Mazzaracchio, C. Gosti, L. Belcastro, F. Arduini, Paper as an ecodesigned and smart material for sample preparation integrated with electrochemical (bio)sensors, *Adv. Sample Prep.* 16 (2025) 100207. https://doi.org/10.1016/j.sampre.2025.100207

[25] N.F.A. Zaini Ambia, S. Hisham, N. Abu Bakar, S.R. Majid, N. Nordin, Smart microfluidic biosensor with polyamine saccharide-GluOx-modified LIG electrodes and real-time ML signal deconvolution for glucose monitoring, *Biosens. Bioelectron.* 289 (2025) 117914. https://doi.org/10.1016/j.bios.2025.117914

[26] G.M. Rani, S.M. Ghoreishian, R. Umapathi, V. Vivekananthan, Y.S. Huh, A biocompatible triboelectric nanogenerator-based edible electronic skin for morse code transmitters and smart healthcare applications, *Nano Energy* 128 (2024) 109899. https://doi.org/10.1016/j.nanoen.2024.109899

[27] M. Gu, X. Zhou, J. Shen, R. Xie, Y. Su, J. Gao, B. Zhao, J. Li, Y. Duan, Z. Wang, Y. Hu, G. Gu, L. Wang, L. Wei, C. Yang, M. Chen, High-sensitivity, ultrawide linear range, antibacterial textile pressure sensor based on chitosan/MXene hierarchical architecture, *iScience* 27 (2024) 109481. https://doi.org/10.1016/j.isci.2024.109481

[28] P. Zhao, D. Gao, Y. Zhou, B. Lyu, J. Ma, Multifunctional integrated flexible triboelectric nanogenerator based on collagen fibers for smart wearable devices, *Chem. Eng. J.* 522 (2025) 167693. https://doi.org/10.1016/j.cej.2025.167693

[29] T. Lu, X. Han, H. Wang, G. Liu, Antibacterial and degradable humidity sensor based on boat-fruited sterculia seeds polysaccharides/silk nanofibers for respiratory monitoring with ultrahigh sensitivity, *Int. J. Biol. Macromol.* 310 (2025) 143053. https://doi.org/10.1016/j.ijbiomac.2025.143053

[30] X. Liu, Y. Yang, S. Song, R. Zhang, C. Zhang, S. Yang, Y. Liu, Y. Song, Lignosulfonate-doped polyaniline-reinforced poly(vinyl alcohol) hydrogels as highly sensitive, antimicrobial, and UV-resistant multifunctional sensors, *Int. J. Biol. Macromol.* 280 (2024) 135959. https://doi.org/10.1016/j.ijbiomac.2024.135959

[31] Y. Fu, S. Wang, Z. Wan, Y. Tian, D. Wang, Y. Ma, L. Yang, Z. Wei, Functional magnetic alginate/gelatin sponge-based flexible sensor with multi-mode response and discrimination detection properties for human motion monitoring, *Carbohydr. Polym.* 324 (2024) 121520. https://doi.org/10.1016/j.carbpol.2023.121520

[32] Z. Xie, Y. Dai, Y. Wen, M. Zhang, M. Tu, F. Sun, Z. An, T. Zhao, B. Liu, Y. Mao, Hydrogel-based flexible degradable triboelectric nanogenerators for human activity recognition, *Sustain. Mater. Technol.* 40 (2024) e00967. https://doi.org/10.1016/j.susmat.2024.e00967

[33] D.M. Roquero, E. Katz, "Smart" alginate hydrogels in biosensing, bioactuation and biocomputing: State-of-the-art and perspectives, *Sens. Actuators Rep.* 4 (2022) 100095. https://doi.org/10.1016/j.snr.2022.100095

[34] J. Shi, Y. Zeng, C. Liu, L. Wang, J. He, Y. Zhang, Y. Zhang, M. Zhang, Biodegradable, fire-safe, and wearable triboelectric nanogenerator enabled by polyphenol-mediated 1D/2D interface architecture, *Chem. Eng. J.* 519 (2025) 164809. https://doi.org/10.1016/j.cej.2025.164809

[35] M. Khakbiz, A. Esteghza, AI-integrated surface topography and experimental investigation of PEO-coated Mg-Zn-Si alloys for next-generation biodegradable medical implants, *Colloids Surf. Physicochem. Eng. Asp.* 726 (2025) 137800. https://doi.org/10.1016/j.colsurfa.2025.137800

[36] S. Ardalan, M. Hosseinifard, M. Vosough, H. Golmohammadi, Towards smart personalized perspiration analysis: An IoT-integrated cellulose-based microfluidic wearable patch for smartphone fluorimetric multi-sensing of sweat biomarkers, *Biosens. Bioelectron.* 168 (2020) 112450. https://doi.org/10.1016/j.bios.2020.112450

[37] Z. Wang, M. Bu, K. Xiu, J. Sun, N. Hu, L. Zhao, L. Gao, F. Kong, H. Zhu, J. Song, D. Lau, A flexible, stretchable and triboelectric smart sensor based on graphene oxide and polyacrylamide hydrogel for high precision gait recognition in Parkinsonian and hemiplegic patients, *Nano Energy* 104 (2022) 107978. https://doi.org/10.1016/j.nanoen.2022.107978
[38] Y. Zhao, S. Mu, B. Hu, J. Hao, Y. Xia, Y. Zhongling, Z. Han, X. Kong, X. Zhu, Z. Xu, W. Hong, P. Wang, S. Wang, X. Guo, Polyvinyl alcohol/silk fibroin/graphene hydrogel-based flexible sensor with self-healing, ultra-stretchable, and eco-friendly properties for advanced wearable electronics and human-machine interaction applications, *Sens. Actuators B Chem.* 445 (2025) 138566. https://doi.org/10.1016/j.snb.2025.138566
[39] Y. Gao, Y. Yu, A rapidly moldable thermogalvanic organohydrogel for self-powered water temperature and level monitoring, *Sens. Actuators Phys.* 394 (2025) 116876. https://doi.org/10.1016/j.sna.2025.116876
[40] Y.J. Lee, S.J. Oh, J.S. Yun, Glutathione-enhanced biodegradable cellulose films for high-performance wearable triboelectric nanogenerators, *Chem. Eng. J.* 522 (2025) 168091. https://doi.org/10.1016/j.cej.2025.168091
[41] Q. Mi, Y. Dong, D. Ge, S. Xie, Y. Tian, F. Zou, H.-Y. Yu, K.C. Tam, Scalable manufacture of efficient, highly stable, and compact 3D imitation skin-based elastic triboelectric nanogenerator for energy harvesting and self-powered sensing, *Nano Energy* 131 (2024) 110283. https://doi.org/10.1016/j.nanoen.2024.110283
[42] B. Deng, F.L. Lasaosa, D. Chen, Y. He, C. Xuan, Y. Cui, M. Doblaré, Transfer learning-enabled viscosity prediction for HAMA/GelMA hybrid hydrogels, *Mater. Today Commun.* 47 (2025) 113018. https://doi.org/10.1016/j.mtcomm.2025.113018
[43] T. Guinovart, G. Valdés-Ramírez, J.R. Windmiller, F.J. Andrade, J. Wang, Bandage-based wearable potentiometric sensor for monitoring wound pH, *Electroanalysis* 26 (2014) 1345–1353. https://doi.org/10.1002/elan.201300558
[44] Md.A. Alam, A. Debnath, Md.T. Uddin, A. Tamanna, S. Kamruzzaman, H.A. Begum, S.K. Ray, S. Fatima, A.A. Khan, Z. Tang, A.K. Mondal, Flexible high electrochemical collagen/lignin composite hydrogel for sensing and supercapacitor applications, *Int. J. Biol. Macromol.* 281 (2024) 136240. https://doi.org/10.1016/j.ijbiomac.2024.136240
[45] K.P. Khadeeja Thanha, P. Ayisha Sana, J.S. Ajesh, P.P. Naseef, H. Tharayil, P. Lubaib, K. Pramod, Advanced smart bioelectronics for wound healing: biosensing, drug delivery, and artificial intelligence, *Int. J. Pharm.* 684 (2025) 126098. https://doi.org/10.1016/j.ijpharm.2025.126098
[46] Z. Bai, X. Wang, M. Huang, Y. Feng, S. Sun, M. Zheng, X. Zou, L. Xie, X. Wang, D. Hao, O. Yue, Y. Chen, X. Liu, Smart battery-free and wireless bioelectronic platform based on a nature-skin-derived organohydrogel for chronic wound diagnosis, assessment, and accelerated healing, *Nano Energy* 118 (2023) 108989. https://doi.org/10.1016/j.nanoen.2023.108989
[47] P. Baniya, M. Tebyani, N. Asefifeyzabadi, T. Nguyen, C. Hernandez, K. Zhu, H. Li, J. Selberg, H.-C. Hsieh, P. Pansodtee, H. Yang, C. Recendez, G. Keller, W.S. Hee, E. Aslankoohi, R.R. Isseroff, M. Zhao, M. Gomez, M. Rolandi, M. Teodorescu, A system for bioelectronic delivery of treatment directed toward wound healing, *Sci. Rep.* 13 (2023) 14766. https://doi.org/10.1038/s41598-023-41572-w
[48] B. Zhao, J. Sun, Y. Xu, J. Li, K. Xie, C. Zhou, B. Su, L. Chen, Programmed temporal modulation by using a magnetoelectric wound dressing for ultra-fast early-stage wound healing, *Biomaterials* 325 (2026) 123569. https://doi.org/10.1016/j.biomaterials.2025.123569
[49] B. Cao, Y. Huang, L. Chen, W. Jia, D. Li, Y. Jiang, Soft bioelectronics for diagnostic and therapeutic applications in neurological diseases, *Biosens. Bioelectron.* 259 (2024) 116378. https://doi.org/10.1016/j.bios.2024.116378

[50] J. Xiao, W. Lu, Z. Li, S. Zhang, X. Zhu, J. Yuan, D. Gan, J. Shen, M. Wang, A photothermal-enhanced thermoelectric nanosheet incorporated with zwitterionic hydrogels for wound repair and bioelectronics, *Acta Biomater.* 200 (2025) 610–628. https://doi.org/10.1016/j.actbio.2025.05.033

[51] D. Lou, Q. Pang, X. Pei, S. Dong, S. Li, W. Tan, L. Ma, Flexible wound healing system for pro-regeneration, temperature monitoring and infection early warning, *Biosens. Bioelectron.* 162 (2020) 112275. https://doi.org/10.1016/j.bios.2020.112275

[52] D. Kaur, R. Purwar, Nanotechnological advancement in artificial intelligence for wound care, in: *Nanotechnological Asp. -Gener. Wound Manag.* Elsevier, 2024: pp. 281–318. https://doi.org/10.1016/B978-0-323-99165-0.00005-8

[53] Md. Sharifuzzaman, A. Chhetry, M.A. Zahed, S.H. Yoon, C.I. Park, S. Zhang, S. Chandra Barman, S. Sharma, H. Yoon, J.Y. Park, Smart bandage with integrated multifunctional sensors based on MXene-functionalized porous graphene scaffold for chronic wound care management, *Biosens. Bioelectron.* 169 (2020) 112637. https://doi.org/10.1016/j.bios.2020.112637

[54] Z. Shi, C. Dai, P. Deng, X. Li, Y. Wu, J. Lv, C. Xiong, Y. Shuai, F. Zhang, D. Wang, H. Liang, Y. He, Q. Chen, Y. Lu, Q. Liu, Wearable battery-free smart bandage with peptide functionalized biosensors based on MXene for bacterial wound infection detection, *Sens. Actuators B Chem.* 383 (2023) 133598. https://doi.org/10.1016/j.snb.2023.133598

[55] K.-Y. Hsu, S.-M. Huang, B.T. Murti, C.-C. Chen, Y.-C. Chao, I.-C. Ha, C.-C. Tsai, C.-Y. Chen, M.-L. Tsai, P.-K. Yang, A flexible, sustainable, and deep learning-assisted triboelectric patch for self-powered interactive sensing and wound healing applications, *Nano Energy* 134 (2025) 110501. https://doi.org/10.1016/j.nanoen.2024.110501

[56] L. Ji, Y. Xiao, K. Xu, X. Wu, O.W. Ojo, L. Diao, K. Mequanint, W. Zhong, P. Zhao, M. Xing, G. Lyu, Smart bandage with multi-sensor system for wound healing and micro-environment monitoring, *Chem. Eng. J.* 507 (2025) 160509. https://doi.org/10.1016/j.cej.2025.160509

[57] C. Li, J.-T. Wang, K. Liu, H. Ding, Q.-F. Li, G. Liang, L. Jin, D. He, Antibacterial and anti-inflammatory synergistic effects of double-layer hydrogel promoting bacterial wound healing, *Chem. Eng. J.* 493 (2024) 152513. https://doi.org/10.1016/j.cej.2024.152513

[58] W. Pan, F.-S. Jiang, S.-T. Huang, G.-X. Li, S.-L. Huang, Z.-H. Zhu, H. Sun, Y.-Z. Long, G.-F. Yu, Low-cost environmental-friendly strain sensor and triboelectric nanogenerator based on waste lyocell fabric, *Sens. Actuators Phys.* 389 (2025) 116541. https://doi.org/10.1016/j.sna.2025.116541

4 Conducting Yarn-Based Smart Wearables

From Lab to Laundry

Anupam Kumari, Angan Sengupta, and Ajay Agarwal

INTRODUCTION

Yarns woven into the fabric for daily life comfort, protection, and expression have become platforms for a new technology in recent years [1]. Smart textiles can sense, compute, communicate, and sometimes even harvest energy. By embedding intelligence into fabrics, smart textiles can change how humans monitor health, interact with machines, and integrate into the Internet of Clothing [2]. Unlike rigid electronic components, conducting yarns provide structural conductivity within the textile. These yarns can be woven, knitted, braided, stitched, or embroidered into fabric to transmit signals, power embedded systems, and sense pressure, temperature, humidity, or biochemical stimuli [3]. Conducting yarn-based wearables are useful for continuous health monitoring (cardiac, respiratory, and hydration), sports and fitness tracking, industrial safety, and fashion technology with flexibility, comfort, and seamless integration into clothing. Conducting yarns maintain electrical pathways even as fabrics bend, stretch, and drape in daily use, making them suitable for wearable devices that feel and behave like ordinary clothing while delivering remarkable functions [4].

Even though much research and many studies have been published, conducting yarn-based smart wearables are still not widely used commercially [5]. The laboratory prototype is more focused on sensitivity, response time, and selection of novel materials. However, when these prototypes are subjected to real-world conditions, such as washing, sweating, abrasion, repeated stretching, and environmental exposure, their performance often degrades rapidly [6]. This "translation gap" between controlled laboratory success and consumer-ready reliability, as shown in Figure 4.1, is the main obstacle to progress in this area. Most studies have provided comprehensive coverage of materials, fabrication techniques, and sensing applications. However, the focus has been largely on what conducting yarns are made of and what they can sense, rather than how they can be made durable, reproducible, and sustainable for long-term use [7]. The aim of this chapter is to take conducting yarn technologies "from lab to laundry" with the following four scopes.

 DOI: 10.1201/9781003507208-4

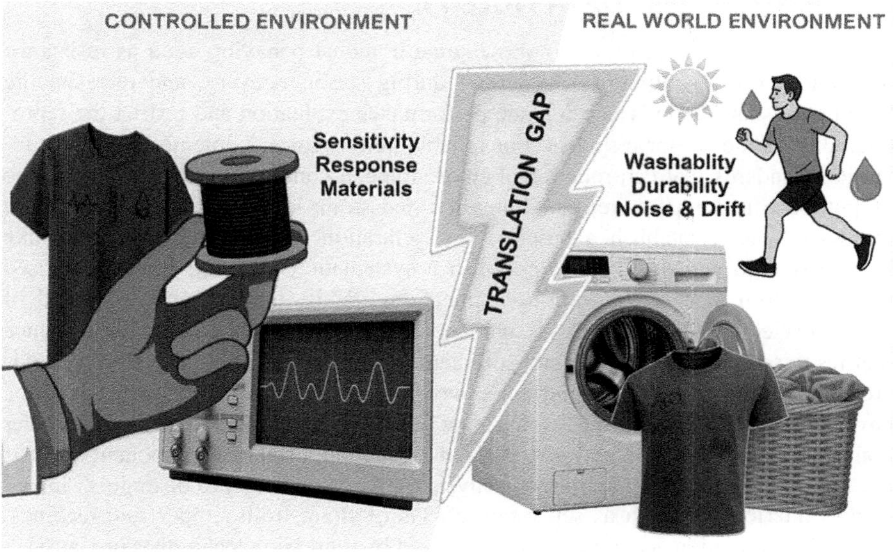

FIGURE 4.1 Schematic illustration of the translation gap between lab-scale performance and real-world durability of conducting yarn-based wearable sensors.

STANDARDS AND RELIABILITY PROTOCOLS

The integration of international testing standards such as IEC 63203-204-1, IPC-8981, and ISO 6330 is critical for the development of conducting yarn-based wearables from research prototypes to commercially viable products. At present, wearable research often suffers from inconsistent durability testing; for instance, "30 washes" in one laboratory may indicate gentle cold-water hand washing, while in another, it may correspond to industrial laundering at 60 °C with aggressive detergents. These inconsistencies make cross-comparison nearly impossible and hinder technology transfer. A universal baseline for testing and evaluation of these smart wearables can be achieved from IEC 63203-204-1 specifying washing durability of e-textiles, IPC-8981 for reliability of e-textile interconnects, and ISO 6330 for household laundering methods. These protocols can be adopted to ensure that the reported results are not only for study purposes but are also consumer ready for commercial deployment. Washability, mechanical durability, and long-term stability can also be checked through some evaluation techniques [8]. Washability test can be conducted with repeated laundry cycles, sweat immersion protocols such as ISO 105-E04, and the effects of detergent on the performance of wearables. Mechanical durability can be evaluated by examining everyday stress on garments, such as cyclic strain, bending, and abrasion test, which are often guided by Martindale or Taber abrasion standards. Long-term stability of smart textiles can be tested by accelerated aging studies like ultraviolet (UV) exposure, humidity, and elevated temperature, to simulate garment lifespan [9]. Laundry-to-failure response and mechanical fatigue studies of garments can be recorded as performance datasets to predict product lifetime and transform lab findings into commercialization of the product.

CALIBRATION AND DATA FIDELITY

Yarn-based sensors are prone to show some nonideal behavior, such as resistance drift under varying humidity, hysteresis during strain recovery, and user-specific baseline offsets. These issues degrade performance evaluation and restrict the reproducibility of the sensor response. For reliable performance, implementation of calibration standards that can reflect real environmental conditions which influence the output, such as temperature, gas, humidity, and strain levels is important. Using a reference sensor, establish a baseline for calibration of performance metrics like hysteresis ratio and drift rate to establish a systematic evaluation for textile-based sensor performance across various applications. While innovation in the field of material selection and structural design can solve the issue of deviating performance metrics to some extent, but it is not sufficient; hence, data-driven correction methods are equally important [8]. Improving wearable sensors' accuracy in a dynamic environment can be achieved by machine learning (ML) frameworks, which offer real-time signal correction and personalized calibration. Principal component analysis (PCA) and independent component analysis (ICA) can be used to distinguish mixed signals interference, such as separating effects of strain from temperature readings. Similarly, calibration accuracy can be improved by regression techniques or Gaussian process models, which establish a predictive relation between sensor response and environmental conditions. Deep learning algorithms or time-series models like long short-term memory networks (LSTMs) or temporal convolutional neural networks (CNNs) capture real-world, time-dependent drift and hysteresis behaviors. The integration of TinyML and edge AI enables lightweight ML algorithms to run directly on low-power microcontrollers embedded within garments, promoting real-time calibration and user-specific adaptation of wearable devices tuned to the individual's physiology and environmental context [10]. Standard calibration techniques and embedded machine intelligence implemented together can lead to reliable and commercially scalable conducting yarn-based smart wearables.

ECO-FRIENDLY AND CIRCULAR DESIGN

To ensure that smart textiles do not evolve into a new e-waste challenge, sustainability considerations must be embedded at the very start of yarn design [11]. This can be achieved by using biodegradable materials like silk, cellulose, polylactic acid (PLA), and chitosan as substrates, which naturally degrade without emitting any harmful residues. For sustainability, the recyclable composites from metals and polymers should be considered [12]. Metals and polymers, components of the conducting yarn, can be separated either mechanically or chemically at the end of the product's life cycle. Apart from material recyclability, evaluation of different types of conducting yarns' effect on environmental footprint, energy consumption, water usage, and toxicity footprints by implementing life-cycle assessment (LCA) frameworks is also important. A notable shift in the field of conducting yarn-based e-textile from just improving sensing performance to focusing equally on environmental sustainability aligns scientific progress with ecological responsibility [13]. End-of-life (EoL) management is still one of the neglected aspects of smart wearables and textile technology development. Various hybrid yarns composed of metallic nanoparticles, carbon fillers, polymers, and multiple layers of coatings cannot be disassembled into reusable components, leading to landfill disposal. To address this challenge, the concept of design-for-disassembly (DfD) has been introduced, which focuses on reversible adhesives or modular encapsulation approaches. It allows materials to be separated through heat or chemical triggers at the end of life. Development of biodegradable encapsulants, such as sustainable alternatives to widely used thermoplastic polyurethane (TPU), will offer protective functions while degrading without releasing toxins. Furthermore, return-and-reuse programs within manufacturing or retail infrastructures can be proposed, where garments are collected and processed to recover valuable conductive fillers such as silver, carbon nanotubes, or MXenes, that would otherwise be wasted. With the rise of eco-labels and global certifications for sustainability, circularity will soon transition from academic ideals to commercial imperatives, making economical and sustainable conducting yarn advantageous in the future marketplace.

APPLICATION TRANSLATION

Most academic prototypes in the field of conducting yarn-based wearables demonstrate what is technologically possible, such as sensing strain, temperature, or wetness [14]. However, relatively few demonstrate what is practically usable. Translation into real-world products requires context-specific design rather than generic proof-of-concept demonstrations [15]. For example, medical sensors must not only function accurately but also survive sterilization processes and comply with biocompatibility regulations. Sports wearables should endure sweat exposure, ultraviolet radiation, and high-elasticity stresses. Safety gears are supposed to be resistant to heat, abrasion, and chemical exposure. Apart from technical robustness, user-centric design is equally vital in the evolution of conducting yarn-based wearable systems. Garments equipped with sensing or computational capabilities must remain comfortable, breathable, and aesthetically acceptable, as even the most robust sensors will be rejected if it lacks wearability. Real-life adoption of these e-textiles is emerging across diverse domains such as healthcare, fitness, safety, and interactive fashion. In healthcare, washable ECG shirts, hydration-monitoring textiles, and wetness sensors integrated into diapers for infant or geriatric care have enabled a new generation of medical textiles [16]. Drift-compensated strain sensors for body movement tracking and compression garments for performance analysis provide real-time insight in the field of sports and fitness. Gas sensors embedded in personal protective equipment (PPE) and flame-resistant conducting yarns in industrial gears promote workplace monitoring and protection. For fashion and interactive clothing, gesture control or augmented reality (AR)/virtual reality (VR) integration within garments helps in human–machine interaction. Designing textile-based smart wearables that meet user needs, must be durable, sustainable, and calibrated as per the requirements for translation of laboratory innovations to consumer-ready products.

MATERIALS AND FABRICATION

Material selection and fabrication methods used in the production of conducting yarns play a major role in determining the reliability and performance of e-textiles.

TABLE 4.1
Fabrication Routes vs. Translation Potential

Fabrication Method	Scale Potential	Control of Properties	Cost	Wash Durability	Industrial Compatibility	References
Electrospinning	Lab scale	Very high	High	Low	Low	[17]
In situ Polymerization	Lab scale	High	Medium	Medium	Medium	[18, 19]
Dip/spray-coating	Pilot scale	Medium	Low	Medium	High	[20, 21]
Knitting/weaving	Industrial	Medium	Low	High	Very high	[22]
Roll-to-roll coating	Industrial	High	Medium	High	Very high	[23, 24]

While academic literature has shown a wide variety of metallic, carbonaceous, polymeric, and hybrid yarns, but the way materials interact with practical factors such as wash durability, comfort, scalability, and environmental impact is still a less explored area. Similarly, the electrical properties of yarns depend upon the fabrication process. Weaving, knitting, and roll-to-roll coating offer scalability, while techniques like electrospinning and in situ polymerization have the advantages of control over material composition, uniform coating, and fine-tuned deposition. This section discusses the deployment of smart wearables in the real world, using conducting yarn materials selection and their fabrication strategies as tools for decision-making (Table 4.1).

Materials Selection

Material composition, such as metallic filaments, carbon-based fillers, conductive polymers, or hybrid combinations, determines the electrical characteristics of conducting yarns. The nonelectrical properties, like wearer comfort, production cost, scalability, recyclability, and wash durability, are equally important for the adoption of wearable technology in real-world applications [25]. Rather than optimizing specific properties in an encapsulated environment, material selection for conducting yarn-based smart wearables must follow a system-level approach for the product to be market-ready [26]. The translational trade-offs, as depicted in Figure 4.2a, are therefore important between the design and adoption of conducting yarn. First is the conductivity versus flexibility trade-off, shown in Figure 4.2c, which can be understood with an example of metallic yarns that provide good conductivity at the cost of fabric comfort, whereas on the other side, polymeric or carbon composites are softer and stretchable but compromise conductivity. Similarly, wash durability with eco-friendliness is difficult to achieve; for example, stainless steel filaments cannot be degraded easily and can withstand 30+ laundry cycles, while biodegradable polymers degrade rapidly but need a protective encapsulation to function well. The protective encapsulation over these yarns is often less eco-friendly. Cost versus performance is another trade-off, where silver-coated yarns are seen to be highly sensitive but are not cost-effective or sustainable, making them unsuitable for mass applications or disposable electronics.

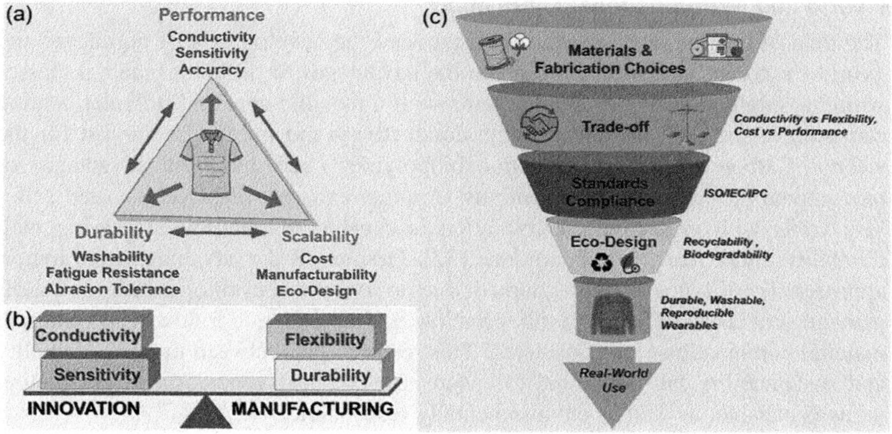

FIGURE 4.2 (a) The interdependence of performance, durability, and scalability parameters. (b) The trade-off between innovation-driven properties and manufacturing-driven factors. (c) The funnel depicting the pathway from material and fabrication choices through trade-offs, standard compliance, and eco-design considerations toward achieving wearable sensors suitable for real-world use.

Lastly, the scalability versus reproducibility trade-off is also difficult for industrial adoption. The nanomaterial-coated yarns are known for good sensitivity but often suffer from response variability across production cycles.

Fabrication: Industrial Methods vs. Laboratory Methods

The fabrication process is the first stage where the electrical and mechanical properties of conducting yarns are decided, as shown in Figure 4.2c. Various prototype devices have been seen at laboratory scale, but industry-scale approaches are necessary for mass adoption [27]. At the laboratory scale, techniques with precise control, such as electrospinning, electroplating, and in situ polymerization, are used, which improve the tunability of properties. These yarns suffer from low throughput and are not suitable for large-scale textile manufacturing. Techniques like spray-coating and dip-coating have poor adhesion and nonuniform coatings. In contrast, industrial fabrication methods like weaving, knitting, braiding, and embroidery use the existing textile machinery, which offer less control over the functionality of yarn but are robust and scalable [28]. Roll-to-roll coating and continuous wet processing have high throughput, reproducibility, and are budget-friendly but may require controlled bath chemistry, deposition parameters, and drying conditions [3]. Low-leakage, adaptive circuitry embedded with conducting yarn systems can improve battery life and make wearable applications energy-efficient. For yarn-based point-of-care applications, accurate biosignal acquisition demands high-input impedance and low-noise amplification circuits for low-power and high common-mode rejection ratio (CMRR) [29–31]. As the laboratory methods deliver high-performance prototypes, while industrial methods work on repeatable, robust, and cost-effective techniques, the need to bridge this gap has become of utmost importance.

Hybrid and Multifunctional Fabrication

The field of conducting yarns is shifting toward the development of multilayer and hybrid structures, designed to combine the advantages of different material classes within a single yarn. For example, yarns with a metallic core and polymer coating can achieve the dual benefits of high conductivity and improved comfort for the wearer. Carbon nanomaterials within biopolymer fibers have the advantages of biocompatibility and sensing sensitivity. Another example of MXene-coated cellulose yarns with protective encapsulation is capable of multimodal sensing with durability under real-world conditions [32]. Despite all the advantages, the hybrid approach is still not widely adopted due to its non-recycling nature and EoL management issues. The circularity principle is challenging to follow when multiple material combinations are considered. This research gap between multifunctionality and sustainability must be filled to ensure that the next generation of conducting yarns is capable, as well as environmentally responsible.

Decision-Making Beyond Materials

Materials and fabrication are starting points in the journey from laboratory prototypes to real-world wearable products. Conducting yarn prototypes do not fail due to a lack of conductivity, but they fail under realistic conditions like inadequate washability, durability, and reproducibility. For performance to align with industry benchmarks, material choices must be guided by standards compliance, such as IEC protocols for washing and durability testing. Fabrication techniques can be examined as per the industrial workflows on the basis of scalability and reproducibility. It should also be ensured that performance cannot be considered at the cost of contribution to future textile e-waste. System-oriented techniques for materials selection and fabrication improve performance and standards. The durability protocols and calibration methods, however, determine whether a yarn sensor survives from lab demonstration to consumer use.

Reliability and Washability

While developing highly sensitive and multifunctional conducting yarns, it is important to remember that if a smart textile cannot survive the washing machine, it cannot survive the market. Reliability and washability are the two most important factors in translating laboratory prototypes into everyday wearable products. Real-world factors that reduce lifespan and degrade the electrical performance of smart garments are repeated laundering, sweat, abrasion, UV exposure, and environmental fluctuations. The reason academic studies are not reliable enough is that they often stop research at initial sensitivity data and durability assessment. This section of the chapter focuses on long-term reliability, guided by international testing standards (Table 4.2).

International Standards and Testing Protocols

Unlike academic tests, industry uses standardized protocols to assess the wash durability and mechanical reliability of wearable electronics and e-textiles. For example, it is pointed out that "after 30 cycles of ISO 6330 domestic laundering"

TABLE 4.2
Common Failure Mechanisms in Wash Testing of Conducting Yarns

Failure Mechanism	Description	Causes	Example Materials at Risk	Mitigation Strategies	References
Mechanical fatigue	Fiber breakage under repeated bending	Stretch cycles	Metal-coated fibers	Encapsulation, core-sheath designs	[33]
Delamination	Conductive coating peels off the base yarn	Poor adhesion, detergents	PEDOT:PSS-coated yarns	Surface pretreatment, chemical coupling	[34, 35]
Oxidation	Electrical loss due to corrosion	Moisture, sweat, UV	Silver-coated yarns, MXenes	Encapsulation, protective coatings	[36, 37]
Abrasion	Physical wear reduces conductivity	Rubbing in laundry	CNT, graphene yarns	Braiding, hybrid composites	[38, 39]
Salt crystallization	Microcracks from sweat crystallization	Perspiration cycles	Carbon-based yarns	Hydrophobic coatings	[38, 40]

is a more meaningful description than simply saying "after 30 washes," since detergent type, water hardness, cycle temperature, mechanical agitation, and drying method also affect the outcome. Standards such as IEC 63203-204-1 (2021) for washing wearables and e-textiles define protocols for mechanical agitation, detergent composition, drying protocols, and reporting of changes in electrical resistance and functionality [41]. Similarly, frameworks such as IPC-8981 (2022) and IPC-8911 (2023) (for e-textile interconnects/conductive-yarn qualification) provide structured approaches around cyclic strain, abrasion, humidity aging, tensile strength, surface resistivity, and wash durability of conductive yarns. Recent reviews underline that one major barrier is the lack of consistent wash protocols in e-textile research, that is, different studies use different machines, detergents, mechanical actions, durations, and temperatures. For example, a 2024 study of textile circuits on various substrates showed performance degradation after up to 100 wash cycles and importantly, identified that mechanical action (substrate swelling, bending of tracks) often dominated damage modes. Incorporating this into yarn-level research means that, when conducting yarn or integrated interconnect in a textile is reported, the test protocol with the industry-style standards should also be aligned (e.g., ISO 6330 for domestic laundering as baseline, plus the upcoming IEC and IPC e-textile standards) so that the results are comparable across studies and closer to real-world use. Aligning with standardized cycles and explicitly reporting machine type, detergent, load size, mechanical agitation, and drying allows meaningful comparison, better reproducibility, and improved industrial relevance.

Common Failure Mechanisms

Common failure mechanisms in conducting yarns, shown in Figure 4.3a, highlight the gap between laboratory performance and real-world durability. Mechanical abrasion is one such issue where yarn surfaces rub against other fabrics or machine drums during wear and laundering, leading to fiber breakage and progressive loss of conductive coatings. Another issue is delamination of conductive coatings (e.g., films of PEDOT: PSS, graphene, or MXene) from polymer substrates when exposed to repeated wet conditions. Inadequate adhesion at the interface can lead to peeling or cracking of the conductive film during moisture or laundering cycles. At the same time, oxidation and corrosion remain major failure modes where metallic yarns

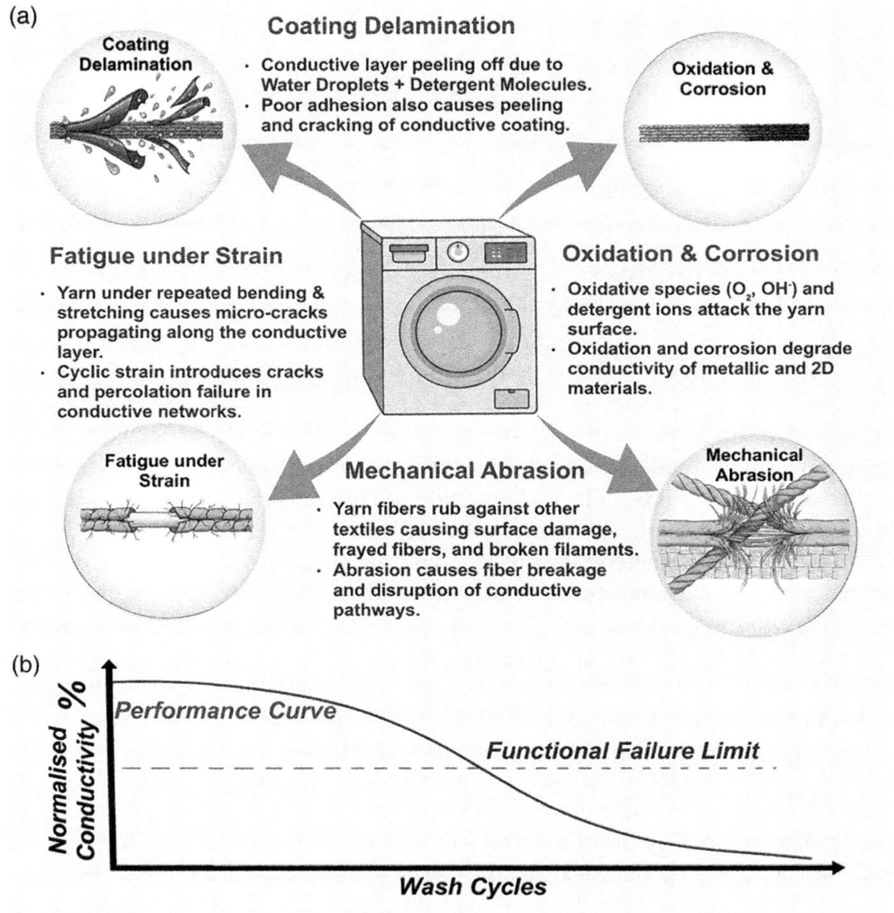

FIGURE 4.3 (a) Illustration of major failure modes affecting conducting yarns. (b) The performance curve shows the decline in normalized electrical conductivity with increasing wash cycles, eventually reaching the functional failure limit.

(such as silver threads) are vulnerable to tarnishing or corrosion in humid or chemically active environments, and MXenes are prone to rapid environmental degradation (e.g., oxidation in humid air) unless suitably protected [32]. Furthermore, mechanical fatigue under cyclic strain (stretching, bending, repeated deformation of yarns, or woven networks) often initiates microcracks in the conductive network, leading to disruption of percolation paths, increased resistance, signal drift, and eventually open-circuit failure. Repeated deformation can degrade connectivity in the conductive layer unless stress-relief architectures or highly conformal coatings are employed. Exposure to detergent chemistry during laundering accelerates degradation, as surfactants, oxidants, and alkali agents in wash cycles promote polymer substrate swelling or breakdown, chemical attack on metal coatings, and moisture penetration, all of which aid in delamination, oxidation, and fatigue mechanisms. The discussed failure mechanisms gradually lead to increases in electrical resistance, drifting or unstable signals, and open-circuit failure of the conductive network embedded in a textile or yarn substrate.

Design Strategies for Robustness

Recent developments have focused on protective strategies to improve the durability of conductive yarns under real-world conditions. One widely adopted method is encapsulation and protective coatings, for example, polymers such as TPU, silicones, or fluoropolymer-based layers that act as barriers against moisture ingress and chemical attack [42]. However, these barrier layers still pose a trade-off: increased protection often comes at the expense of breathability and comfort (air and moisture permeability may decline), so wearable integration must carefully balance protection and comfort. Innovation with yarn architectures makes the yarn robust, like a core-sheath structure (conductive core wrapped by a protective polymer sheath) or twisted yarn provide mechanical reinforcement that reduces delamination and enhances strain tolerance [43]. Stainless steel threads based on low-cost braided e-yarns embedded in heat-shrink tubing increase wash durability and mechanical flexibility for smart textiles. As research in the field of hydrophobic and self-healing treatments is emerging, polymer-based hybrid yarns are now capable of reforming conductive pathways after cracks and damage in the conducting part, offering long-term electrical reliability. A wearable conductive-yarn system must balance durability with wearer comfort, unlike a yarn with thick silicone encapsulation might survive 100 laundry cycles but become too stiff, reduce breathability, or compromise stretchability. For a smart wearable technology, designing a protective system is crucial, as it requires improved wash life and reliability, without compromising softness, flexibility, or breathability.

Laundry-to-Failure Testing and Standard MIWTE Framework

Figure 4.3b demonstrates the laundry-to-failure curve, which is a useful translational metric for assessing the durability of conducting yarns. This approach involves plotting normalized conductivity, or sensor sensitivity, against the number of standardized wash cycles. Functional failure of the wearable is then reported as the cycle at which performance falls below a defined application-specific threshold. Such a standard paired with mechanical and environmental stress tests provides a fair basis to make inter-laboratory comparisons meaningful and offers valuable data for forecasting garment lifespan to the industry [13]. Sweat simulation tests, as described in EN ISO 105-E04, can be carried out where yarns are exposed to acidic and alkaline perspiration to replicate physiological conditions with UV aging tests, involving accelerated light exposure for outdoor garments, where sunlight-driven degradation can be studied. Quantitative abrasion testing (Martindale, Taber) is the industry standard for surface durability, where instrumental end-point detection and image-analysis methods have improved reproducibility and reduced operator bias. These standard metrics altogether create a complete system for predicting performance and translating laboratory prototypes into reliable, consumer-ready smart textiles. The durability of wearable textiles is one of the most under-reported aspects in academic publications, which makes it difficult to compare results across studies or align academic findings with industrial performance standards. To address this gap, the adoption of a Minimum Information for Wearable Textile Experiments (MIWTE) framework has been proposed, where authors would be required to report the material composition (including fiber type, conductive elements, coatings, and yarn linear density), fabrication and integration methods (such as weaving, knitting, printing, or lamination), yarn's linear density (tex/denier), and the method employed for fabric integration [44]. The MIWTE framework would simplify performance comparisons across different studies via a data reporting and transparency process.

Calibration, Drift, and Data Reliability

For real-case scenarios, wearable sensors undergo environmental fluctuations, mechanical fatigue, user-to-user differences, and hysteresis, leading to electrical instability. This performance deterioration due to nonlinearity, noise, and signal drift creates trust issues that restrict its clinical and industrial applications. To overcome this issue, in-depth material optimization is necessary from both signal calibration and data reliability perspectives. As shown in Figure 4.4, this section discusses calibration protocols, ML, and TinyML-based approaches for correcting drift and variability.

Calibration Strategies

Conducting yarns are heterogeneous in nature, and their structural complexity depends upon material optimization [23]. Due to the complex structure, the electrical performance may vary over time. Factors such as fiber orientation, coating uniformity, and environmental exposure contribute to signal instability [3, 45]. Hysteresis occurs when resistance does not return to its original baseline after strain is released, resulting in memory effects in the signal. Baseline drift is another problem where conductivity changes due to oxidation, fatigue, or fluctuations in humidity. Body geometry difference causes user-to-user variability (shown in Figure 4.4a) as movement patterns, temperature, and sweat composition differ from individual to individual. As a result, calibration and drift compensation must be taken care of for stable, accurate, and personalized output. One of the calibration strategies is the use of a reference component. The reference element such as a metallic yarn is integrated along with the active sensor to provide a baseline for correction. Another technique is to use an array of sensors. Averaging signals from multiple yarns reduces

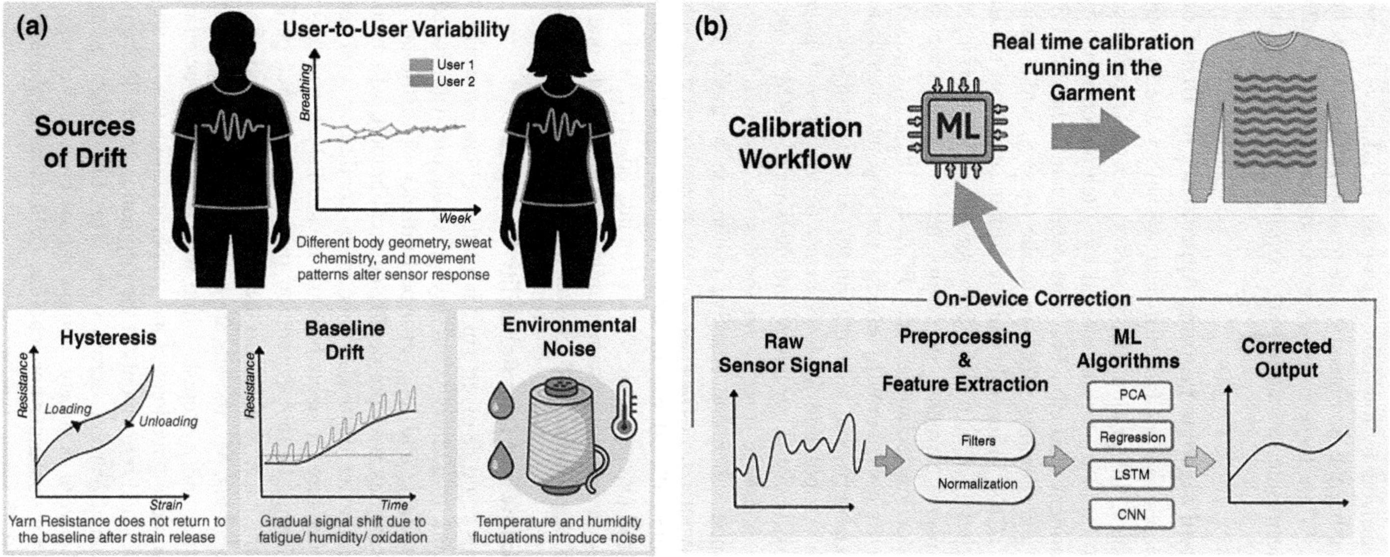

FIGURE 4.4 (a) Sources of drift. (b) Calibration workflow to produce a corrected output for accurate and reliable sensing performance.

the impact of local inconsistencies. Signal processing method is used to suppress noise and motion artifacts by low-pass, high-pass, or adaptive filtering. Improving the signal by correcting the nonlinear strain–resistance impact can be achieved by polynomial fitting. While these methods might be capable of handling short-term variability, they are unable to handle long-term drift and the complexity of multimodal signals.

Sources of Drift and Machine Learning Approaches for Drift Compensation

Conducting yarn-based sensors face drift and variability due to both material and integration challenges. Integration effects distort electrical responses, generating nonlinear and non-stationary signals caused by stitching tension, fabric strain heterogeneity, and garment elasticity. For time-dependent behavior, LSTM networks capture hysteresis and memory effects in textile sensors, while temporal convolutional networks (TCNs) offer efficient, real-time sequence modeling. Domain adaptation approaches such as transfer learning and online learning enable models to adjust to new users, garments, or environments without extensive retraining. Dimensionality reduction techniques like PCA and ICA isolate dominant signal components (e.g., strain vs. humidity effects) and separate overlapping biosignals, improving interpretability and reducing noise. At the hardware level, TinyML supports on-device computation for low-latency performance. Lightweight, quantized neural networks can run on microcontrollers (e.g., ARM Cortex-M, ESP32), enabling real-time inference directly on the textile. For reliability, ML provides powerful drift compensation and adaptive calibration methods. Regression-based models (linear, polynomial, or Gaussian Process Regression) map sensor outputs to reference values, with GPR offering uncertainty-aware predictions for self-correcting calibration. These techniques enable self-calibrating, adaptive, and autonomous textile sensor systems, capable of sustaining high performance despite environmental variations, material degradation, and long-term use (Figure 4.5).

Calibration Workflow and Data Reliability Strategies

As conducting yarns could be diverse in nature, the calibration workflow should be a multilayer framework to address variability in conducting yarn-based sensors. It is important for the reliable and long-term performance of textile-based smart wearables. The process begins with pre-deployment calibration, where conducting yarn behavior is studied under controlled laboratory conditions with restricted strain, humidity, and temperature. A stable baseline is established, which is then followed by user-specific tuning. In this calibration protocol, a wearer performs standard movements while the garment adapts to their body geometry and movement pattern. For practical applications, ML models are employed for real-time correction, which compensates for drift and hysteresis while maintaining data accuracy in dynamic environments. In some applications, cloud synchronization models are implemented, which allow updates and refinements at the population level. This enables garments to benefit from shared learning across users. This hierarchical system provides a sensible balance between accuracy, computational cost, and energy consumption, making it suitable for consumer applications. Calibration implementation for every study should specify the strategies, whether it is classical or ML-based, to promote

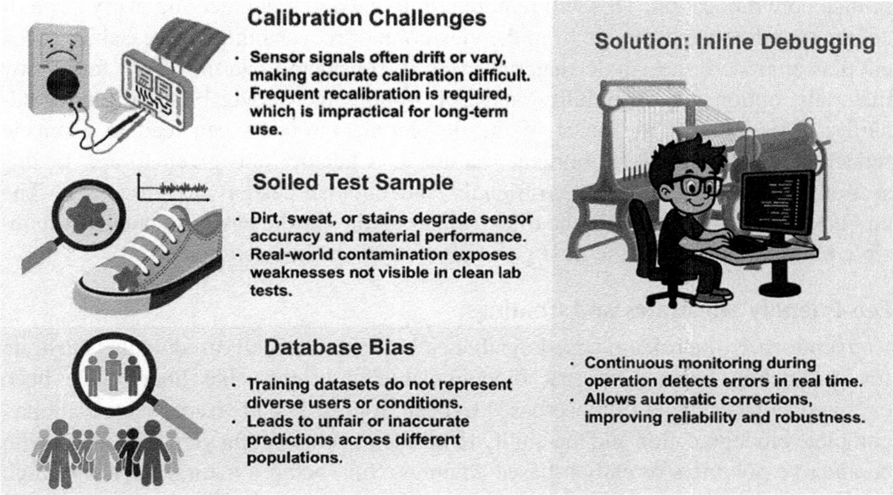

FIGURE 4.5 Calibration, contamination, and data bias challenges in wearable sensors and potential solutions.

reproducibility and comparability across studies. Similarly, information on the sampling rate, dataset size, and preprocessing steps that are used to train or validate models should be recorded in the public domain. Parameters such as drift error, hysteresis ratio, and calibration stability can be included in performance reporting to capture long-term reliability in a better way. Also, it should be mentioned if the calibration strategy is a one-time adjustment or requires continuous learning and processing during use. Such measures can prevent the adoption of non-reproducible and non-standard demonstrations. It is now evident that material and fabrication alone cannot ensure a reliable smart garment, but calibration strategies and data reliability are equally important.

ECO-FRIENDLY AND CIRCULAR DESIGN

With the emerging relevance of e-textiles, sustainability becomes equally relevant alongside the functionality and reliability. The textile industry is already a big contributor to landfills and uses a large amount of water and chemicals during its production. Approximately 20% wastewater is contributed by the textile industry globally. With the incorporation of conducting yarn-based e-textiles, these concerns are poised to grow even more. Conducting yarns introduce metals, polymers, and nanomaterials embedded into garments. If not environmentally accounted for, these will result in "embedded e-waste" that will be inseparable from the garments, leading to other complexities. Recent reports have emphasized the adoption of circular design principles in the development of e-textiles. Sustainable materials scientists are advocating for eco-friendly substrates, biodegradable coatings, and recyclable conductive composites for a shift toward viable pathways of innovation. It invigorates the rethinking of the materials of conducting yarns used to enable sensing, heating, or data communication. This will require rethinking and reengineering every stage of conducting the yarn life cycle from the viewpoint of recyclability and reusability. In a circular approach, the whole material cycle is considered, starting from fewer raw materials, option for repairability, and EoL options like recyclability or biodegradability. This LCA at the onset of the development process can lead to a drastic reduction in environmental burden, leading to a healthy and green future. In this section, different aspects of eco-friendly and circular designs are discussed. The emphasis is on understanding the role of smart textiles as the environmentally responsible building blocks for the next generation of wearable electronics.

Eco-Friendly Substrates and Coatings

Conventionally, petroleum-based synthetics have been widely used as the substrate for conducting yarns. Different biodegradable and renewable fibers have been explored and reported. Cellulose-based Lyocell, Ioncell-F, and viscose have exhibited complete biodegradation and the ability to be used as conducting yarn by coating with conductive polymers or carbon-based coatings. Silk, being a natural fiber with high tensile strength, biocompatibility, and controlled biodegradability, is also a viable option for implantable e-textiles. PLA and chitosan-based fibers, with very good biodegradability, are a promising alternative to be explored in future works. PLA is widely used as a thermoplastic material in additive manufacturing, exhibiting adequate flexibility and strength. Chitosan, a naturally derived polysaccharide, shows antimicrobial activity and can enhance multifunctionality if used in blended yarns. Water-soluble PEDOT:PSS-based coatings are gaining popularity as an alternative to toxic organic solvents. Some reports have explored biomass-derived conductive additives, such as lignin-based carbon, bio-sourced graphene, and carbon black alternatives, as eco-friendly alternatives to conventional nanomaterials. Some recent efforts to synthesize MXenes and other 2D nanomaterials using fluorine-less processes highlight the feasibility of sustainable fabrication of conducting yarns [32]. The combination of these efforts showcases critical steps toward circular and sustainable design of conducting yarns.

Sustainable Fabrication Pathways

In many cases, although the materials may be eco-friendly and sustainable, the synthesis process itself can contribute some pollutants. The conductive coatings are generally based on solvents that release volatile organic compounds (VOCs). A sustainable approach can be the adoption of water-based solvents for dip coatings. In some cases, roll-to-roll deposition of conductive inks is mooted as an economical and sustainable alternative. Another time and energy-saving approach is combining the conductive treatment with the garment dyeing process. Traditional silver and copper conductive inks can be replaced with carbon-filler-based blends. Such efforts and techniques can reduce the environmental impact of the fabrication/synthesis process of conducting yarns.

End-of-Life and Circularity

Even after the EoL of e-textiles, their proper extraction and disposal are serious issues in ensuring their sustainability [12]. The challenge is even aggrieved by the physical nature of conductive yarns being quite like the garments. The yarn may be embedded, intertwined, or knitted in the garment itself. In such cases, if the embedded electronics fail, the conducting yarn may be misclassified as part of the garment and end up in landfills. Such a scenario has the potential for creating a new waste stream "embedded e-waste". To mitigate this challenge associated with the EoL of e-textiles, circular-economy strategies from design to disposal can prove to be effective. In general, conducting yarns have components like textile fibers, conductive parts, and encapsulation layers, each with different material properties and unique recyclability challenges. The matter is more complicated when heavy metals, nanomaterials, or rare earth minerals are involved. Circular-economy strategies can be implemented as DfD, material recovery, and closed-loop systems. The concept of using modules and detachable parts that can be easily extracted at EoL – repaired or refurbished, is labeled as DfD. It has been attempted in some works such as the incorporation of knitting channels that allow for the easy removal and recovery of the conducting yarn part, the use of reversible adhesives and mechanical fastenings that replace permanent glues or laminations, avoiding the combination of incompatible materials, like metal parts fixed on elastic straps to facilitate easy separation for recycling [46, 47]. Another useful strategy is the implementation of collection systems for EoL e-textiles and refurbishing them for reuse. It will decrease the probability of them ending up as waste. Additionally, embedding life-cycle information and providing instructions on how to properly dispose of items at the EoL has the potential to further promote proper handling and recycling. The adoption of circular-economy principles, illustrated in Figure 4.6, is essential for the sustainable use of conducting yarn-based e-textiles. With proactive design and regulations, these innovations will not only be high functioning but also environmentally responsible.

Life-Cycle Assessment (LCA)

LCA is a standard method (ISO 14040/14044) for assessing the environmental impacts of a product, process, or service throughout its life cycle. Using LCA, hotspots in the production and life cycle of a conducting yarn can be evaluated in the context of e-textiles. The replacement of petroleum-based fibers with biodegradable ones, as well as the switch from solvent-based coatings to aqueous or bio-carbon-based coatings, substantially reduces greenhouse gas and VOC emissions [48]. Durability is another very important factor, as a garment that maintains conductivity over 50 laundry cycles will have approximately 70% less environmental impact compared to one that survives only five wash cycles. Although the LCA of conducting yarns can provide critical insights, its application is limited by data gaps for novel materials and uncertainties in recycling and biodegradability. The studies at present seldom include circular design or EoL recovery. Inclusion in LCA will further reduce the environmental impact. Future LCAs should prioritize durability, recyclability, and low-impact fabrication as key design criteria.

Eco-Compliance and Regulation

Agencies and governments throughout the world are trying to formulate and implement regulations to achieve their sustainability goals, in line with their commitments.

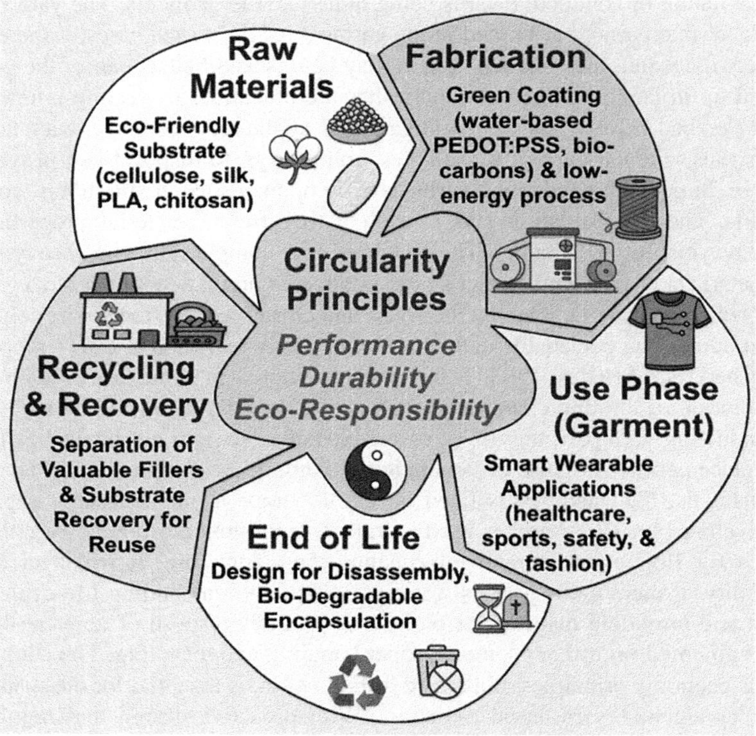

FIGURE 4.6 An illustration of the circular life cycle of sustainable smart textiles.

Smart textiles are not an exception in this. The EU, under ESPR 2024 (Ecodesign for Sustainable Products Regulation), has brought textiles under the eco-design rules mandating reparability, recyclability, durability, and non-use of hazardous substances. Policies like REACH and RoHS regulate the use of certain heavy metals and phthalates in textiles and electronic parts. The EU's new EPR (Extended Producer Responsibility) scheme is expected to put the onus of collection and recycling of textiles on manufacturers. For the success of conducting yarns, researchers need to adopt the principles of regulation. The first step in this direction can be the adoption of voluntary certifications such as OEKO-TEX®, Global Recycled Standard (GRS), and ISO 14040/14044. These regulations demand complete and transparent documentation across the supply chain and encourage the adoption of sustainable processes and materials. Collectively, adoption of these frameworks has the potential to push conductive-yarn development toward circular design. It will ensure that smart textiles achieve both performance and environmental responsibility and be competitive in the market too [13].

Translational Applications

In general, sensors and embedded electronics are described either in terms of their sensing target (humidity, temperature, gas, pressure, or tactile) or the mechanism (thermionic, piezoelectric, chemiresistive, or capacitive). The same is the case with conducting yarn-based e-textiles reported in the literature. These end-application or mechanism-based classifications are well-suited for material benchmarking, but it fails to capture the aspects that lead to end-user adoption. E-textiles, in addition to being electronic systems, are part of daily wear and must be evaluated from that perspective, too. For a user, the factors that influence his/her decisions are comfort, reliability, washability, and worthwhileness in day-to-day chores [41]. As such, wearable sensors need to be designed contextually, taking into account the specific needs and challenges of the application it is designed for. For context, aesthetics are designed for fashion, sweat, and UV exposure for sportswear or sterilization and laundry for medical wear. The question to be answered by these e-textile systems shifts from "What can the yarn sense?" to "Can the garment survive 30 washes, deliver reliable data, and feel like normal clothing?". This section will look beyond the proof-of-concept demonstrations and focus on the conducting yarns specifically engineered for key domains of occupational safety, fitness, healthcare, and interactive fashion, described in Figure 4.7. Critical factors like calibration, comfort, washability, and sustainability that enables the yarn-based smart textiles to cross the barrier from lab scale and enter consumer's arena will be pondered upon.

FIGURE 4.7 Applications of smart wearable textiles across four major sectors: Healthcare, Sports & Fitness, Industrial Safety Gears, and Fashion & Interactive Clothing.

Healthcare and Medical Textiles

The field of healthcare is known for tight specifications due to its critical nature. The same stringent specifications, such as high accuracy, wide range of operation, biocompatibility, and resilience to the sterilization process, apply to medical garments as well [49]. The conducting yarn embedded in the patient robe is likely to repeatedly interact with sebum, sweat, and body fluids and has to go through frequent laundry cycles. It has been observed that silver-coated conducting yarns lose their conductivity in less than 30 washes due to the effect of bleaching agents used. Care specifications to reduce this fatigue and prolong the life of conducting yarns have been presented but are not expected to always be adhered to. On the other hand, techniques such as encapsulation, composite fiber structure, and embedding stainless steel in the core have shown to be effective in increasing the life to around 100 laundry cycles. In the wetness sensing systems for neonatal diapers, the conducting yarn is supposed to be in contact with body fluids for a prolonged period. Its conductivity degrades rapidly, as well as biocompatibility is a concern in this case. PEDOT:PSS or fluoropolymer coatings have been demonstrated to effectively increase the operational life, acting as a moisture-resistant coating while preserving comfort. An important concern with e-textiles is the signal fidelity due to the inter-patient differences in sweat composition, skin impedance, and microenvironmental conditions [50]. To mitigate these calibration challenges, ML model-based implementations have been explored and demonstrated to reduce diagnostic errors. The environmental impact of medical wearables is increasingly scrutinized [51]. Studies have been exploring a plethora of biodegradable materials and trying to use natural fibers for conducting yarns. Zinc and magnesium, as the conducting materials, have even been explored to facilitate environmentally safe disposal.

Sports and Fitness Wearables

Sportswear e-textile is expected to be lightweight, breathable, and able to withstand multiple laundry cycles. It endures intense sweat and continuous mechanical strain. Knitted electrodes are reported to provide conformity but suffer resistance drift over repeated use. Braided and core-sheath structures of conducting yarn have demonstrated high gauge factor, elastic recovery, and long cycle and wash endurance. Although encapsulation is the general technique for increasing the life of conducting yarns, but they increase the bulk of the yarn, an undesirable factor for sportswear. In-fiber conductions by functionalizing the fiber with materials like MXene and carbon nanotube in elastic and helical structures have been demonstrated. These innovations claim to maintain the lightweight feature of yarns while providing durability [52]. In this case, sweat is a common modulator of the conducting yarn's resistance. Recent works have reported ML-assisted online drift compensation, adaptive regression/RNNs, and domain adaptation to correct hydration-driven signal shifts. Collectively, the state-of-the-art sports e-textiles are converging on translational factors like breathability, lightweight, sweat resistance, and in-fiber conduction without encapsulation.

Industrial and Occupational Safety

E-textiles in an industrial setup are expected to sense gases, temperature, and perform physiological monitoring. Their robustness and physical properties must be at par with the industrial garments, as they are evaluated based on sector-specific standards too [53]. In firefighting, the conducting yarn ought to follow the same standard as a firefighter's dress does. Many conducting yarn-based gas sensors (polyaniline-coated, electro-spun 1D nanofibers, and MXene functionalized) have been demonstrated to detect gases like NH_3 and CO [32, 46]. But their electromechanical studies reveal low mechanical stretchability and poor fatigue resistance. Recent work on copper-coated polyamide yarns has shown promise as a thermally durable, electrically and mechanically robust alternative in an industrial setup. Practical reliability of these sensors is equally important in such scenarios, as a false alarm can lead to panic, as well as economic costs due to abrupt disruptions [54]. This factor has been shown to be mitigated by deploying redundant sensors and on-garment edge processing. Ultimately, sustainability and being eco-friendly are equally important in this sector. Biodegradable substrates and coatings showcase the efforts in this arena. Together, these efforts are steering conducting yarn e-textiles from lab prototypes toward garments capable of continuous hazard monitoring in industrial setups.

Fashion, Lifestyle, and Interactive Wearables

In fashion and consumer e-textiles, the specifications are not that strict, but still, the garments are expected to endure regular laundry cycles. In this case, visual appearance and aesthetics are more important than functional performance. The integration of conducting yarns is a preferred method, as they seamlessly fit into existing textile technology and can be easily woven, knitted, or braided. A primary design challenge in this sector is the seamless integration of sensors, interconnects, and small electronic components into textile structures, with a lower reliance on encapsulations or metallic sheaths unless these are explicitly incorporated as part of the aesthetic design [55]. From a functional perspective, user adoption of fashion-oriented wearable systems depends on the accuracy of sensing, calibration stability, rapid response time, and gesture or motion recognition capabilities. For applications such as AR, VR, music-responsive garments, or gesture-controlled interfaces, the e-textile system should perform well under various wear and wash conditions [2, 56]. With the rapid innovation in the fashion-tech hybrid sector and the high turnover of apparel products, sustainability concerns are a central issue. This is mitigated by efforts like modular electronics, EoL reuse strategies, and recycling. The translation of conducting yarn-based electronics in healthcare, sports, safety, and fashion depends on factors like staying calibrated for reliable data, withstanding laundry cycles, and remaining comfortable and visually appealing [55]. An e-textile sensor with equitably good sensing parameters will not prevail if garments feel stiff or lose functionality after a few washes. Hence, e-textiles should be evaluated beyond "what they sense". Sustainability, recyclable and biodegradable elements, is also becoming an important factor in setting future standards. Ultimately, e-textiles must prove that they can endure laundering, deliver consistent data, and stay comfortable and eco-friendly to move from lab prototypes to everyday use (Table 4.3).

TABLE 4.3
Key Requirements vs. Gap in Translational Application

Domain	Application	Performance Requirement	Current Gaps	References
Healthcare	ECG, glucose monitoring	Biocompatibility, washability, accuracy	Drift, poor durability, lack of regulatory standards	[57]
Sports and fitness	Strain, hydration, posture	Stretchability, sweat resistance, washability	Signal drift in sweat, calibration issues	[8]
Industrial safety	Gas, heat, stress sensing	Flame resistance, abrasion resistance, long life	Harsh environment durability, certification gap	[2]
Interactive fashion and VR	Gesture, haptic, gaming	Comfort, responsiveness, aesthetics	Power supply, consumer acceptance, washability	[56]

Future Roadmap: Toward Standards, Intelligence, and Market Adoption

It has been demonstrated that conducting yarns can be used to sense temperature, humidity, strain, etc. The adoption aspect of conducting yarn-based wearables depends upon the practical considerations of reliability, adaptability, and sustainability. Advancing material choices and improving the conventional sensing parameters alone cannot result in commercial adaptation of the conducting yarns. It has become important to fix the standards of test protocols, edge intelligence, robust calibration, and material circularity for this purpose. In this section, the roadmap of e-textiles, focusing on the adaptation of international reliability standards, integration of edge intelligence, and sustainable design, has been discussed and depicted in Figure 4.8. For an innovation to be meaningful, aligning technological innovation with regulatory frameworks and consumer/industry expectations is important.

Toward Reliable, Standardized, and Smart Yarn-Based Wearables

For years, e-textile research has been focused on achieving higher sensitivity, faster response, and lower detection limits. While these metrics are important, real-world deployment prioritizes reliability, repeatability, and durability, focusing on how well sensors perform after repeated washing, across production batches, and over months of use [42]. Studies on conductive material degradation are currently gaining popularity. Under the same wash counts, washability outcomes are seen to vary widely depending on substrate, circuit design, and integration method. One of the major barriers is the lack of standardized terminology and test protocols for durability and reliability. There are some e-textile-specific standards like IEC 63203-204-1 (washability test methods) and IPC 8981 (quality and reliability of e-textiles) that can be adapted [41]. Beyond durability, embedded intelligence in yarn-based wearables is gaining traction, where instead of researching new sensors, the focus is shifting toward making sensors smart. Emerging technologies such as TinyML, on-garment processing, federated learning, and digital twin frameworks enable on-device

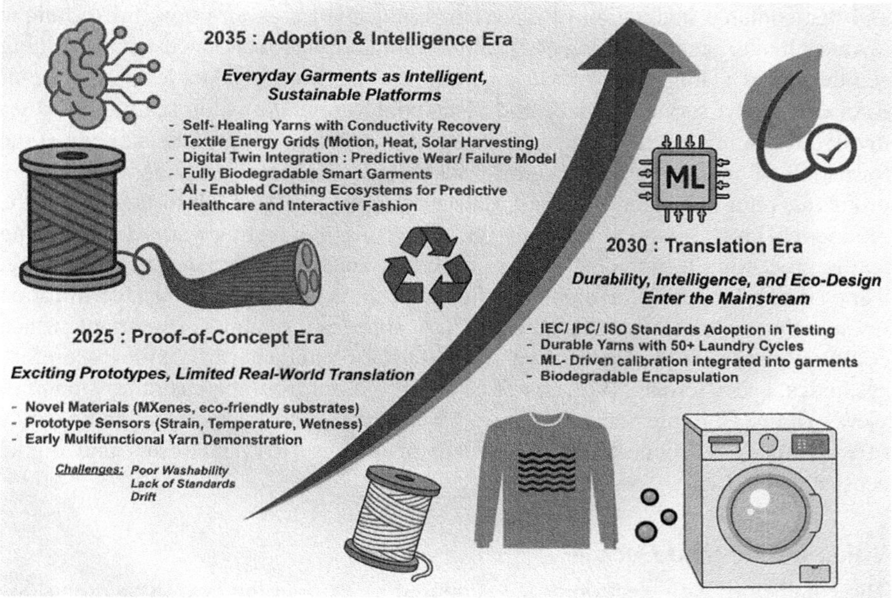

FIGURE 4.8 A roadmap illustrating the evolution of smart textiles from 2025 to 2035.

computation [58]. It reduces reliance on external hardware, solves privacy concerns, and enhances responsiveness. A globally accepted reporting framework would be beneficial for comparability and reproducibility, somewhat similar to biomedical research guidelines like CONSORT or PRISMA. Cross-lab consistency can be addressed by the proposed MIWTE, outlining essential details as a global reporting standard. The field of wearable technology can align with industrial standards and user expectations in terms of reliability and reproducibility.

Sustainability and Circularity as Market Gatekeepers

As markets demand eco-friendly products, eco-labeling has become important for the e-textile industry, which can cover recyclability, life-cycle carbon footprint, toxicity level, and other environmental performances. The study shows sustainability as a major issue caused by mixed materials, embedded electronics, and limited EoL options [59]. Therefore, materials for developing conducting yarns should have biodegradable encapsulants or recyclable cores, so that products can be returned to the biosphere if not reused. Designing conducting yarn should not only reduce disposal but also address EoL recovery or recycling to cut down on environmental burden. Apart from electrical performance metrics, the LCA reporting is equally important to know the environmental impact of the production, use, and disposal phases. LCA studies for wearable smart textiles highlight material and manufacturing obligations and the need for circular-economy strategies [60]. DfD principles for e-textiles, on the other hand, focus on constructing smart garments such that sensors, conductive yarns, and textile substrates can be separated at EoL, enabling recovery of valuable metals, printed electronics, and fibers [61]. Literature has shown modular and demountable integration of electronics into textiles as an impactful technique toward circular smart clothing [62]. Some literature has mentioned smart clothing eco-design as a four-pillar transformation: shifting from product-level to system-level circularity, service models, and dematerialization. Basically, scaling e-textiles involves bridging research, manufacturing, and end-user needs. One strategy is the formation of industry-academia consortia that decide, share, and approve testing protocols, pilot manufacturing, and translate technology from lab to factory. At the production level, adoption depends on integrating conducting yarns into existing textile processing technology, such as weaving, knitting, embroidering, or printing, using standard industrial equipment rather than some new setups. Then, certification in emerging fields to the CE-mark or FDA approvals is utmost important, which could validate safety, reliability, and regulatory compliance of textiles. Comfort, aesthetics, and affordability must not be compromised from the customer's point of view. The sensor integration needs to be seamless into everyday clothing while offering intuitive connectivity to Internet of Things (IoT) platforms and digital ecosystems [63, 64].

Vision for the Next Decade

The conducting yarn electronics are supposed to grow in the next decade, pushing the scope of wearable technology. By 2035, the sector of smart and interactive wearables is expected to cross the mark of 65 billion USD (a ten times leap from 6.6 billion USD in 2025). The key milestone in this trend will be laundry-proof wearables [60]. Commercially viable garments must sustain performance parameters through 50 or more wash cycles, even with abrasions and detergents. Concurrently, material scientists have been working on self-healing yarn systems that can repair microcracks and restore conductivity post-wash, thereby extending e-textiles' lifespan and reliability. Another important aspect is the powering of the e-textiles by lightweight non-conventional textile-based energy sources that harvest from sources such as body heat, motion, solar exposure, or their combinations [47]. This aspiration is well reflected in the market trends, too. With the broader textile industry already focusing on sustainability and circular-economy models, it will become a defining dimension [59]. Biodegradable smart garments with fully decomposable parts and mechanisms to recover the reusable parts are gradually becoming a reality. The design of smart wearables with EoL in mind is a necessity to keep up the pace with trends. Finally, on the intelligence front, garments will evolve into AI-enabled ecosystems learning wearer patterns, adapting sensor calibrations, predicting health or risk states, and interacting with IoT platforms seamlessly.

CONCLUSION

In recent years, the conducting yarns have developed from conductive filaments to multi-structured and multifunctional yarns that are capable of sensing, energy harvesting, and communication. The central challenge is no longer achieving electrical conductivity under controlled laboratory conditions but maintaining stable and reliable performance under real-world situations such as washing, abrasion, perspiration, and repeated mechanical deformation. ML and TinyML have aided in adaptive calibration to maintain long-term signal stability. Meanwhile, a trend of shift is being seen for circular design from voluntary eco-friendly policies to regulatory necessity for sustenance. This chapter also discussed the factors essential for scaling e-textile technologies toward practical adoption. The material and fabrication trade-offs demonstrate that conductivity alone is insufficient. Components like durability, comfort, and environmental compatibility must be considered as design priorities. The discussion highlights the importance of standard reliability frameworks, including IEC 63203-204-1, IPC 8981, and ISO 6330. A future roadmap for the adoption of conducting yarn-based e-textiles addressed manufacturability, scaling, user acceptance, and ecosystem design. The idea behind conducting yarn-based wearable sensors is an adaptive, active, and communicative interface to enhance human interaction with the environment. To achieve this, a balance for the translation of conducting yarns from laboratory innovation to everyday application is needed.

REFERENCES

[1] A. Lund et al., "Roll-to-Roll Dyed Conducting Silk Yarns: A Versatile Material for E-Textile Devices," *Adv Mater Technol*, vol. 3, no. 12, p. 1800251, 2018, doi:10.1002/admt.201800251

[2] H. Liu, Y. Shi, Y. Pan, Z. Wang, and B. Wang, "Sensory Interactive Fibers and Textiles," *npj Flexible Electronics*, vol. 9, no. 1, p. 23, Mar. 2025, doi: 10.1038/s41528-025-00398-4

[3] J. Marten, N. Gaukel, Y. Hu, Y. Wang, G. Yuan, and N. Willenbacher, "High-Throughput Production of Electrically Conductive Yarn (E-Yarn) for Smart Textiles," *Adv Electron Mater*, vol. 11, no. 7, May 2025, doi: 10.1002/aelm.202400700

[4] K. Pattanarat, N. Petchsang, T. Osotchan, Y.-H. Kim, and R. Jaisutti, "Wash-Durable Conductive Yarn with Ethylene Glycol-Treated PEDOT: PSS for Wearable Electric Heaters," *ACS Appl Mater Interfaces*, vol. 13, no. 40, pp. 48053–48060, Oct. 2021, doi: 10.1021/acsami.1c13329

[5] A. Kumari, A. Sengupta, and A. Agarwal, "The Role of Conducting Yarns in Shaping Next-Generation Wearable Sensors: A Review," *ACS Appl Mater Interfaces*, vol. 17, no. 30, pp. 42541–42567, Jul. 2025, doi: 10.1021/acsami.5c09050

[6] B. K. Dejene, "The Future of Fabric: A Comprehensive Review of Self-Powered Smart Textiles and their Emerging Applications," *Energy Reports*, vol. 14, pp. 898–943, Dec. 2025, doi: 10.1016/j.egyr.2025.07.002

[7] F. Huang, J. Hu, and X. Yan, "Review of Fiber- or Yarn-Based Wearable Resistive Strain Sensors: Structural Design, Fabrication Technologies and Applications," *Textiles*, vol. 2, no. 1, pp. 81–111, Feb. 2022, doi: 10.3390/textiles2010005

[8] L. Liza, M. H. Kabir, L. Jiang, S. Jerrams, and S. Chen, "The Technology of Wearable Flexible Textile-Based Strain Sensors for Monitoring Multiple Human Motions: Construction, Patterning and Performance," *Sensors & Diagnostics*, vol. 2, no. 6, pp. 1414–1436, 2023, doi: 10.1039/D3SD00194F

[9] E. Bakker, "Wearable Sensors," *ACS Sens*, vol. 8, no. 4, pp. 1368–1370, Apr. 2023, doi: 10.1021/acssensors.3c00437

[10] F. Wen et al., "Machine Learning Glove Using Self-Powered Conductive Superhydrophobic Triboelectric Textile for Gesture Recognition in VR/AR Applications," *Advanced Science*, vol. 7, no. 14, pp. 1–15, 2020, doi: 10.1002/advs.202000261

[11] M. G. Tadesse, M. T. Abate, J. Felix Lübben, and M. Rauch, "Recycling and Sustainable Design for Smart Textiles: A Review," *Adv Sustain Syst*, vol. 9, no. 8, Aug. 2025, doi: 10.1002/adsu.202401072

[12] M. Dulal et al., "Sustainable, Wearable, and Eco-Friendly Electronic Textiles," *Energy & Environmental Materials*, vol. 8, no. 3, May 2025, doi: 10.1002/eem2.12854
[13] F. Wentz, M. Mohammadi, K. Tybrandt, M. Berggren, R. Arvidsson, and A. Rahmanudin, "Integrating Environmental Assessment into Early-Stage Wearable Electronics Research," *J Mater Chem C Mater*, vol. 13, no. 39, pp. 19983–19999, 2025, doi: 10.1039/D5TC02280K
[14] A. Kumari, A. Agarwal, A. Sengupta, and Y. Garg, "Conducting Yarn-based Capacitive Humidity Sensor," in *2023 IEEE Applied Sensing Conference (APSCON)*, IEEE, Jan. 2023, pp. 1–3, doi: 10.1109/APSCON56343.2023.10101251
[15] X. Wu, X. Ma, Z. Liu, X. Yu, and H. Yu, "Porous Core–Shell Yarn in Wearable Electrochemical Sensors for Real-Time Sweat Monitoring," *ACS Appl Mater Interfaces*, vol. 16, no. 48, pp. 66727–66737, Dec. 2024, doi: 10.1021/acsami.4c15353
[16] Z. Xu et al., "Smart Textiles for Personalized Sports and Healthcare," *Nanomicro Lett*, vol. 17, no. 1, p. 232, Dec. 2025, doi: 10.1007/s40820-025-01749-6
[17] K. R. S. Dinuwan Gunawardhana et al., "The Potential of Electrospinning to Enable the Realization of Energy-Autonomous Wearable Sensing Systems," *ACS Nano*, vol. 18, no. 4, pp. 2649–2684, Jan. 2024, doi: 10.1021/acsnano.3c09077
[18] N. Joseph, J. Varghese, and M. T. Sebastian, "In Situ Polymerized Polyaniline Nanofiber-Based Functional Cotton and Nylon Fabrics as Millimeter-Wave Absorbers," *Polym J*, vol. 49, no. 4, pp. 391–399, Apr. 2017, doi: 10.1038/pj.2016.121
[19] O. Ala, B. Hu, D. Li, C.-L. Yang, P. Calvert, and Q. Fan, "Conductive Textiles via Vapor-Phase Polymerization of 3,4-Ethylenedioxythiophene," *ACS Appl Mater Interfaces*, vol. 9, no. 34, pp. 29038–29046, Aug. 2017, doi: 10.1021/acsami.7b06771
[20] S. Babaeipour et al., "Thin Multifunctional Coatings for Textiles Based on The Layer-By-Layer Application of Polyaromatic Hybrid Nanoparticles," *Mater Adv*, vol. 5, no. 15, pp. 6114–6131, 2024, doi: 10.1039/D4MA00085D
[21] B. Abdi, A. Tarhini, H. Baniasadi, and A. R. Tehrani-Bagha, "Developing Graphene-based Conductive Textiles Using Different Coating Methods," *Adv Mater Technol*, vol. 9, no. 4, Feb. 2024, doi: 10.1002/admt.202301492
[22] S. Y. Park, J. Lee, and S. Lee, "Structural Influence of Knitting Patterns on Mechanical, Electrical and Durability Characteristics of Conductive Fabrics," *Sci Rep*, vol. 15, no. 1, p. 36945, Oct. 2025, doi: 10.1038/s41598-025-20780-6
[23] S. P. Beeby et al., "Heterogeneous E-Textiles: Materials, Manufacturing and Sustainability," *Adv Mater Technol*, vol. 10, no. 3, Feb. 2025, doi: 10.1002/admt.202400844
[24] A. Khan, T. Alam, M. Rashid, S. R. Mir, and G. Hossain, "Roll to Roll Triboelectric Fiber Manufacturing for Smart-Textile Self-Powered Sensor and Harvester," *Nano Energy*, vol. 111, p. 108378, Jun. 2023, doi: 10.1016/j.nanoen.2023.108378
[25] L. Fang, Y. Zhou, and Q. Huang, "Ionic Conductive Textiles for Wearable Technology," *Advanced Materials*, Jun. 2025, doi: 10.1002/adma.202502140
[26] T. Ding et al., "Scalable Thermoelectric Fibers for Multifunctional Textile-Electronics," *Nat Commun*, vol. 11, no. 1, p. 6006, Nov. 2020, doi: 10.1038/s41467-020-19867-7
[27] J. S. Marion, N. Gupta, H. Cheung, K. Monir, P. Anikeeva, and Y. Fink, "Thermally Drawn Highly Conductive Fibers with Controlled Elasticity," *Advanced Materials*, vol. 34, no. 19, May 2022, doi: 10.1002/adma.202201081
[28] B. Fang et al., "Scalable Production of Ultrafine Polyaniline Fibres for Tactile Organic Electrochemical Transistors," *Nat Commun*, vol. 13, no. 1, p. 2101, Apr. 2022, doi: 10.1038/s41467-022-29773-9
[29] A. Kumari and V. Pandey, "Non-Dynamic Power Reduction Techniques for Digital VLSI Circuits: Classification and Review," in *2020 International Conference on Computational Performance Evaluation (ComPE)*, IEEE, Jul. 2020, pp. 579–583, doi: 10.1109/ComPE49325.2020.9200156

[30] A. Kumari and T. Sharan, "CCII+ Cell Based Instrumentation Amplifier Using BD-CM Gain Enhanced OTA," *SSRN Electronic Journal*, 2020, doi: 10.2139/ssrn.3517362
[31] V. Pandey and N. K. Singh, "BER performance simulation in a multi user MIMO TAS/MRC Nakagami-m channel using BPSK, QPSK and QAM," in *2014 International Conference on Communication and Signal Processing*, IEEE, Apr. 2014, pp. 1494–1498, doi: 10.1109/ICCSP.2014.6950098
[32] K. A. S. Usman et al., "The Future of MXene Fibers," *Advanced Materials*, vol. 37, no. 36, Sep. 2025, doi: 10.1002/adma.202506437
[33] T. Busolo, P. K. Szewczyk, M. Nair, U. Stachewicz, and S. Kar-Narayan, "Triboelectric Yarns with Electrospun Functional Polymer Coatings for Highly Durable and Washable Smart Textile Applications," *ACS Appl Mater Interfaces*, vol. 13, no. 14, pp. 16876–16886, Apr. 2021, doi: 10.1021/acsami.1c00983
[34] S. Devaraju, A. K. Mohanty, D. Won, and H. Paik, "One-Step Fabrication of Highly Stable, Durable, Adhesion Enhanced, Flexible, Transparent Conducting Films Based on Silver Nanowires and Neutralized PEDOT: PSS," *Mater Adv*, vol. 4, no. 7, pp. 1769–1776, 2023, doi: 10.1039/D3MA00031A
[35] M. P. Moniz et al., "Electrospray Deposition of PEDOT: PSS on Carbon Yarn Electrodes for Solid-State Flexible Supercapacitors," *ACS Appl Mater Interfaces*, vol. 15, no. 25, pp. 30727–30741, Jun. 2023, doi: 10.1021/acsami.3c03903
[36] S. Park, H. Kim, and S. Lee, "Changes in Characteristics of Silver Conductive Fabrics Owing to Perspiration and Washing," *RSC Adv*, vol. 13, no. 41, pp. 28444–28461, 2023, doi: 10.1039/D3RA04276F
[37] A. M. Eagleton et al., "Fabrication of Multifunctional Electronic Textiles Using Oxidative Restructuring of Copper into a Cu-Based Metal–Organic Framework," *J Am Chem Soc*, vol. 144, no. 51, pp. 23297–23312, Dec. 2022, doi: 10.1021/jacs.2c05510
[38] J. Lv et al., "Printable Elastomeric Electrodes with Sweat-Enhanced Conductivity for Wearables," *Sci Adv*, vol. 7, no. 29, Jul. 2021, doi: 10.1126/sciadv.abg8433
[39] S. Afroj et al., "Engineering Graphene Flakes for Wearable Textile Sensors *via* Highly Scalable and Ultrafast Yarn Dyeing Technique," *ACS Nano*, vol. 13, no. 4, pp. 3847–3857, Apr. 2019, doi: 10.1021/acsnano.9b00319
[40] W. Fan et al., "An Antisweat Interference and Highly Sensitive Temperature Sensor Based on Poly(3,4-ethylenedioxythiophene)–Poly(styrenesulfonate) Fiber Coated with Polyurethane/Graphene for Real-Time Monitoring of Body Temperature," *ACS Nano*, vol. 17, no. 21, pp. 21073–21082, Nov. 2023, doi: 10.1021/acsnano.3c04246
[41] S. Rotzler, M. von Krshiwoblozki, and M. Schneider-Ramelow, "Washability of e-textiles: Current Testing Practices and the Need For Standardization," *Textile Research Journal*, vol. 91, no. 19–20, pp. 2401–2417, Oct. 2021, doi: 10.1177/0040517521996727
[42] H. Kim et al., "Waterproof and Conductive Tough Fibers for Washable e-textile," *npj Flexible Electronics*, vol. 9, no. 1, p. 28, Mar. 2025, doi: 10.1038/s41528-025-00399-3
[43] A. Agarwal and A. Kumari, "Absorbent Pads," INA202511048246, May 19, 2025
[44] G.-H. Lee et al., "Meter-Scale Heterostructure Printing for High-Toughness Fiber Electrodes in Intelligent Digital Apparel," *Nat Commun*, vol. 16, no. 1, p. 4320, May 2025, doi: 10.1038/s41467-025-59703-4
[45] Y. Wang et al., "Hierarchically Core-Dual Sheath Stretchable Conductive Yarns for On-Demand Thermoregulation and Wearable Strain Sensing," *Adv Funct Mater*, Oct. 2025, doi: 10.1002/adfm.202518220
[46] S. Zou, D. Li, C. He, X. Wang, D. Cheng, and G. Cai, "Scalable Fabrication of an MXene/Cotton/Spandex Yarn for Intelligent Wearable Applications," *ACS Appl Mater Interfaces*, vol. 15, no. 8, pp. 10994–11003, Mar. 2023, doi: 10.1021/acsami.2c18425
[47] I. Marriam, Y. Wang, and M. Tebyetekerwa, "Polyindole Batteries and Supercapacitors," *Energy Storage Mater*, vol. 33, pp. 336–359, Dec. 2020, doi: 10.1016/j.ensm.2020.08.010

[48] V. Pandey, A. Kumar, A. S. Razeen, A. Gupta, S. Tripathy, and M. Kumar, "Pd/AlGaN/GaN HEMT-Based Room Temperature Hydrogen Gas Sensor," *IEEE Sens J*, vol. 24, no. 24, pp. 40409–40416, Dec. 2024, doi: 10.1109/JSEN.2024.3487877
[49] Z. Wang et al., "Lead-Free Piezoelectric Composite Based on a Metamaterial for Electromechanical Energy Conversion," *Adv Mater Technol*, vol. 7, no. 12, Dec. 2022, doi: 10.1002/admt.202200650
[50] Y. Wang, C. Sun, and D. Ahmed, "A smart Acoustic Textile for Health Monitoring," *Nat Electron*, vol. 8, no. 6, pp. 485–495, May 2025, doi: 10.1038/s41928-025-01386-2
[51] Y. Li et al., "Influence of Halides on the Interactions of Ammonium Acids with Metal Halide Perovskites," *ACS Appl Mater Interfaces*, vol. 15, no. 20, pp. 24387–24398, May 2023, doi: 10.1021/acsami.3c01432
[52] H. Souri et al., "Wearable and Stretchable Strain Sensors: Materials, Sensing Mechanisms, and Applications," *Advanced Intelligent Systems*, vol. 2, no. 8, Aug. 2020, doi: 10.1002/aisy.202000039
[53] H. Liu, Z. Wang, and G. Zhou, "Wearable Sensors for Monitoring Workplace Chemical Exposures in Occupational Health Management," *Analytical Methods*, vol. 17, no. 39, pp. 7863–7889, 2025, doi: 10.1039/D5AY00961H
[54] E. Svertoka et al., "Wearables for Industrial Work Safety: A Survey," *Sensors*, vol. 21, no. 11, p. 3844, Jun. 2021, doi: 10.3390/s21113844
[55] W. Gong et al., "Scalable and Reconfigurable Green Electronic Textiles with Personalized Comfort Management," *ACS Nano*, vol. 16, no. 8, pp. 12635–12644, Aug. 2022, doi: 10.1021/acsnano.2c04252
[56] K. Yao et al., "A Fully Integrated Breathable Haptic Textile," *Sci Adv*, vol. 10, no. 42, Oct. 2024, doi: 10.1126/sciadv.adq9575
[57] Md. M. Hossain, B. M. Li, B. Sennik, J. S. Jur, and P. D. Bradford, "Adhesive Free, Conformable and Washable Carbon Nanotube Fabric Electrodes for Biosensing," *npj Flexible Electronics*, vol. 6, no. 1, p. 97, Dec. 2022, doi: 10.1038/s41528-022-00230-3
[58] G. Loke et al., "Digital Electronics in Fibres Enable Fabric-Based Machine-Learning Inference," *Nat Commun*, vol. 12, no. 1, p. 3317, Jun. 2021, doi: 10.1038/s41467-021-23628-5
[59] M. Dulal et al., "Toward Sustainable Wearable Electronic Textiles," *ACS Nano*, vol. 16, no. 12, pp. 19755–19788, Dec. 2022, doi: 10.1021/acsnano.2c07723
[60] P. Luoma, E. Penttinen, P. Tapio, and A. Toppinen, "Future Images of Data in Circular Economy for Textiles," *Technol Forecast Soc Change*, vol. 182, p. 121859, Sep. 2022, doi: 10.1016/j.techfore.2022.121859
[61] A. Komolafe et al., "Integrating Flexible Filament Circuits for E-Textile Applications," *Adv Mater Technol*, vol. 4, no. 7, p. 1900176, Jul. 2019, doi: 10.1002/admt.201900176
[62] S. Rotzler, M. von Krshiwoblozki, C. Kallmayer, and M. Schneider-Ramelow, "Washability of E-Textiles: Washing Behavior of Textile Integrated Circuits Depending on Textile Substrate, Circuit Material and Integration Method," *Adv Funct Mater*, vol. 35, no. 11, Mar. 2025, doi: 10.1002/adfm.202417344
[63] V. Pandey, A. Gupta, S. Tripathy, and M. Kumar, "IoT-Enabled hBN/AlGaN/GaN High Electron Mobility Transistor for Carbon Dioxide Monitoring," in *2024 IEEE Sensors*, IEEE, Oct. 2024, pp. 1–4, doi: 10.1109/SENSORS60989.2024.10784655
[64] K. Chandnani et al., "A Novel Smart Baby Cradle System Utilizing IoT Sensors and Machine Learning for Optimized Parental Care," *Sci Rep*, vol. 15, no. 1, p. 19080, May 2025, doi: 10.1038/s41598-025-02691-8

5 Textile-Based Wearable Sensors

Advances in Multimodal Sensing for Continuous Human Health Monitoring

Seyed Majid Ghoreishian and Hubert Bryan Riley

INTRODUCTION

Textiles, frequently described as the "second skin" of the human body, have accompanied humankind for approximately 27,000 years [1]. To date, the textile industry is among the most advanced industries through the integration of nanotechnology, high-technology materials, and electronics, particularly in artificial intelligence systems [2].

Modern textiles integrate a wide range of advanced functionalities, including electrical conductivity, ultraviolet (UV) shielding, flame retardancy, electrochromism, thermoregulation, self-cleaning properties, antimicrobial activity, solar-energy harvesting, photonic behavior, and even catalytic performance [3]. These advancements mainly rely on sophisticated tools and methodologies, including multiscale simulations of material properties and processing parameters, advanced mechatronics, robotics, machine-vision systems, and self-regulating or machine-learning-enabled processes, to achieve efficient, single-stage production of complex, multilayered, three-dimensional (3D) structures [4].

Among modern textiles, "Smart Textiles" constitute an indispensable component of emerging technology field. This term mainly refers to a broad field of studies and products that extend the functionality and usefulness of common fabrics [5]. According to the International Organization for Standardization (ISO), smart textiles are distinguished from conventional fabrics by intrinsic functional properties that extend far beyond mechanical protection and aesthetics [4].

Smart textiles can integrate electronic devices into yarns, fabrics, or garments and respond to a wide range of external stimuli. Owing to their inherent mechanical flexibility, compatibility with established knitting and weaving processes, and excellent wearer comfort, smart textiles represent a highly promising platform for next-generation wearable electronic systems [6]. Although smart textiles have emerged over 20 years, in recent times, the true promise of smart textiles lies in

DOI: 10.1201/9781003507208-5

their ability to evolve everyday garments into multifunctional platforms that simultaneously serve as thermal regulators, energy harvesters, and power-storage units, thereby enabling a new generation of truly intelligent and autonomous clothing [4, 7].

Of all the functional segments in the smart textiles market, the sensing category is expected to deliver the strongest growth, contributing an additional USD 2.14 billion to global annual revenue by 2026. According to a report published by Straits Research, the global electronic-type smart textile market was valued at USD 2.82 billion in 2022 and is projected to reach USD 17.94 billion by 2031, representing a compound annual growth rate (CAGR) of 22.8 % over the forecast period. Similarly, a Euratex survey estimates that the European electronic textiles sector alone will attain a market value of approximately €1.5 billion by 2025. To accelerate innovation in this field and shorten the time-to-market for high-potential prototypes, the European SmartX Community initiative was established, running from 2019 to 2022, with dedicated support for the development and commercialization of smart textile solutions [4].

Among the different types of smart textiles, high-performance functional wearable sensors are the most widely used, with applications across industries such as military, defense, healthcare, sports, and fitness [8, 9]. Wearable sensors enable active responses to environmental or physiological stimuli, marking a fundamental shift from passive garments to interactive and adaptive wearable systems. These systems are fully body-worn electronic platforms integrating a broad spectrum of capabilities, including embedded biosensors for monitoring pressure, temperature, humidity, strain, and chemical analytes; on-board data processing and wireless communication; as well as actuation, thermal regulation, energy harvesting, storage, and even aesthetic responsiveness (Figure 5.1) [4, 11, 12].

Textile-based sensors include energy harvesting and thermoelectric power generation from body heat, fully wireless human–machine interfaces, inertial and strain sensors for precise tracking of human-motion direction and velocity, advanced biosensors for real-time healthcare diagnostics, and aesthetically integrated smart fabrics within the fashion and apparel industries [4, 8]. However, current wearable sensors still face several critical limitations:

- The introduction and sealing processes required for durability often employ non-breathable materials, which compromise long-term wearer comfort and skin health.
- Integration into garments typically relies on rigid or cumbersome attachment techniques that restrict natural movement and limit practical wearability.
- Fabrication routes remain complex, time-consuming, and costly, hindering scalability and broad adoption.

These technical challenges collectively impede the transition of textile sensing platforms from laboratory prototypes to reliable, everyday wearable systems [6].

From this perspective, researchers are actively pursuing wearable sensors that combine high sensitivity, self-powering capability, and additional performance-enhancing features. To realize these multifaceted requirements, substantial efforts have focused on incorporating advanced materials into sensor components and

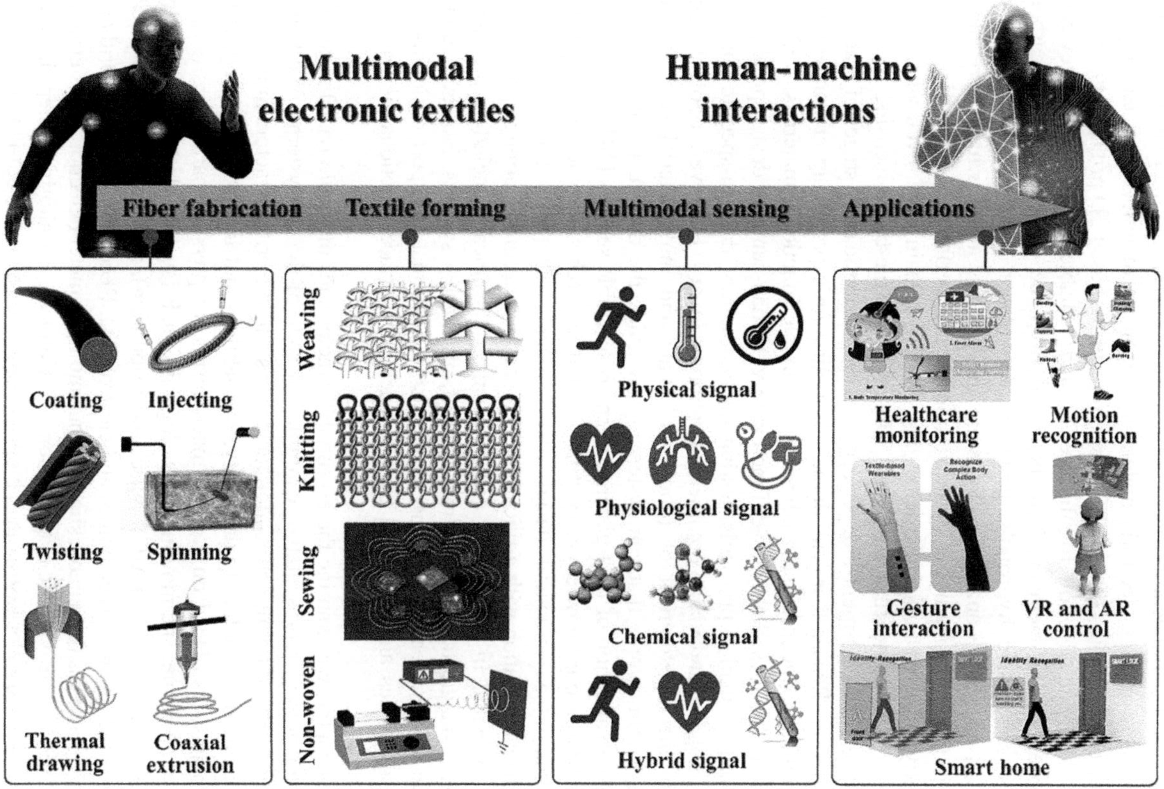

FIGURE 5.1 Schematic overview of modern electronic textiles designed for intelligent human–machine interfaces, including fiber fabrication, textile forming, multimodal sensing, and application for human–machine interfaces.

(This figure was adapted with permission from Ref. [10]. Copyright 2023, Soft Science.)

refining fabrication technologies [13]. Accordingly, these multifunctional textiles can be combined with non-textile components, such as shape-memory alloys, electroactive polymers, thermoelectric modules, or flexible energy-storage devices [4]. In addition, notable examples include the integration of multidimensional carbon nanofibers, hybrid nanocomposites, liquid-metal alloys, and nanomaterial-based conductive networks as electrode materials. Advanced manufacturing techniques, such as selective laser sintering (SLS) and photocuring-based 3D printing, have also gained widespread adoption for the precise and scalable production of devices. Furthermore, sensing elements have benefited from the extensive use of carbon nanotubes, conductive hydrogels, graphene derivatives, as well as textile- and paper-based capacitive and ionic-electronic transducers, all of which have been explored in numerous studies to optimize transduction performance [6, 13].

Currently, most wearable microelectronic devices remain rigid, even when integrated into flexible textile substrates. However, textile-based sensing systems capable of seamless, garment-like performance remain in the early stages of practical implementation. Research in this domain is both highly promising and inherently multidisciplinary, encompassing advances in flexible functional materials, scalable fabrication techniques, device-level functionality, human-centered design principles, standardized evaluation protocols, testing methodologies, domain-specific application requirements, and viable commercialization pathways. However, a notable gap persists in the development of robust multiscale and multimaterial science and engineering frameworks that can systematically guide the design, fabrication, and performance validation of next-generation smart wearable sensors [14].

To the best of our knowledge, existing original research and review articles have primarily addressed structural design and fabrication strategies for textile-based sensors, their integration with artificial intelligence and machine-learning approaches, and their potential in sustainable and personalized healthcare, including wearable point-of-care diagnostics and continuous health-monitoring systems. However, comprehensive reports specifically focusing on smart textiles that incorporate multimodal sensing capabilities and advanced human–machine interface functionalities remain scarce.

In this chapter, we examine the essential characteristics of textile-based wearable sensors and present representative examples of state-of-the-art systems. We begin with an overview of textile fabrication techniques and the critical factors influencing sensor performance, including durability, washability, and long-term stability. Particular emphasis is placed on sensing mechanisms for physical parameters (strain, pressure, temperature, and motion), physiological signals (such as electrocardiography (ECG), electromyography (EMG), pulse, and respiration), and chemical/biochemical analytes (including sweat metabolites and electrolytes). Recent progress in hybrid and multimodal platforms that enable simultaneous multiparameter detection is systematically reviewed, along with their applications in healthcare monitoring, sleep disorder diagnosis, cardiovascular assessment, human-motion recognition, sign-language translation, and human–machine interfaces. Finally, we discuss the principal challenges remaining in the field—ranging from mechanical robustness and energy autonomy to scalability, biocompatibility, and sustainability—and outline promising directions for future research. This analysis aims to provide a comprehensive framework for advancing next-generation smart textile systems toward practical, real-world implementation in continuous health monitoring and personalized medicine.

TEXTILE FABRICATION TECHNIQUES

In general, textiles are formed by the controlled interlacing or bonding of fibers and yarns through several established techniques, including weaving, knitting, braiding, and nonwoven production (Figure 5.2). Each approach offers specific benefits and trade-offs. Woven fabrics, for instance, permit precise control of dimensions and generally exhibit high tensile strength, but they typically possess limited extensibility. In contrast, knitted structures provide excellent stretch and recovery yet tend to have lower stiffness (Young's modulus). Braided textiles are valued for their robustness and resistance to shear deformation, although they usually display reduced porosity relative to other constructions. 3D textile architectures—particularly 3D

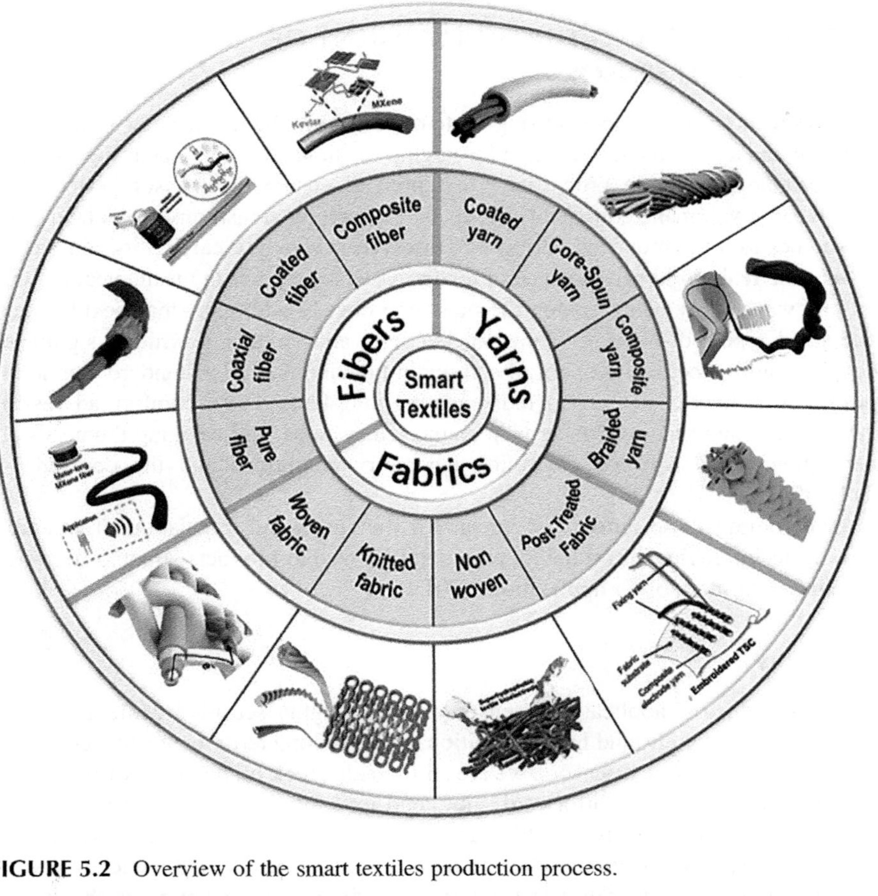

FIGURE 5.2 Overview of the smart textiles production process.

(This figure was adapted with permission from Ref. [15]. Copyright 2025, Springer Nature.)

knitted and woven spacer fabrics—can deliver superior mechanical performance and are especially well suited for the seamless integration of electronic sensors. Their inherent cushioning effect enhances wearer comfort, while the spaced construction allows tailored adjustment of thickness, compressibility, and permeability [16, 17].

In recent years, the entire production chain—from fiber spinning and yarn formation to the assembly of multilayered composite fabrics was systematically investigated. The resulting textiles are suitable for both traditional garments and advanced smart textile platforms with sensing capabilities [18]. In many reports on wearable smart textiles designed for mood regulation, however, the choice of base textile materials and fabric constructions remains poorly defined. The emphasis instead tends to fall solely on sensor integration and electrical conductivity. To advance the field, future work should place greater priority on the deliberate engineering of fibers and fabric architectures, given that these substrates are in direct and prolonged contact with the skin and thus play a critical role in achieving reliable, mood-responsive performance [19].

FACTORS AFFECTING THE DEVELOPMENT OF TEXTILE-BASED SENSORS

The fabrication of wearable sensing systems represents both a frontier in academic research and a formidable challenge for large-scale industrial manufacturing. Achieving scalable production processes is essential to ensure that sensor performance and reliability remain consistent as sensor technology transitions from laboratory prototypes to mass-produced garments. However, several critical factors continue to govern the real-world performance of textile-based sensors, with mechanical durability and washability being among the most pressing. In addition, smart textiles must withstand repeated cycles of abrasion, bending, and stretching without significant degradation of electrical properties, and retain functionality after standard laundering procedures. Other challenges include ensuring long-term wearer comfort, addressing data privacy and security concerns in connected systems, and bridging the persistent performance gap between controlled laboratory conditions and the demands of daily life.

Overcoming these hurdles is crucial for the successful translation of wearable sensing technologies into reliable, commercially viable products suitable for widespread health-monitoring applications [20].

DURABILITY

For most industrial applications, sensors must be highly sensitive while also being robust enough to withstand harsh conditions over the long term. Unfortunately, when it comes to textile-based sensors, we are still mostly stuck in the world of wearables. The main reason is straightforward: they tend to deteriorate after repeated use and simply do not last. Because textiles are inherently soft and flexible, developing a sensor that remains functional after numerous washes, bends, and stretches—while still allowing air or water to pass through—is a significant challenge. In the case of printed textile sensors, the major challenges remain wash fastness, flexibility under bending, and stretch resistance without compromising performance [20]. The fabric needs to be soft and smooth—no texture or pile—so the print comes out crisp. It is worth adding a thin interface (or barrier) layer on top; that is the key to obtaining sharp, high-quality results every time. This layer protects the conductive ink or paste from water during washing, preventing circuit degradation. As for the conductive material itself, it must remain flexible and stable through repeated bending and laundry cycles—that is what gives you real, long-lasting durability [21].

Washing Stability

Ever since the first wearable sensors started being sewn or laminated into clothing, the question designers and researchers hear most often is a simple, practical one: "Can I wash it?" For smart textiles to become truly wearable, their electroconductive elements must withstand repeated laundering or dry cleaning without degrading performance. One effective strategy is to protect these functional layers by laminating them, which shields the conductive substrates from mechanical stress and creasing during washing cycles. [22].

Several groups have explored methods to enable wearable sensors to withstand repeated washing cycles. Among the approaches that have shown promise are encapsulating the sensing elements with polyurethane (PU) [23], pretreating the textile with bovine serum albumin (BSA) to block nonspecific binding sites [24], coating the finished sensor with polydimethylsiloxane (PDMS) [25], and introducing a screen-printed PU interface layer directly on the fabric surface [26]. In this context, Schäl *et al.* [27] fabricated the sensor by applying a graphite/PU composite coating (with graphite loadings of 25–33 wt%) onto 100% cotton plain-weave fabric, then securing an adhesive tape layer on top. The resulting sensing system was subjected to ten standard washing cycles at 40°C and 1400 rpm. Remarkably, after these ten washes, no residual graphite flakes could be detected on the fabric surface.

Furthermore, it was reported that after washing, the cotton fabric surface appeared noticeably smoother than in its original state; however, a few minor defects—small scuffs and faint abrasions—had appeared, most likely from repeated contact with coarser items in the wash.

In another study, Yan *et al.* [28] developed a highly sensitive, flexible strain sensor by combining thermoplastic polyurethane (TPU) with carbonized woven fabric made from polyacrylonitrile nanofiber yarn. The resulting sensor shows impressive performance, delivering an average gauge factor (GF) of 77.3 across a 12 % strain range while maintaining excellent durability and stability—even after more than 1000 stretching cycles at 5% strain.

Aging/Storage Effect

The long-term stability of electronic textiles is significantly influenced by aging processes, which ultimately govern their practical shelf-life. To enable reliable deployment in wearable applications, the effects of aging on electrical conductivity under varying storage conditions and over extended periods must be systematically evaluated and benchmarked.

To date, limited data are reported and available on the impact of prolonged storage on the electrical performance of conductive textile-based sensors. Systematic studies are therefore required in which samples are stored for several weeks (or months) under controlled atmospheric conditions (e.g., different combinations of temperature, relative humidity, and ambient gases), with conductivity measured at regular intervals. Weekly or biweekly monitoring of resistance/conductivity is recommended to quantify the kinetics of any degradation.

Such experiments will reveal the rate and mechanisms of performance loss, allowing direct correlation with environmental parameters. Establishing the magnitude and temporal profile of these changes is essential for defining realistic shelf-life specifications and for implementing appropriate storage protocols or protective coatings for next-generation conductive textile sensors [20].

APPLICATIONS OF TEXTILE SENSORS

As highlighted previously, textile-based electronics, alongside other flexible electronic platforms, have garnered considerable attention for their versatility in wearable sensing applications. These sensors can play a pivotal role in advancing continuous health monitoring, human-motion tracking, and personalized medical management, ultimately serving as indispensable tools for promoting individual well-being, as shown in Figure 5.3.

In general, textile-based sensors are classified as passive or active. Passive sensors depend on an external power source for operation, whereas active sensors directly transduce input stimuli—such as mechanical strain or thermal energy—into an electrical potential difference. To date, the majority of reported textile sensors fall into the passive category. Regardless of type, all textile-based sensing systems require a reliable power supply [30–32].

Given the inherently portable nature of wearable devices, batteries remain the predominant energy source. Early battery designs were rigid and incompatible with full integration into soft, conformable textiles. Consequently, the development of flexible—and, more recently, stretchable—energy-storage devices—often combined with energy-harvesting mechanisms such as triboelectric, piezoelectric, thermoelectric, and magnetoelastic effects—has emerged as a critical enabling technology for truly integrated, self-powered smart textiles and wearable sensing platforms. Furthermore, the wireless connectivity feature allows direct coupling of these systems to smartphone platforms or the Internet of Things (IoT) networks, eliminating the need for cumbersome wiring [33–35]. In the following sections, we examine recent progress in textile-based wearable sensors across a broad range of operational parameters evaluated in Section (Applications of Textile Sensors), encompassing physical signals (strain, pressure, temperature, and full-body motion recognition), physiological signals (ECG, EMG, pulse wave, respiration, and sleep-related patterns), chemical and biochemical signals (sweat metabolites, electrolytes, pH, and gas detection), as well as advanced hybrid and multimodal platforms that enable simultaneous multiparameter sensing for applications in healthcare monitoring, cardiovascular assessment, sleep disorder diagnosis, sign-language translation, and human–machine interfaces.

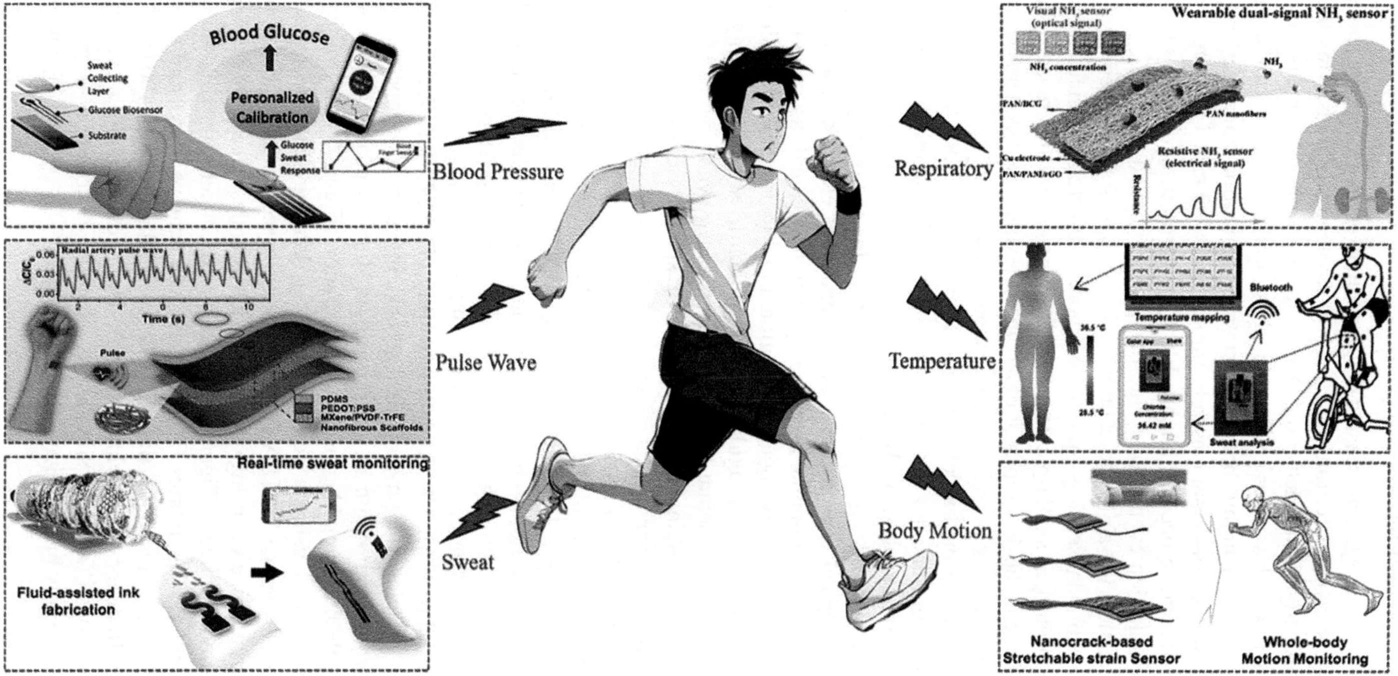

FIGURE 5.3 Recent advances in textile-based wearable sensors for various applications.

(This figure was adapted with permission from Ref. [29]. Copyright 2023, MDPI.)

Physiological Signal Sensing

As mentioned earlier, integrating sensing devices into everyday garments and fabrics enables wearable sensors to continuously track physiological signals. This seamless approach enables early disease detection, remote health monitoring, and personalized healthcare in real-world settings. Wearable physiological sensors can monitor a range of parameters—including body temperature, respiration rate, heart rate, and various bioelectric signals—noninvasively through skin contact. Over the past few years, these devices have become remarkably lighter, softer, and more stretchable, enabling secure adhesion even during vigorous physical activity.

In clinical practice, the ability to capture multiple signals simultaneously—often referred to as multimodal sensing—provides a more complete and reliable picture of a patient's condition than single-parameter measurements. For this reason, the field has seen a rapid rise in the development of advanced textile-based systems capable of simultaneously monitoring pulse, breathing patterns, and other vital signs. Notably, it has been reported that the biopotential signals captured by such sensors typically span frequencies from about 1 Hz to several kHz, with amplitudes ranging from microvolts to approximately 1 V. Because these signals are often extremely weak, conventional rigid electrodes are no longer adequate. Instead, next-generation wearable electrophysiological sensors must exhibit excellent electrical conductivity, a high signal-to-noise ratio, minimal measurement variability, and fast response times to ensure reliable, high-fidelity signal acquisition and downstream processing [13].

Importantly, humans spend roughly one-third of their lives asleep. Quality sleep is essential for physical and mental health, yet sleep apnea syndrome (SAS) poses a serious threat to both. Continuous monitoring of respiratory effort, pulse, or heart rate patterns during sleep provides an effective means of detecting and managing this condition. Furthermore, by analyzing pulse transit time (PTT) derived from these signals, clinicians can pinpoint the specific type of SAS affecting the patient and tailor treatment accordingly. Beyond sleep apnea, smart textiles enable the diagnosis of a range of other conditions—for instance, cardiovascular diseases through continuous pulse-wave monitoring, COVID-19 or other respiratory infections via breathing patterns, and more [10]. In this context, Fan *et al.* [36] prepared a machine-knitted triboelectric all-textile sensor array (TATSA) for precise, noninvasive monitoring of epidermal physiological signals, integrating conductive stainless steel/Terylene yarns with nylon in a full cardigan stitch to enable high-sensitivity detection of subtle pressures like pulse and respiration (Figure 5.4a). In this study, TATSA was fabricated using a computerized flat knitting machine for scalable production. The developed TATSA features a loop-based structure that optimizes the triboelectric contact area, using conductive yarn (stainless steel core twisted with Terylene) and nylon thread (Figure 5.4b) to ensure breathability, comfort, and compatibility with various fabrics. This sensor could achieve a pressure sensitivity of 7.84 mV Pa^{-1}, a response time of 20 ms, stability over 100,000 cycles (Figure 5.4c), a frequency bandwidth up to 20 Hz, and washability exceeding 40 cycles, validated by COMSOL simulations of force distribution (Figure 5.4d). Experimental tests show linear voltage–pressure response ($R^2 = 0.99$) and negligible

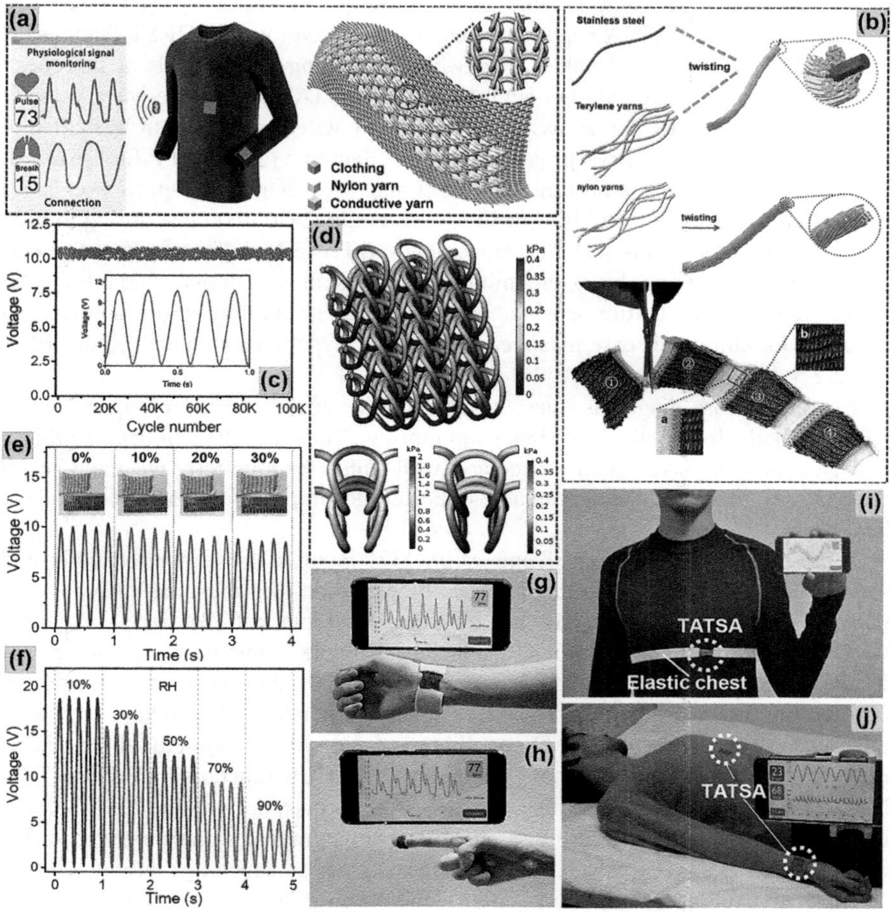

FIGURE 5.4 (a) Two TATSAs integrated into a shirt for the monitoring of pulse and respiratory signals in real time. (b) Structure of the conductive yarn, nylon thread, and multiple TATSAs were knitted together through a conventional plain stitch. (c) Durability test of a TATSA under a pressure of 1 kPa. (d) Simulation result of the force distribution of a TATSA under an applied pressure of 0.2 kPa using the COMSOL software. The output voltages of the sensor (e) at various relative humidity from 10 to 90% and (f) at different tensions in the wale direction. Measurement of the pulse at the (g) wrist and (h) fingertip ankle. (i) Photograph showing the display of the TATSA placed on the chest for measuring the signal in the pressure associated with respiration. (j) Photograph showing two TATSAs placed on the abdomen and wrist for measuring respiration and pulse, respectively, during sleep.

(This figure was adapted with permission from Ref. [36]. Copyright 2020, American Association for the Advancement of Science.)

degradation under tension (Figure 5.4f), twisting, or humidity (10–90% RH) (Figure 5.4e). The TATSA captures detailed pulse waveforms (PQRST features) at neck, wrist, fingertip, and ankle; meanwhile, the respiratory signals at the chest/abdomen enable the extraction of augmentation index (AIx) and pulse-wave velocity (PWV) for cardiovascular assessment. Integrated wireless Bluetooth system into clothing like shirts and socks provides real-time data to a mobile APP for CAD risk evaluation (AIx > 0.3 indicating high risk) and SAS diagnosis (apnea events via respiration patterns), with a high signal-to-noise ratio (>20 dB) even during motion (Figure 5.4g and h). This scalable, biocompatible TATSA platform advances wearable health monitoring by overcoming limitations in sensitivity, durability, and integration of prior textile sensors, offering superior comfort, breathability, and accuracy for chronic disease management, such as hypertension and sleep disorders (Figure 5.4i and j).

Beyond disease diagnosis, these smart textiles can remain in prolonged contact with the body for continuous signal monitoring. Their excellent biocompatibility with human skin makes them particularly well-suited for long-term wearable health applications, thereby creating valuable opportunities in preventive and personalized healthcare.

With respect to this topic, Dong *et al.* [37] reported an ultra-stretchable and superhydrophobic nonwoven textile-based bioelectrode for robust self-cleaning and personal health monitoring. By integrating carbon black (CB)/carbon nanotube (CNT) hybrid (CB/CNT) into a conductive network containing perfluorooctyltriethoxysilane (PFOTES)–TiO_2 nanoparticles (NPs), the outer layer could be enabled to show a high-performance sensing in harsh environments, including underwater and corrosive conditions (Figure 5.5a). In this system, the Arduino platform was interfaced with two textile-based PTCCS (anchored PFOTES-TiO_2 NPs on CB/CNT (CCS) textiles), as electrodes, placed over the belly of the biceps brachii and near the elbow joint. Electromyographic signals were recorded during three distinct biceps contraction tasks—fist clenching, elbow flexion (lifting), and forearm supination—after standard amplification and filtering steps. When the same conductive fabric was used as a strain sensor, it produced highly repeatable changes in resistance that directly reflected the reorganization of its internal conductive network during deformation. Technically, PTCCS electrodes were fabricated via electrospinning by blending styrene–ethylene–butylene–styrene (SEBS) and Poloxamer (F127) to enhance hydrophilicity and fiber elasticity, followed by anchoring CB/CNT and spray-coating with PFOTES–TiO_2 to achieve microtopographic superhydrophobicity (Figure 5.5b and c). The PTCCS strain sensor demonstrates an adjustable ultrabroad sensing window up to 1050% with tunable GFs exceeding 1130 depending on the CB/CNT ratio (Figure 5.5d), rapid response and recovery times of ~102 and ~151 ms ensuring real-time signal acquisition (Figure 5.5e), excellent signal stability under static holding at large strains up to 180% (Figure 5.5f), and a robust deformation-tolerant sensing mechanism enabled by adaptive CB/CNT conductive networks that maintain percolation under extreme stretching, as revealed by scanning electron microscopy (SEM) observations (Figure 5.5g). This wearable strain sensor monitors subtle skin motions and large deformations with high sensitivity, while as a bioelectrode, it detects electrocardiogram

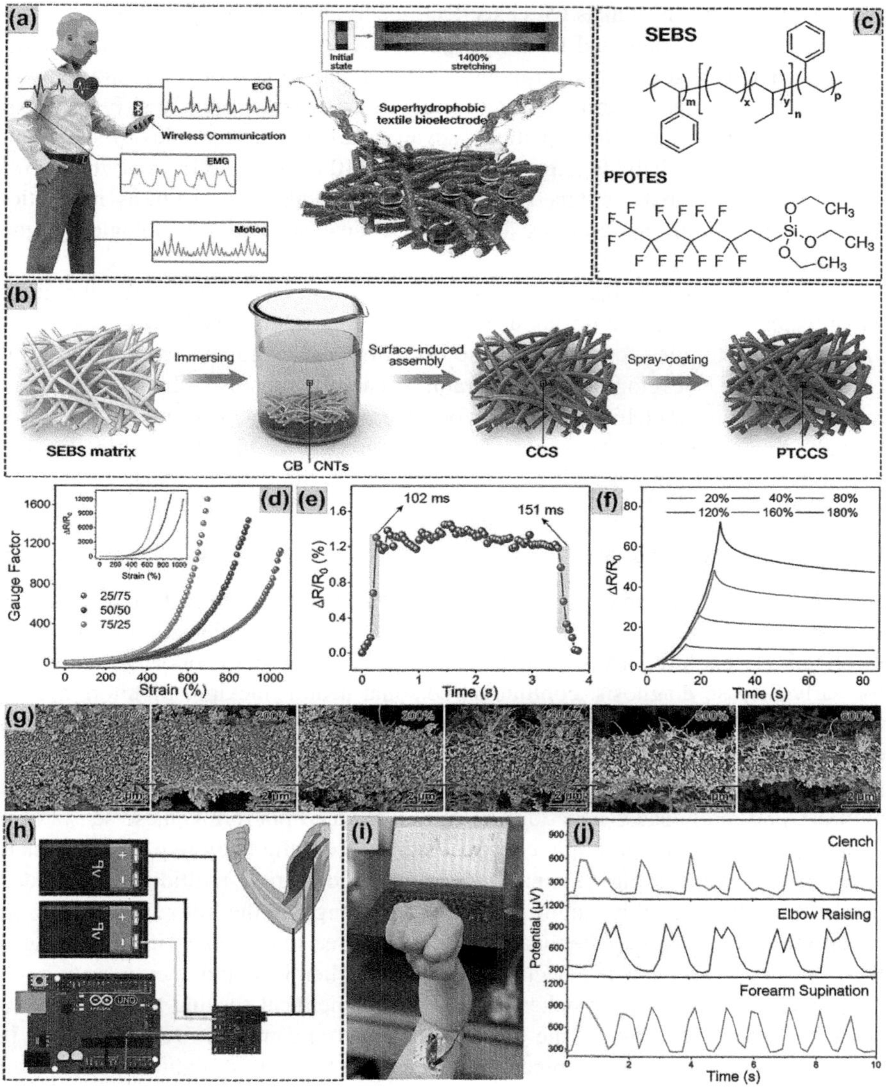

FIGURE 5.5 (a) Schematic presentation of ultrahigh stretchable and superhydrophobic bioelectrodes for personal health monitoring. (b) The fabrication process of the ultra-stretchable and superhydrophobic textile by anchoring CB/CNT hybrids onto electrospun SEBS. (c) Chemical structure of SEBS and PFOTES. (d) Gauge factor versus strain of PTCCS strain sensor associated with CB/CNT volume ratios of 25/75, 50/50, and 75/25 (inset: plots of $\Delta R/R_0$ versus strain). (e) Response/recovery times of the PTCCS strain sensor with CB/CNT volume ratios of 25/75. (f) Static response of PTCCS under 20%, 40%, 80%, 120%, 160%, and 180% strains and holding for 80s. (g) SEM images of PTCCS fiber under gradually increased strains of 100–600%. (h, i) Schematic setup and photograph of EMG device for measuring bioelectricity of bicipital muscle, and (j) acquired EMG signals of clenching, elbow raising, and forearm supination.

(This figure was adapted with permission from Ref. [37]. Copyright 2022, Elsevier.)

and electromyography signals with excellent fidelity, maintaining performance during clenching, elbow raising, and forearm supination (Figure 5.5h and i). Other demonstrations include real-time human-motion tracking and bioelectrical monitoring, with antifouling properties ensuring long-term stability in dynamic scenarios (Figure 5.5j). This cost-effective and scalable platform advances textile bioelectronics by surpassing prior sensors in stretchability (e.g., GF = 268.4 at 110% strain or GF = 71.76 at 120% strain), multifunctional resistance to harsh environments, and seamless integration for wearable health applications like motion analysis and physiological signal detection.

In conclusion, textile-based sensors can quietly track key physiological signals. At the same time, people go about their daily lives undisturbed, thereby creating valuable opportunities for early disease detection and ongoing health monitoring. Looking ahead, developing truly durable wearable sensors—able to withstand repeated washing, stretching, and real-world wear—will be a major research priority. Equally exciting is the integration of fabric sensors with the IoT, a direction that promises to transform wearable health technology in the coming years.

Chemical Signal Sensing

Beyond the physical and physiological sensors already discussed, another major branch of wearable sensors focuses on multimodal chemical detection—one of the most actively explored areas in smart textiles today. These sensors are widely used for early disease diagnosis, continuous personal health tracking, monitoring emotional states via sweat chemistry, and detecting hazardous gases in the surrounding air. Because they are built directly into everyday textiles—such as T-shirts, wristbands, masks, or headbands—they feel natural against the skin, provide excellent comfort, and rarely cause irritation. As fully integrated platforms, these sensors can simultaneously capture multiple chemical signals from the body (e.g., sweat metabolites, pH, electrolytes) and the environment, providing rich, multidimensional data needed for reliable health insights and timely warnings. In the following section, we group these textile-based chemical sensors into three main categories according to their operating mechanisms: electrochemical, optochemical, and chemiresistive.

Among the various approaches to developing chemical-sensing textiles, electrochemical detection is by far the most investigated and most widely deployed. Its popularity stems from a combination that is hard to beat: straightforward design, robust performance, and the ability to translate analyte concentration directly into an easily measured electrical signal—either a voltage or a current.

These electrochemical sensing platforms generally fall into two broad categories: conventional electrochemical sensors (potentiometric, amperometric, or voltammetric) and organic electrochemical transistors (OECTs).

Technically, standard electrochemical sensors operate according to the Nernst equation: they measure electrode potential or current, which varies predictably with analyte concentration, thereby translating chemical information into an electrical signal. This signal is then amplified and processed by the readout circuitry. In this context, integrating multiple selective electrodes into a single textile platform enables simultaneous monitoring of multiple analytes.

From this perspective, Zhao *et al.* [38] developed a core–shell structured gold nanorods (AuNRs@Au) embedded in a thread-embroidered fabric-based microfluidic device for *ex situ*, multifunctional sweat sensing, enabling simultaneous surface-enhanced Raman scattering (SERS)-based detection of glucose and lactate alongside colorimetric assessment of local sweat loss (Figure 5.6a). This sensor was fabricated by embroidering mercerized cotton threads into hydrophobic cotton fabric for detection sites and viscose threads for microfluidic channels, followed by deposition of AuNRs@Au SERS tags. The flexible thread-embroidered sensing band (Figure 5.6b and c) conforms well to highly curved body regions such as the head, elbow, and wrist, ensuring stable skin contact during motion (Figure 5.6d). The comparative on-body measurements show that sweat glucose levels remain nearly position-independent at ~51.86–67.19 μM and correspond to blood glucose values of ~5.1–6.4 mM after 45 min of exercise (Figure 5.6e), in contrast to sweat lactate concentrations that vary with body location, ranging from 3.50 to 10.34 mM due to differences in local muscle activity (Figure 5.6f), alongside position- and environment-dependent sweat loss quantified at ~12.5–100 μL and correlated with infrared (IR) thermal imaging under indoor and outdoor conditions (Figure 5.6g). In this analysis, colorimetric sweat loss monitoring uses active dyes (active red 3B, orange SNE, yellow 4GL, green 6G, blue 4B, and purple K-3R) in a thread-embroidered/fabric sensing band (Figure 5.6c). This integrated, cost-effective textile platform advances noninvasive health monitoring by combining SERS sensitivity with microfluidic efficiency, outperforming prior sensors in multiplexed metabolite detection, portability, and real-time athletic/medical applications like fatigue assessment and diabetes tracking.

Healthcare Monitoring

In recent years, wearable sensors have been integrated directly into fabrics, enabling the development of advanced smart textiles through continuous, real-time health monitoring. Among the many threats to human health, cardiovascular diseases have remained the leading cause of death worldwide for decades. Recording a person's ECG does far more than flag sudden cardiac events—it reveals a wealth of information about overall physiological state simply through the shape and timing of the waveforms.

In textile-based ECG systems, working and reference electrodes are seamlessly integrated into the garment and positioned to make stable contact with the wearer's upper arms. This simple configuration yields clean, high-signal-to-noise-ratio tracings that remain reliable even during prolonged wear.

Beyond ECG, today's ultrasensitive textile mechanical sensors can detect the subtle pulse wave at the wrist with high fidelity. From these recordings, clinicians—and even automated algorithms—can extract telling clues about age, gender, and early signs of conditions such as hypertension [39]. In this regard, Takamatsu *et al.* [39] developed a simple and reliable technique for direct patterning of organic conductors on knitted textiles, creating flexible electrocardiography electrodes for long-term, noninvasive heart monitoring with low-impedance skin contact (Figure 5.7a). This sensing platform was fabricated by dip-coating knitted polyester

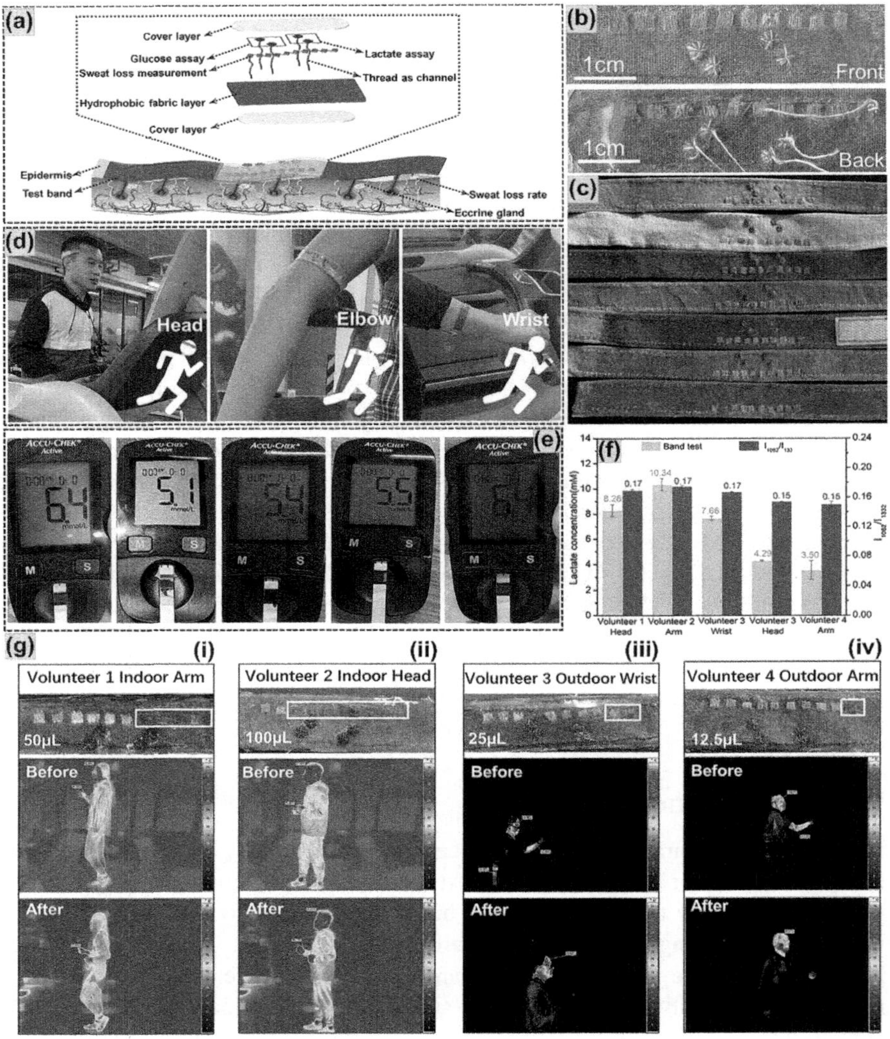

FIGURE 5.6 (a) Exploded view illustration of a thread-embroidered/fabric sensing band and its interface with the skin. (b) Front view and back view of the produced band. (c) Optical images of flexible, wearable thread-embroidered/fabric microfluidic devices for sweat sensing and monitoring. (d) Deployed wearable device on regions of the human body with different curvatures and shapes. (e) Test results of the blood glucose from the four volunteers using the commercial blood glucose meter. (f) The peak intensity ratio of I_{1082}/I_{1332} and the value of the lactate concentration calculated through the fitting function $g(x)$ ($g(x)$ is the concentration, and x is I_{1082}/I_{1332}). (g) Local sweat loss tested by the thread-embroidered/fabric sensing band (top) and infrared thermal imaging of the subjects before and after 45 min exercise in the gym (indoor, (i) and (ii)) and in the playground (outdoor, (iii) and (iv)).

(This figure was adapted with permission from Ref. [38]. Copyright 2022, Elsevier.)

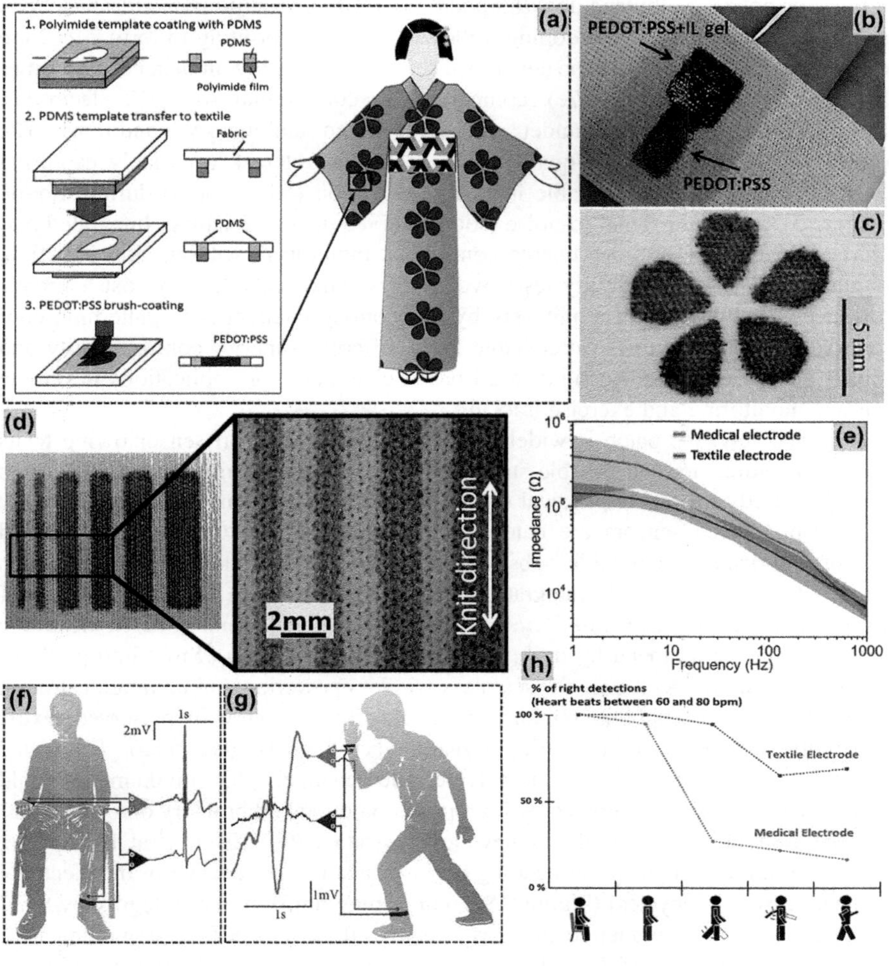

FIGURE 5.7 (a) Process flow for the patterning of PEDOT:PSS on textiles, inspired by the Japanese kimono dying method. (b) Photograph of a textile wristband with a PEDOT:PSS electrode. (The electrode has an area of $1\,cm^2$, which is also coated with an ionic liquid (IL) gel, indicated by the PEDOT/PSS + IL gel arrow.) (c) Flower pattern of PEDOT:PSS on a textile. (d) A "lines and spaces" test pattern with lines of 1, 2.5, 3, and 4mm width, with a magnified view of the 1 and 2.5mm lines. (e) Impedance spectra of a textile and a medical electrode of similar area measured in the 1Hz to 1kHz frequency range in contact with skin from three different volunteers. (f, g) ECG recordings performed with the PEDOT:PSS textile electrode (lower trace) and a medical Ag/AgCl electrode (upper trace) ((f) from a volunteer sitting at rest and (g) during movement). (h) Percentage of accuracy of heartbeat detection during different types of activity (sitting, standing up, leg moving, arm moving, walking) with medical and textile electrodes during a 50s epoch.

(This figure was adapted with permission from Ref. [39]. Copyright 2015, Springer Nature.)

fabrics in poly(3,4-ethylenedioxythiophene):poly(styrene sulfonate) (PEDOT:PSS) solution and ethylene glycol doping, followed by screen-printing to form 2 × 2 cm electrodes (Figure 5.7b–d). The developed sensors achieve low impedance (2.5 times lower at 10 Hz) (Figure 5.7e), comparable to commercial Ag/AgCl electrodes, and maintain conductivity under real clinical and ambulatory conditions. The electrodes demonstrate high-quality ECG recordings at rest, accurately capturing PQRST waveforms, and reliable heart rate detection (60–80 bpm) during motion (Figure 5.7f–h), with signal-to-noise ratio preserved under stretching. Integrated into shirts, the textile electrodes enable continuous monitoring without skin irritation, confirmed by biocompatibility tests over 3 days. This scalable, low-cost approach advances wearable health monitoring by overcoming challenges in patterning conductive polymers on thick, stretchable knits, offering superior conformability and signal quality compared to rigid metal-based electrodes for applications in remote cardiac monitoring and exercise tracking.

Alternatively, the patch is widely used as a wearable smart sensor owing to its ease of manufacture and flexible structure. In healthcare monitoring, patch sensors are often used to measure physical signals (e.g., stress and temperature). In general, film-shaped patch sensors are categorized into two types: traditional film-based sensors fabricated on flexible substrates, such as PDMS, and textile-based patch sensors that can be directly integrated into fabrics to enhance wearability and conformability. For instance, Lin *et al.* [40] developed digitally embroidered electronic textiles using liquid metal by infiltrating galinstan ($Ga_{68.5}/In_{21.5}/Sn_{10}$) into perfluoroalkoxy alkane (PFA) tubing for seamless wireless powering and communication in wearable systems, enabling nonintrusive health monitoring with preserved textile flexibility and permeability during dynamic activities (Figure 5.8a). Fabricated liquid-metal fibers into UV-curable polymer microchannels (200 μm diameter) could form stretchable fibers, followed by high-precision digital embroidery onto substrates (Figure 5.8b and c), the textiles achieve low resistance ($0.02\,dB\,cm^{-1}$ at 13.56MHz) (Figure 5.8d), and high cyclic bending performance ($<$ 1% change in the electrical resistance over 10^4 cycles) (Figure 5.8e), confirming uniform fiber integrity by SEM analysis. Optimized inductive patterns support efficient near-field communication (NFC) at 13.56 MHz (~25% ideal power transfer efficiency), maintaining connectivity under treadmill exercise (0–9.2 km h^{-1}) and tissue phantoms mimicking implants (Figure 5.8f and g). Integrated into an athletic shirt, the system wirelessly powers a temperature-sensing node via a smartphone relay, enabling continuous axillary temperature monitoring with GPS-tracked activity data, validated against clinical thermometers (Figure 5.8h and i). It was found that this customizable embroidery pattern can transmit data from the temperature sensor embedded in the clothing to smartphones via the NFC function of smartphones or other devices. Briefly, this scalable, biocompatible platform advances smart textiles by resolving mechanical/thermal/chemical durability challenges, outperforming rigid or coated alternatives in comfort, washability, and real-time IoT integration for cardiovascular and fitness applications.

In another study, He *et al.* [41] reported a multifunctional triboelectric nanosensor textile fabricated via a simple dip-coating method with PEDOT:PSS on cotton and paired with polytetrafluoroethylene (PTFE), enabling energy harvesting, strain sensing, and CO_2 detection for wearable healthcare applications (Figure 5.9a). As shown in Figure 5.9b, the PEDOT:PSS-functionalized textile displays uniform

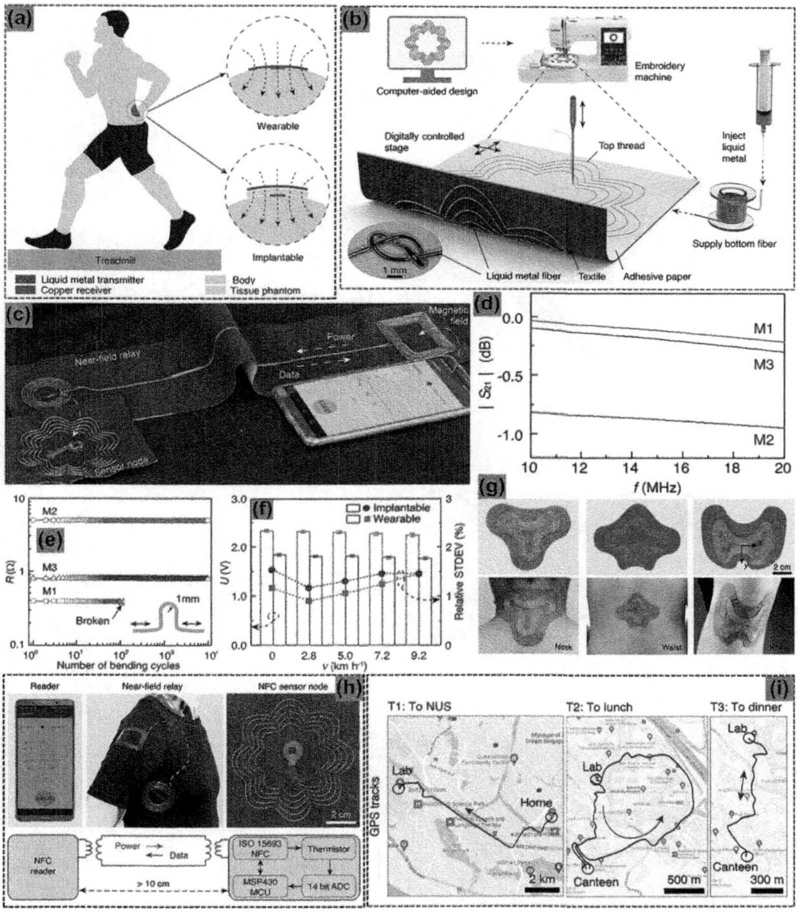

FIGURE 5.8 (a) Schematic of the wearable wireless system. (The receiver is placed underneath the textile transmitter with/without a tissue phantom between them, which simulates the implantable and wearable configuration.) (b) Illustration of the digital embroidery process. (Liquid-metal fibers consist of PFA tubing infiltrated with galinstan.) (c) An embroidered wireless system comprising a relay for long-range NFC between a smartphone and a textile sensor node. (d) Measured scattering parameter $|S_{21}|$ from embroidered 10 cm length transmission lines. (e) Resistance of the spiral inductors as a function of the number of bending cycles. (f) Images of skin patches fabricated using digital embroidery of liquid-metal fibers attached to the neck, waist, and knee. (The skin patches are used as wearable transmitters connected with a signal generator for wireless powering of the underneath wearable and implantable devices during daily activities.) (g) Comparison of received voltage and its relative standard deviation (STDEV) during different activities. (h) Textile thermal monitoring system comprising a sensor node connected with a smartphone reader via a near-field relay integrated on a shirt. (The top row shows images of the system components, while the bottom row shows the block diagram. The distance between the reader and sensor node is more than 10cm.) (i) Map of the GPS position of the subject.

(This figure was adapted with permission from Ref. [40]. Copyright 2022, Springer Nature.)

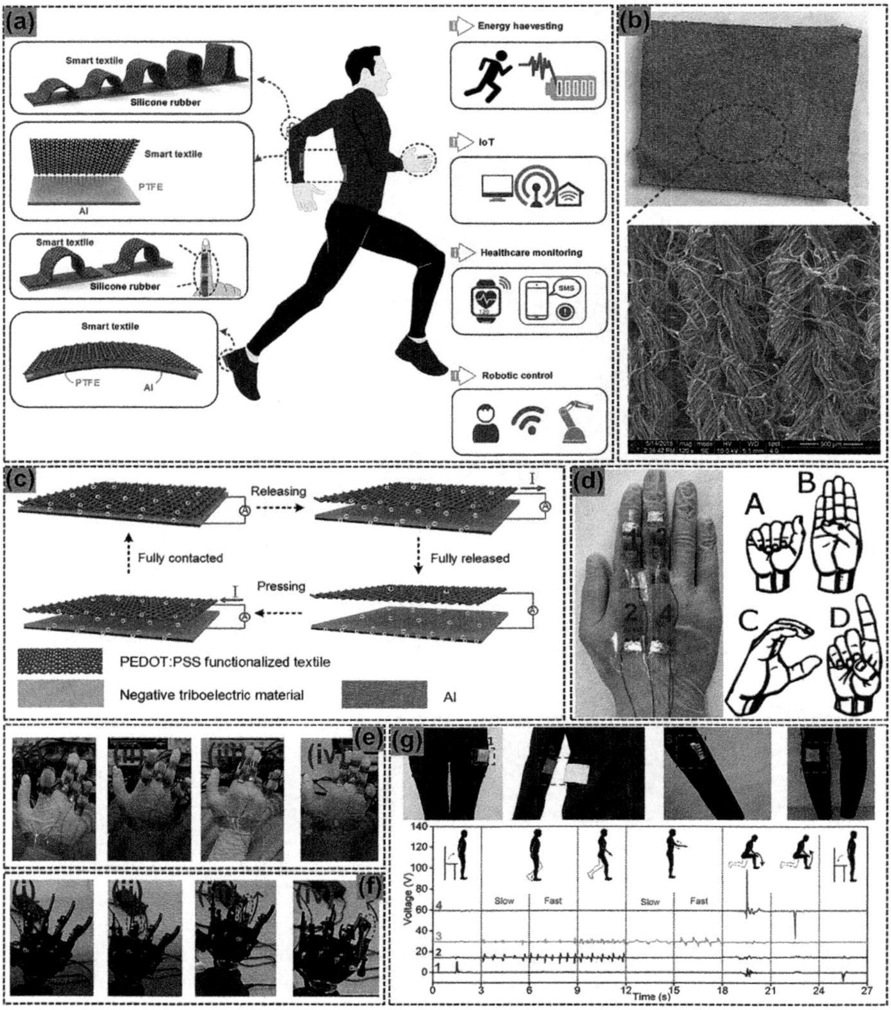

FIGURE 5.9 (a) Schematic diagrams showing the 3D structures of the versatile textile-based TENGs and their applications. (b) Photograph of the PEDOT:PSS-functionalized cotton textile (up) and SEM image of functionalized textile (down). (c) Working mechanism of PEDOT: PSS-based TENGs. (d) Hand gesture capture and robotic hand control with the finger-bending sensors. (Left, the photographs of the two-arch finger-bending sensors attached to the index finger and the middle finger. The four arches are labeled from 1 to 4 in accordance with channels 1 to 4, and right, illustration of the hand gestures representing the alphabet from "A" to "D" in American Sign Language.) Photographs of (e) bending one of the fingers at 90 degrees and (f) the robotic hand gestures corresponding to the four different hand motions. (g) A simple three-layer TENG based on the smart textile attached to an arched PET confinement structure on the hip, sleeve, and underneath the arm, elbow, and knee. (The associated output voltages of the four devices when the person is standing up, slowly walking, fast walking, running, slowly bending the arm, fast bending the arm, falling down, getting up, and sitting down.)

(This figure was adapted with permission from Ref. [41]. Copyright 2019, Elsevier.)

conductive coating on fibers with preserved porous microstructure, while Figure 5.9c illustrates the contact-separation triboelectric mechanism, enabling efficient charge generation at the PEDOT:PSS/PTFE interface for energy harvesting and self-powered sensing applications. Figure 5.9d demonstrates the efficient performance of the developed nanosensing device in real-time American Sign Language recognition and robotic hand control through self-powered gesture monitoring; meanwhile, Figure 5.9e and f depicts stable, repeatable voltage outputs from multi-arch strain sensors during finger bending. From Figure 5.9g, the developed smart textile shows outstanding CO_2 sensing performance after polyethyleneimine modification, with triboelectric voltage decreasing linearly as CO_2 concentration rises due to altered surface charge and gas interaction, indicating strong potential for self-powered wearable air quality monitoring. In conclusion, this scalable, self-powered textile platform advances wearable electronics by integrating energy harvesting and multifunctional sensing, outperforming prior triboelectric nanogenerators (TENGs) in versatility, comfort, and environmental robustness for IoT-enabled health monitoring.

On the other hand, wearable devices have now reached a relatively mature stage, with everyday examples including fitness bands, smart bracelets, and smart glasses. Thanks to steady improvements in the underlying technology, the sensors in these devices can communicate reliably and consistently with smartphones or other personal gadgets. Among these, wrist-worn health monitors are most commonly used to track heart rate. Their primary strengths are convenience and compactness. Many of these bands are also designed to integrate seamlessly with traditional watches, combining classic style with modern functionality in a single device.

For instance, Meng *et al.* [42] produced a wireless textile-based sensor (TS) system for self-powered personalized health care, featuring a highly flexible, scalable triboelectric sensor integrated into everyday clothing for real-time, noninvasive monitoring of pulse waves and diagnosis of obstructive sleep apnea-hypopnea syndrome (OSAHS) with enhanced wearing comfort (Figure 5.10a). This sensing device was fabricated via scalable sewing and weaving techniques using silver-coated polyester fabric as the base/electrode layer and embossed flower-shaped polyester–metal hybrid fibers (three-ply twisted) as the sensing superstructure (Figure 5.10b–d). Figure 5.10e presents COMSOL simulations of mechanical deformation in the TS at 0.95 and 3.2 kPa, respectively, showing intensified deformation and increased contact area in the flower-shaped hybrid fiber superstructure at the higher pressure, thereby enhancing triboelectric charge generation for sensitive pulse-wave detection. From these figures, it can be concluded that the deformation regions and magnitudes within the hybrid fiber superstructure of the triboelectric textile sensor are highly dependent on the applied external excitation. Moreover, Figure 5.10f schematically depicts the triboelectric mechanism in the TS, in which pulse pressure induces periodic relative motion and contact-area variation between the hybrid fiber and the silver-coated fabric, thereby driving charge transfer and electrostatic induction to produce self-powered electrical signals proportional to arterial pulsation. Figure 5.10g and h demonstrates that the wireless biomonitoring system has the capability for wearable pulse-wave acquisition from non-wrist sites in elderly users, reflecting the device worn on the temple of a 73-year-old woman with the acquired signal (Figure 5.10g) and on the arm of a 75-year-old woman with the

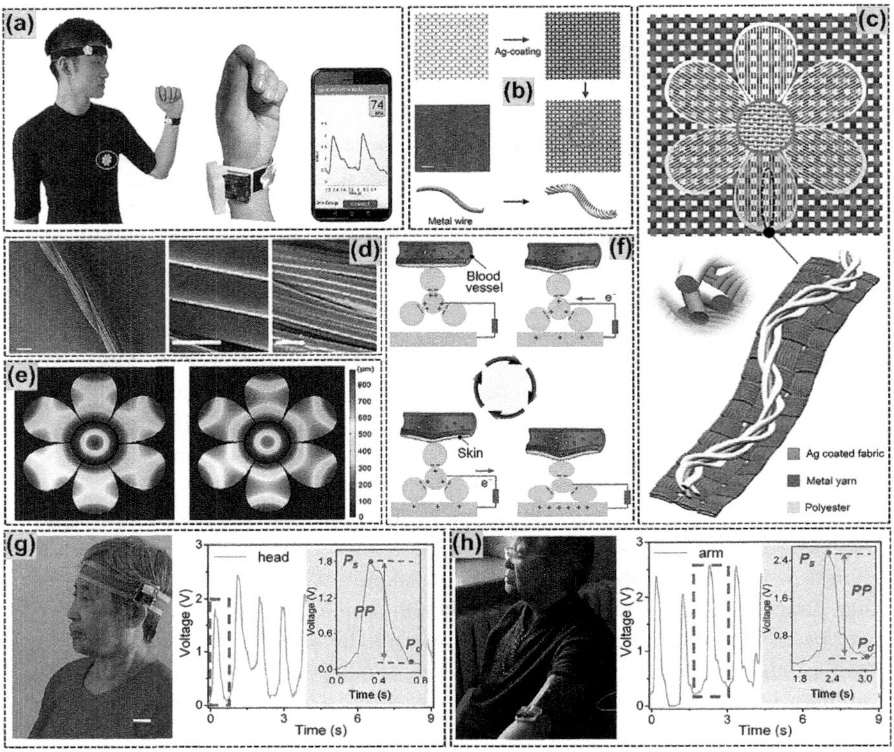

FIGURE 5.10 (a) A photograph of a wearable TS-based wireless biomonitoring system and its wide range of applicability, which can be worn on various body parts for health monitoring. (b) Schematic illustration showing the fabrication procedure for the silver-coated base as an electrode, as described in the text. (c) A schematic illustration of the TS as a flower shape and a basic unit of the TS. (d) SEM images with different magnifications. (The scale bars are 100, 20, and 10 μm, respectively.) (e) Simulation results of the distribution of TS deformation under different applied pressures. (f) Schematic diagrams showing a cycle of electrical signal generation of the basic unit in the TS in response to the human pulse. (g) Photograph and acquired temple pulse-wave signals obtained when the TS-based wireless biomonitoring system was worn on a 73-year-old woman (scale bar: 2 cm). (h) Photograph and acquired arm pulse-wave signals obtained when a 75-year-old woman wore the TS-based wireless biomonitoring system.

(This figure was adapted with permission from Ref. [42]. Copyright 2020, Elsevier.)

corresponding waveform (Figure 5.10h). In these figures, enlarged insets (left side graphs) reveal clear characteristic points (Ps, PP, Pd) indicative of reliable cardiovascular monitoring even in weak or aged individuals. This cost-effective, breathable, and biocompatible platform advances wearable health care by overcoming limitations in sensitivity, comfort, and motion artifacts of prior devices, establishing a body area network for cardiovascular and sleep disorder diagnostics in the IoT era.

Interestingly, wearable sensors demand materials that can bend and stretch with the body, stay gentle on the skin, and still deliver reliable energy harvesting or sensing performance even when the wearer is sweating, moving, or tugging at the fabric. The magnetoelastic effect—long confined to stiff, bulky metals—becomes far more promising once we reimagine it in soft, fiber-based forms that can be woven or knitted directly into everyday clothing without anyone noticing they are there. In this context, Zhao *et al.* [43] developed soft fibers exhibiting magnetoelasticity for wearable electronics (Figure 5.11a), demonstrating stronger magnetomechanical coupling (8.28×10^{-8} T Pa^{-1}) than traditional rigid metals through a wavy chain model based on magnetic dipole–dipole interactions and demagnetizing factors. This sensor was fabricated by extruding neodymium–iron–boron (NdFeB) nano-magnets (50, 75, and 83 wt%) in porous silicone polymer via mechanical stirring (Figure 5.11b) and impulse magnetization (2.65 T). The produced fiber (83 wt%, as optimum sample) achieve Young's modulus of 630 kPa, 180% stretchability, and up to 18 mT magnetic field variation under 450 kPa, validated by SEM analysis (inset of Figure 5.11b), DMA stress–strain curves (Figure 5.11c), and SQUID hysteresis loops (Figure 5.11d). The textile magnetoelastic generator (MEG), woven with silver-coated conductive yarns on a loom (Figure 5.11e–g), generates 0.63 mA cm^{-2} short-circuit current density, 16 mW peak power at 180 Ω internal impedance, and intrinsic waterproofness confirmed by immersion (7 days) and perspiration tests (Figure 5.11h). As a wristband, the MEG detects arterial pulses with a 0.05 kPa limit underwater, extracting parameters like heart rate (HR), K value (which indicates the mean arterial blood pressure), and PWV via Bluetooth cell-phone application for telemedicine (Figure 5.11i–k). This biocompatible MEG platform (97.6% fibroblast viability) advances soft-matter electronics by enabling humidity-resistant biomechanical energy harvesting and sensing, outperforming triboelectric/piezoelectric textiles in output, durability, and comfort for healthcare, robotics, and IoT applications.

Wearable Temperature Sensor

Body temperature is a direct reflection of metabolic activity and is a fundamental requirement for normal physiological function; it influences virtually every process in the human body. Importantly, real-time monitoring of core temperature has therefore become clinically vital, as sudden deviations can signal life-threatening events such as myocardial infarction. Beyond medical settings, precise temperature tracking is equally important for athletes and military personnel, where intense physical exertion pushes the body to its limits, and performance (as well as safety) depends heavily on maintaining thermal balance [13].

Moreover, the COVID-19 pandemic has made it clear that we need innovative strategies to reduce infection risk, and smart textiles are a natural place to apply them. One particularly appealing direction is the development of textile-based sensors based on functionalized carbon-based materials with built-in biocidal properties. Beyond simply shielding the wearer from pathogens, these materials could be designed to be genuinely biocompatible—avoiding unwanted immune responses and potentially accelerating wound healing when in direct contact with skin.

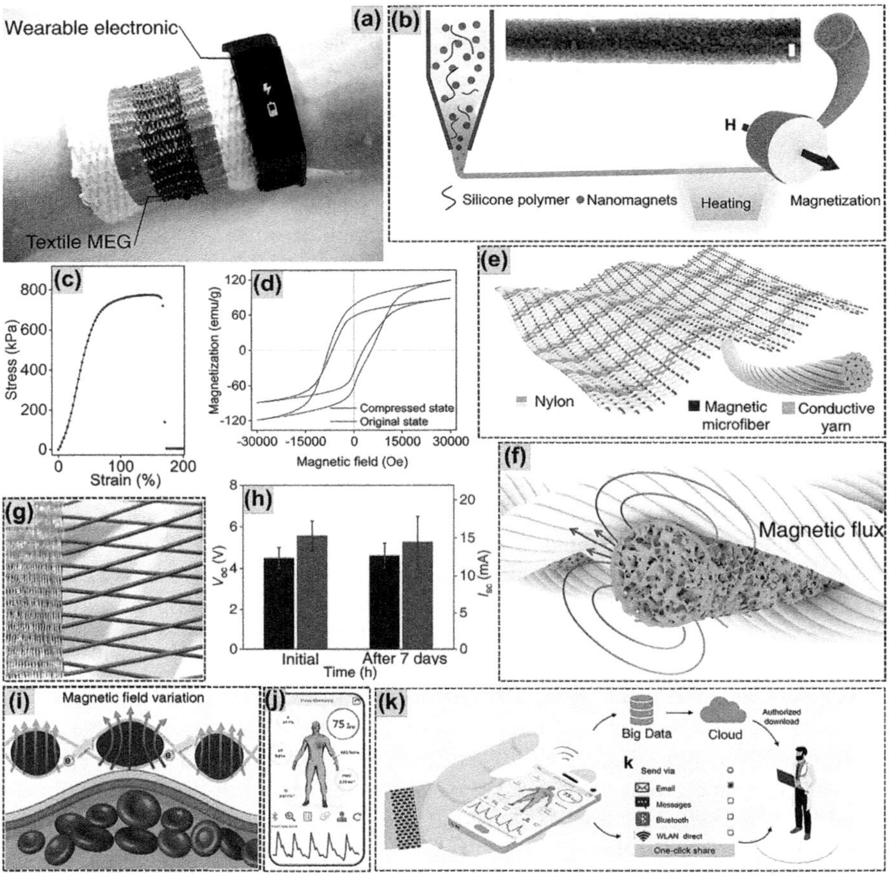

FIGURE 5.11 (a) Textile MEG mixed with wool fibers is used to drive a wearable biosensor system. (b) Schematic of the fabrication process and micro-CT (micro-computed tomography) image of the soft magnetic fiber. (c) Stress–strain curves of the soft magnetic fibers with 83 wt% magnetic concentration. (d) Magnetic hysteresis loop of the soft magnetic fiber in the original state and in the compressed state. Schematic (e) design and (f) working mechanism of the textile MEG. (g) Photograph of the weaving process using a loom (scale bar: 1 cm). (h) Electric outputs of the textile MEG before and after being submerged in water for 7 days (data are mean ± SD; $n = 5$). (i) Schematics of the working principle of the textile wristband. (j) The developed user interface on a cell phone application. (k) Schematic of a fully integrated cardiovascular monitoring system. (The personal health data could be processed via two pathways: directly sending to the physicians for immediate clinical diagnosis, or uploading into the cloud to conduct big data analysis for data-driven diagnosis.)

(This figure was adapted with permission from Ref. [43]. Copyright 2021, Springer Nature.)

Another promising avenue is the use of these materials to design advanced wearable sensors that monitor subtle changes in skin temperature during drug delivery. Such real-time feedback could help patients (and clinicians) maintain optimal therapeutic conditions while giving pharmaceutical researchers richer data on side effects and physiological responses to new compounds. Technically, temperature sensing in textiles is not just a technical upgrade; it is a versatile platform that could touch everything from infection control and personalized medicine to mental health monitoring and smarter nutrition [3].

Various recent studies have explored the development of textile-based temperature sensors using thermoelectric materials [44]. For instance, Afroj *et al.* [45] introduced a highly scalable and ultrafast yarn dyeing technique for engineering graphene flake-based wearable textile sensors, enabling flexible, washable, and bendable yarns integrated into knitted structures for wireless temperature monitoring via self-powered radio frequency identification (RFID) or low-power Bluetooth (Figure 5.12a and b). The development process of this sensor was optimized by reducing graphene oxide (rGO) dispersions with ascorbic acid or sodium hydrosulfite (SH) reduction (24 h at 1:10 GO:polymer ratio yielding I_D/I_G ~1.73 and C/O ~6.59), followed by batch dyeing of cotton yarns at 60°C for 30 min and curing at 150°C for 3 min. The produced rGO-coated yarn samples (Figure 5.12c and d) achieve low resistance (~42.7 kΩ cm^{-1} after three coating cycles) and exponential temperature-dependent conductance with a negative coefficient (stable from 20°C to 60°C, repeatable 25°C–55°C cycles) (Figure 5.12e). Morphological and chemical analyses via SEM (uniform flake wrapping) (Figure 5.12f and g), XPS (residual oxygen groups for bonding), and Raman confirmed enhanced reduction and adhesion, supporting wash durability (>10 cycles with minor resistance increase) and mechanical robustness (> 1000 bending/compression cycles, no creasing) (Figure 5.12h). Demonstrations include real-time body temperature tracking in smart garments, with potential industrial scale-up to ~1000 kg h^{-1}. This eco-friendly approach advances smart textiles by addressing the scalability, cost, and environmental challenges of metal-based methods while outperforming them in flexibility, breathability, and integration for healthcare and fitness applications.

In another study, Rahmani Charvadeh *et al.* [46] investigated the influence of fever on textile permittivity to design a wearable temperature sensing antenna, enabling noninvasive, real-time monitoring via changes in fabric relative permittivity (ϵ_r) from human body sweat evaporation in the 35–41°C range (Figure 5.13a–c). Using a custom setup with a Keysight N1501A-101 dielectric probe and vector network analyzer to measure ϵ_r and tan δ at 12 GHz (Figure 5.13d), four commercial fabrics (linen, cashmere, crepe, cotton) were evaluated in a sealed container simulating sweat moisture, identifying cashmere as optimal with ~0.02 ϵ_r change per 2°C step and tan δ decreasing from 0.1287 to 0.1219 (Table 5.1). A rectangular microstrip patch antenna (dimensions: L = 58.7 mm, W = 40 mm, h = 1.3 mm) fabricated on cashmere with copper tape exhibits a resonant frequency shift of 60–80 MHz per 2°C (measured) or ~35 MHz per 1°C, validated by simulation (at 300 MHz total shift) using CST Studio Suite 2021 software and measurements showing good agreement despite minor discrepancies from fabrication and temperature errors (Figure 5.13e).

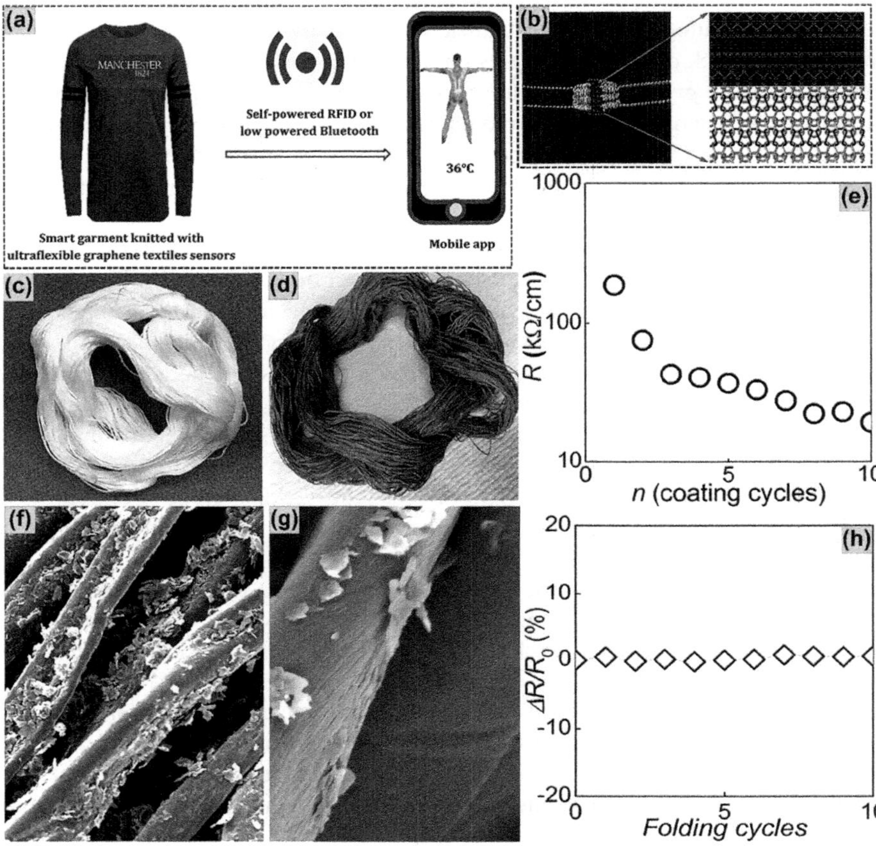

FIGURE 5.12 (a) Concept smart garment knitted with ultraflexible graphene textile sensors for monitoring of physiological conditions of the human body and send data to a mobile app via a self-powered RFID or low-powered Bluetooth device. (b) Fabrication and characterization of a knitted temperature sensor with graphene-coated yarn. (c) Undyed hank of scoured–bleached control cotton yarn. (d) Hank of rGO-dyed (coated) cotton yarn. (e) Resistance of rGO (SH)-coated and dried (at 100°C) yarn versus the number of coating cycles. SEM images rGO flake dyed (coated) cotton yarn (f) before and (g) after 10 washing cycles. (h) The electrical resistance variation of graphene yarn sensors degrades after 10 folding–releasing cycles.

(This figure was adapted with permission from Ref. [45]. Copyright 2019, American Chemical Society.)

On-body simulations with a 44 mm phantom (dry skin, fat, muscle properties from Table 5.2) confirm minimal S_{11} impact and safe the specific absorption rate (SAR) (<0.0878 W kg^{-1} at 0.5 W input) (Figure 5.13f), recommending axillary placement for accuracy and comfort. This passive, flexible antenna advances fever detection by leveraging the relative permittivity of textile (REPOT) sensitivity, outperforming prior methods in integration, cost, and IoT compatibility for COVID-19 and other disease monitoring.

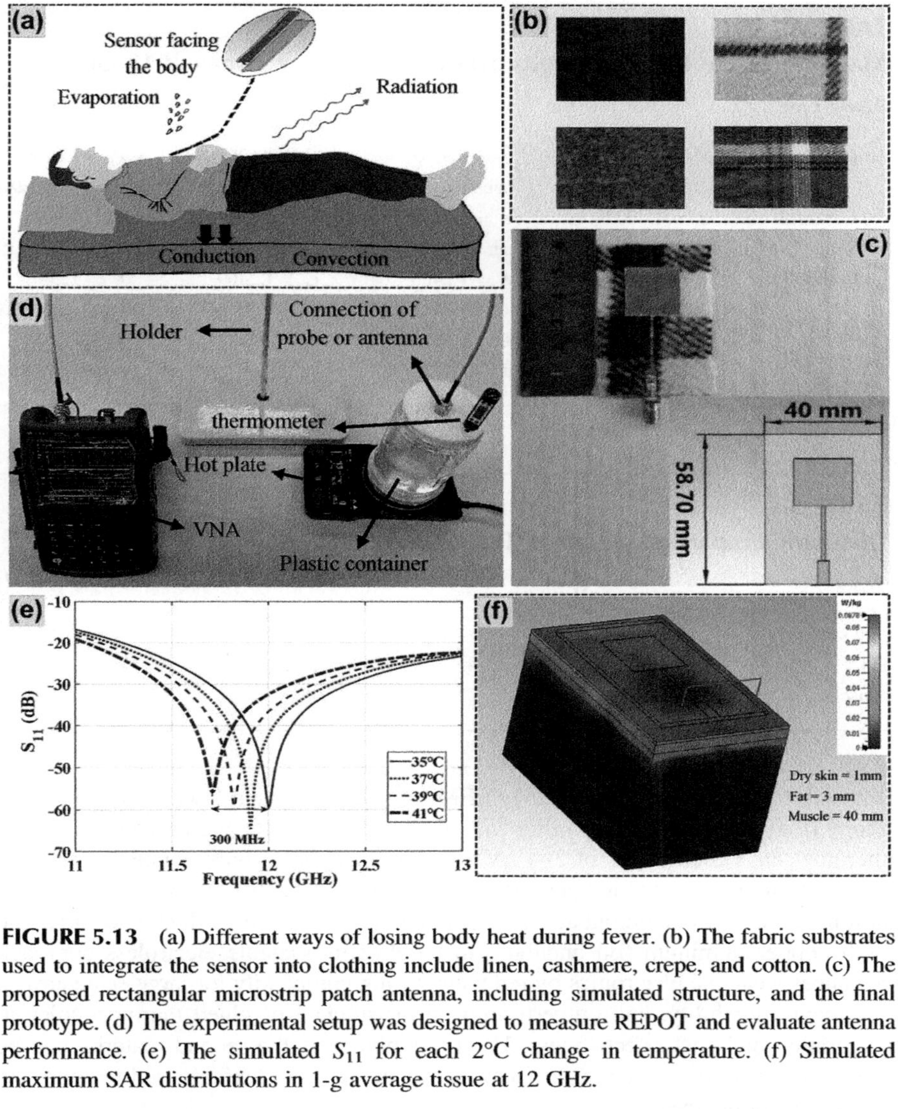

FIGURE 5.13 (a) Different ways of losing body heat during fever. (b) The fabric substrates used to integrate the sensor into clothing include linen, cashmere, crepe, and cotton. (c) The proposed rectangular microstrip patch antenna, including simulated structure, and the final prototype. (d) The experimental setup was designed to measure REPOT and evaluate antenna performance. (e) The simulated S_{11} for each 2°C change in temperature. (f) Simulated maximum SAR distributions in 1-g average tissue at 12 GHz.

(This figure was adapted with permission from Ref. [46]. Copyright 2022, Elsevier.)

Landsiedel *et al.* [47] produced a multipoint flexible temperature sensor array and thermoelectric generator using electroless copper-coated lyocell cellulose textiles, enabling direct integration of sensors and conductors while preserving textile flexibility and bendability. This system was fabricated via surface activation with tin chloride, silver seeding, and copper deposition (Figure 5.14a), and the conductive fabrics achieved a uniform copper-layer thickness of ~260 nm, with a copper content of 147 mg g^{-1} of fabric. The developed sensor shows low internal resistance (708 Ω)

TABLE 5.1
Measurement of REPOTs Results over a Temperature Sweep at 12 GHz

Fabric Type	35	37	39	41
(a) Linen	$\epsilon_r = 1.6261$	$\epsilon_r = 1.6263$	$\epsilon_r = 1.6265$	$\epsilon_r = 1.6269$
	$\tan\delta = 0.1284$	$\tan\delta = 0.1284$	$\tan\delta = 0.1284$	$\tan\delta = 0.1284$
(b) Cashmere	$\epsilon_r = 1.4207$	$\epsilon_r = 1.4419$	$\epsilon_r = 1.4590$	$\epsilon_r = 1.4823$
	$\tan\delta = 0.1287$	$\tan\delta = 0.1265$	$\tan\delta = 0.1243$	$\tan\delta = 0.1219$
(c) Crepe	$\epsilon_r = 1.2446$	$\epsilon_r = 1.2466$	$\epsilon_r = 1.2469$	$\epsilon_r = 1.2469$
	$\tan\delta = 0.0323$	$\tan\delta = 0.0322$	$\tan\delta = 0.0322$	$\tan\delta = 0.0322$
(d) Cotton	$\epsilon_r = 1.4851$	$\epsilon_r = 1.4875$	$\epsilon_r = 1.4895$	$\epsilon_r = 1.4919$
	$\tan\delta = 0.0697$	$\tan\delta = 0.0696$	$\tan\delta = 0.0695$	$\tan\delta = 0.0694$

TABLE 5.2
Dielectric Properties of Human Tissues at 12 GHz

Tissue	ϵ_r	$\tan\delta$	σ (s m^{-1})
Dry skin	29.327	0.52797	10.337
Fat	4.4570	0.24462	0.7278
Muscle	40.101	0.50616	10.355

and stable morphology confirmed by SEM (Figure 5.14b) and laser scanning microscopy (Figure 5.14c). After pairing with aluminum strips, the thermocouples exhibit thermoelectric coefficients of 3–4 μV K^{-1}, comparable to bulk copper–aluminum, with outstanding voltage-temperature response (Figure 5.14d). A six-junction generator produces 0.4 mV output at a 71 K gradient (Figure 5.14e), which is comparable with copper/aluminum foil and bulk material. This scalable electroless approach advances wearable electronics by reducing wiring complexity for large-scale arrays, enhancing durability, and outperforming prior stitched or glued thermocouples in integration and cost-effectiveness. Using aluminum as the second material can be considered a sufficient technical factor for manufacturing textile-based temperature measurement devices.

Motion Recognition

In sports and exercise healthcare, accurate motion recognition plays a critical role in injury prevention. Accordingly, real-time monitoring of joint angles, body posture, and subtle movements enables the system to assess an athlete's physical state in real time and issue timely warnings when form breaks down or fatigue sets in.

The same technology has clear applications in medical training, particularly for first-aid and rehabilitation scenarios, where precise feedback on technique can make

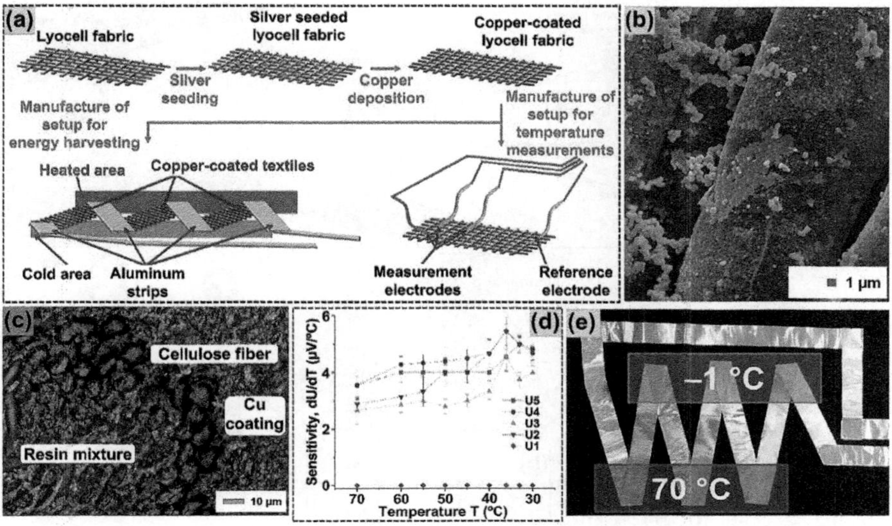

FIGURE 5.14 (a) Manufacture of textile-based thermocouples usable for temperature measurements and energy harvesting, made from copper-coated lyocell fabrics. (b) Scanning electron microscopy reveals the growth of various structures on the surface. (c) Laser scanning microscopy image of a cross-section, which indicates that almost 50% of the fibers of one yarn are coated with copper. (d) Temperature measurements were performed using two setups, each comprising five measuring terminals and one reference terminal, placed on a copper-coated lyocell fabric. (e) Thermoelectric output voltages of arrangements in series of three thermocouples using strips of copper and aluminum foils.

all the difference. In general, traditional rigid sensors have long been a bottleneck: their mechanical bulk makes it difficult to create truly lightweight, compact devices that people can wear comfortably for hours. However, textile-based sensors sidestep this limitation efficiently—they are flexible, soft, and seamlessly integrate into clothing, enabling practical, everyday use.

Most daily human activities ultimately revolve around the movement of major joints—knees when we run, elbows when we reach or lift, shoulders when we throw. By focusing on these key joints, a well-designed system can reliably capture and interpret the vast majority of natural motions. In this context, Wu *et al.* [48] developed highly integrated, scalable liquid alloy/silicone rubber core/shell fibers (LCFs) for strain sensing and self-powered smart textiles, enabling textile triboelectric nanogenerators (t-TENGs) with robust performance amid mechanical and environmental stresses. This study reports a simple, one-step, all-in-one fabrication method (Figure 5.15a and b) for producing LCFs using a coaxial needle (Figure 5.15c) that enables coordinated yet independent flow of the liquid-metal core and silicone rubber shell. The resulting LCFs are uniform, continuous, pore-free fibers with a nearly circular cross-section and well-defined core–shell structure

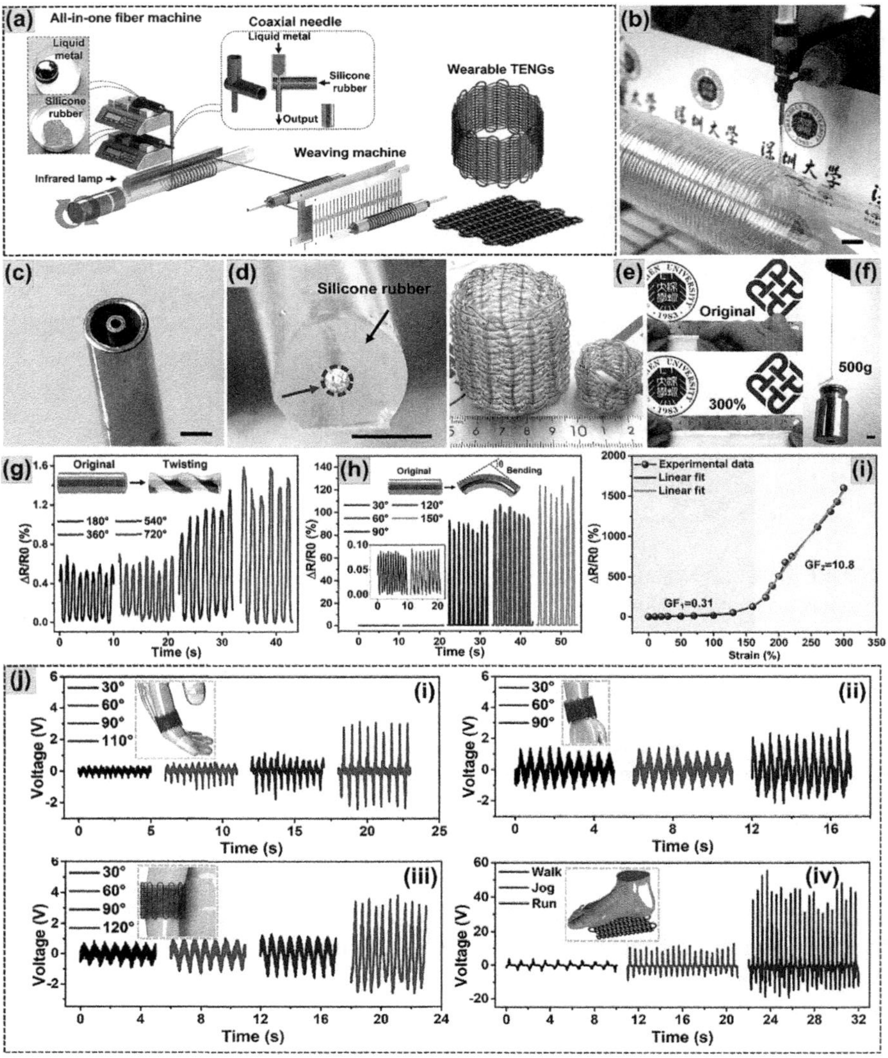

FIGURE 5.15 (a) Schematic illustration and fabrication procedure of the LCFs. Photograph of (b) fabrication procedure of the LCFs (the scale bar was 1 cm), (c) the output of the coaxial needle (the scale bar was 1 mm), and (d) cross-sectional morphology of the LCFs (the scale bar was 1 mm). (e) Ring weaving t-TENGs based on the LCFs (the scale bar was 1 cm). (f) Tensile performances of the LCFs (the scale bar is 1 cm). (g) Resistance change rate under different twist angles (180°, 360°, 540°, and 720°) during the twisting-releasing cycle. (h) Resistance change rate under different flexion angles (30°, 60°, 90°, 120°, and 150°) during the bending–releasing cycle. (i) Resistance change rate versus tensile strain via linear fit. (j) Application of motion monitoring by the ring and plain t-TENGs, and the corresponding V_{oc} under different bending angles or frequencies ((i) finger, (ii) wrist, (iii) elbow, and (iv) foot).

(Figure 5.15d), offering a highly integrated, automated, and scalable alternative to conventional multistep assembly techniques. Fabricated LCF yarns could be woven into t-TENGs with different shapes (Figure 5.15e) and exhibit exceptional stretchability (300% strain under 500 g load) (Figure 5.15f), promising pliability with stable resistance changes under twisting (up to 720°) (Figure 5.15g) and bending (up to 150°) (Figure 5.15h), validated through maintaining output over 1800 cycles. Interestingly, Figure 5.15i shows the relative resistance change ($\Delta R/R_0$) of LCF as a function of tensile strain, revealing a low GF ($GF_1 = 0.31$) in the small-strain regime (0–130%) transitioning sharply at an inflection point to a high GF ($GF_2 = 10.8$) in the large-strain regime (up to 150–300%), with $\Delta R/R_0$ reaching ~1595% at 300% strain, driven by progressive thinning and elongation of the liquid alloy core that sharply increases resistance at higher deformations. Notably, the developed multifunctional TENGs exhibit real-time motion-monitoring capability when attached to various body parts, yielding distinct, repeatable open-circuit voltage (V_{oc}) patterns. As can be seen in Figure 5.15j, this sensing setup could clearly discriminate joint bending angles (30°–120°) at the elbow (i), wrist (ii), and knee (iii), as well as gait phases during walking, jogging, and running (iv), enabling accurate, self-powered human activity recognition for healthcare and human–machine interface applications. This all-in-one LCF platform advances wearable electronics by overcoming assembly complexities, enhancing washability and durability, and outperforming prior methods in mass-producibility and practical reliability.

In recent decades, human–computer interaction and personalized medicine have offered flexibility and integration into everyday clothing. However, achieving high sensitivity, a wide sensing range, and durability in these textiles remains challenging, often requiring complex fabrication methods. In this regard, Pan *et al.* [49] introduced a highly stretchable and sensitive PU/CB/MXene/Ecoflex (PCME) composite yarn for wearable smart textiles, utilizing a continuous impregnation strategy to create a multidimensional conductive structure with zero-dimensional (0D)–two-dimensional (2D) nanomaterials for enhanced performance. As shown in Figure 5.16a, by immersing PU yarn in CB ink and MXene dispersion, followed by Ecoflex encapsulation, the final structure was validated by SEM–energy-dispersive X-ray spectroscopy (EDS) (Figure 5.16b–d). The PCME yarn achieves a GF of 59.99, a wide sensing range (0–200%), fast response (289 ms), and recovery (575 ms), along with high durability over 8000 s, confirmed according to tensile test results that show outstanding elastic recovery and stable resistance (Figure 5.16e and f). As shown in Figure 5.16g, PCME enables real-time monitoring of joint motions (finger, elbow, knee) and gesture recognition ("A," "E," "TWO"). In addition, by integrating the PCME composite yarn into a glove, the developed system demonstrates excellent finger-bending detection capability. It produces distinct, repeatable $\Delta R/R_0$ patterns corresponding to thumb, index, middle, ring, and little finger movements at different bending degrees (Figure 5.16h). Notably, Figure 5.16i and j dedicates the practical smart control system, where PCME signals from finger flexion are converted, transmitted via Wi-Fi to a relay module, and used to remotely switch household appliances like a fan on/off, showcasing the yarn's high sensitivity, wide detection range, and promising application in human-motion recognition and intelligent home control. This low-cost, scalable PCME platform advances smart textiles by overcoming

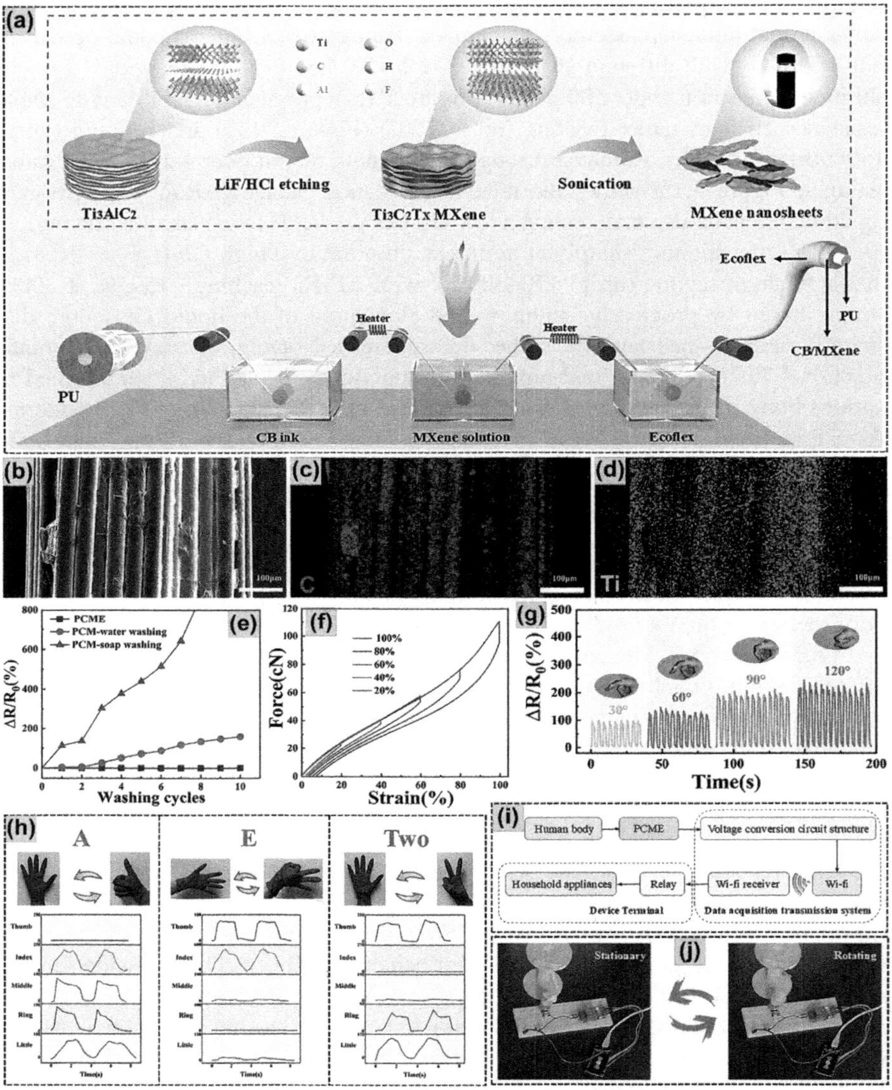

FIGURE 5.16 (a) Schematic illustration of the preparation process of the PCME (preparation of (up) $Ti_3C_2T_X$ MXene and (down) PCME. (b) SEM image of PCM and its corresponding (c) C and (d) Ti EDS mappings. (e) Stress–strain curves of PCME for constant elongation elastic recovery tests with increased maximum strain from 20 % to 100 %. (f) The relative change in the resistance of PCM and PCME during washing. (g) Resistance change of the index finger at different bending angles (30°, 60°, 90°, and 120°). (h) Monitoring three gestures using a glove integrated with five PCME. (i) Block diagram of PCME-based intelligent control. (j) Demonstration of the PCME in controlling an electric fan.

(This figure was adapted with permission from Ref. [49]. Copyright 2025, Elsevier.)

trade-offs in sensitivity and range, outperforming prior sensors in washability, cyclability, and multifunctionality for healthcare and smart home applications.

From a technical perspective, wearable sensors are generally classified into four main categories based on their transduction mechanism: capacitive, piezoelectric, piezoresistive, and triboelectric. Among these, piezoelectric sensors have attracted particular attention in recent years because they directly convert mechanical deformation into electrical signals in real time. This inherent energy-harvesting capability paves the way for truly self-powered devices that operate without any external power source or frequent battery replacement [50–52]. Ahn *et al.* [53] developed a wearable multilocal strain sensor (MLS sensor) based on a 3D textile structure with prestrained monofilament. Figure 5.17a schematically illustrates the multimodal sensing capability of the MLS sensor across various human body regions, categorizing signal detection into vibration mode (50 Pa–2 kPa) for subtle biosignals such as voice, carotid pulse, respiration, and fingertip tactile input, and tensile/compression mode (2 kPa–200 kPa) for larger mechanical stimuli, including neck/elbow/knee bending and gait. Using the Raschel knitting machine (Figure 5.17b), the developed 3D textile structure with prestrained monofilaments could amplify multilocal strain and enhance piezoelectric output in polyvinylidene fluoride (PVDF) film, enabling comprehensive, self-powered wearable health monitoring (Figure 5.17c and d). Demonstrations include precise vocal cord vibration detection for speech recognition ("advance" and "sensor") (Figure 5.17e) and elbow/knee bending tracking (Figure 5.17f and g). This air-permeable MLS sensor advances wearable electronics by leveraging macro-3D structures to enhance piezoelectric performance, surpassing prior nanostructured PVDF in multimodal sensitivity, breathability, and low-pressure detection.

Recently, it was confirmed that, when the MLS sensor is used for gait measurement, it can act as a piezoelectric energy harvester. The sensor can also detect gait changes as early indicators of dementia, enabling prediction before diagnosis. Li *et al.* [54] developed a highly stretchable knitted strain sensor integrated into leggings using weft-knitted nylon/nylon-wrapped spandex/silver-coated yarn for human-motion detection. This sensing system proposes a novel 3D curved-surface bursting test method that more accurately simulates real-world body deformations than traditional 2D tensile testing (Figure 5.18a(i)). Fabricated sensor composed of a single circular knitting machine with sensor zones (5 × 1.5 cm) (Figure 5.18a(ii)) positioned around the knee joint for seamless wearability (Figure 5.18a(iii)). In this study, to simultaneously record the resistance and strain data of the developed sensor under cyclic loads, a resistance testing hardware unit (MCU) was utilized (Figure 5.18b), while for the uniaxial tensile test, the knitted fabric samples were stretched in the longitudinal direction, as can be seen in Figure 5.18c. The sensor exhibits a 2D tensile sensing range of 60% with GF decreasing from 33.3 to 25 (Figure 5.18d). Gait detection demonstrations captured distinct resistance patterns for walking, running (higher frequency), climbing (larger amplitude), and descending steps, enabling eigenvector-based multivariate analysis (Figure 5.18e). In 3D bursting tests per ASTM D3787 (Figure 5.18f and g), the sensor achieves a doubled sensing range of 120% with linear resistance-strain response, faster 350 ms response time, and negligible speed effects (100–600 mm min^{-1}), validated by area increase rate calculations and monotonicity analysis up to 51% strain. The developed sensor

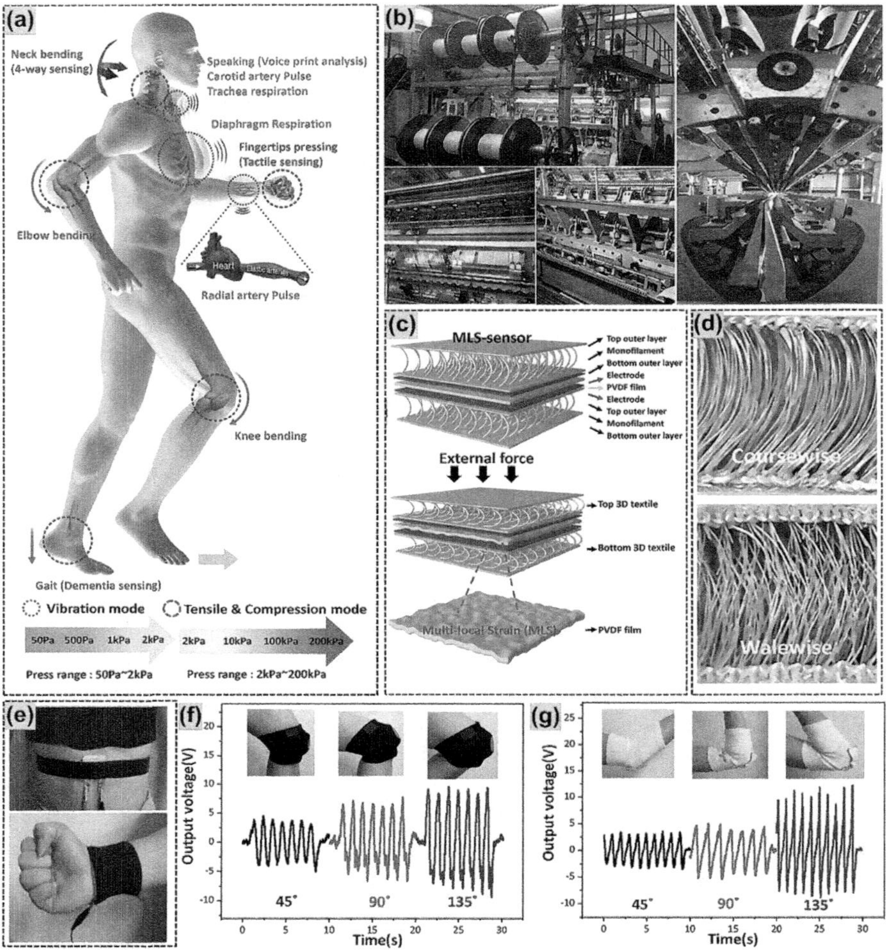

FIGURE 5.17 (a) The location and pressure range at which the MLS sensor is applied to the body are determined by classifying body signals as Vibration Mode (circle with dashed border) or Tensile and Compression Mode (circle with dotted border). (b) A Karl Mayer double-needle bar raschel machine is used for 3D textiles (left), and a 3D textile weaving method using a double-needle bar raschel machine is shown (right). (c) MLS-sensor structure (up) and deformation of 3D textile and multilocal strain in PVDF films when external force is given (down). (d) The course-wise and wale-wise optical microscope images of a 3D textile. (e) MLS sensors applied to the human body (respiratory movements (up) and arterial pulse pressure (down)). Measurement of (f) knee and (g) elbow bending movements at 45°, 90°, and 135°angles.

(This figure was adapted with permission from Ref. [53]. Copyright 2020, Elsevier.)

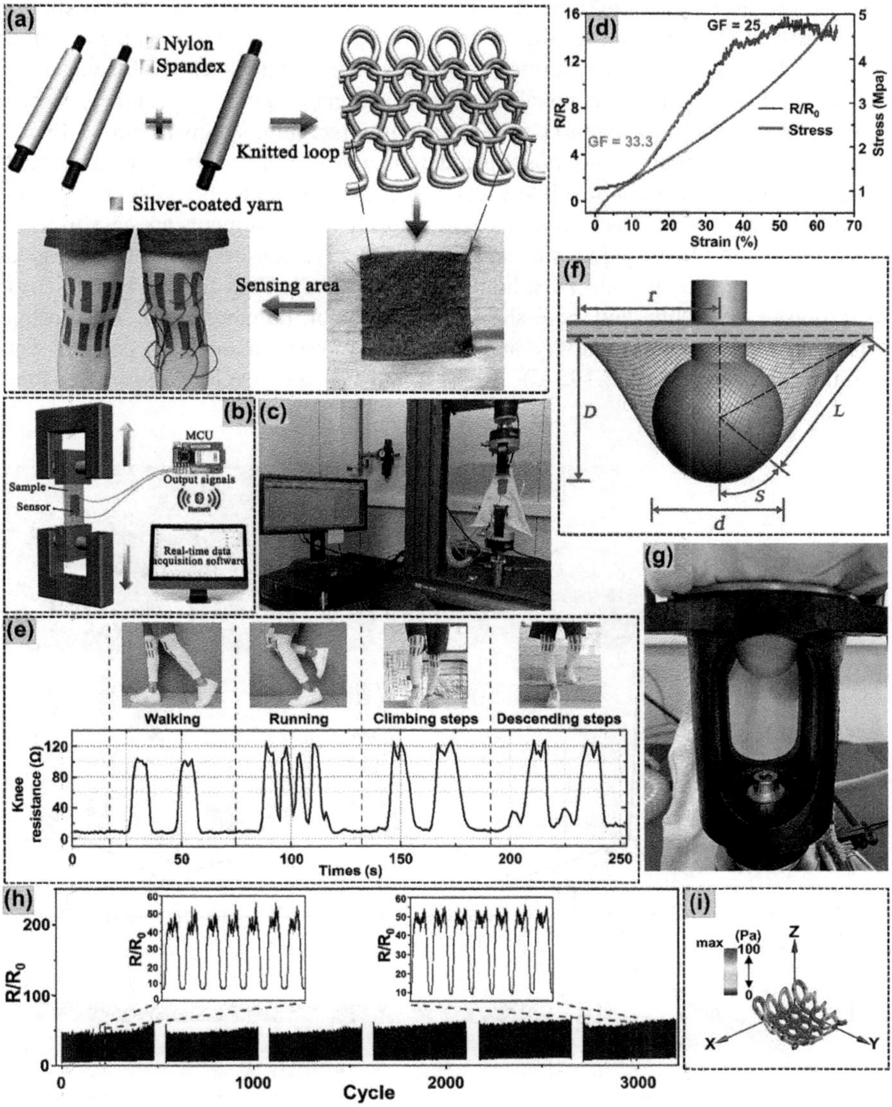

FIGURE 5.18 (a) A high-stretch knitted sensor based on spandex/silver-coated yarn. (b) The process of collecting resistance signal through the tensile instrument and (c) 2D longitudinal tensile test. (d) Enlarged diagram of resistance change in the first stage. (e) Diagram of resistance changes of legs under four kinds of sports. (f) 3D deformation of the fabric during the bursting procedure. (g) Photograph of the bursting test setup to analyze the flexible knitted sensor. (h) Long-term sensor stability and repeatability, and (i) loop surface stress analyses.

(This figure was adapted with permission from Ref. [54]. Copyright 2021, Elsevier.)

shows a response time of 400 ms, and durability over 3000 cycles at 350 mm min^{-1}, as confirmed by Instron tensile tests showing three-stage resistance changes (Figure 5.18h). This integrated knitted sensor advances wearable textiles by enhancing practical reliability, multidirectional sensitivity, and gait monitoring accuracy, outperforming prior 2D-limited methods in simulating human joint strains (Figure 5.18i).

Beyond the obvious cues of posture and deliberate gestures, subtler signals—such as the rhythm of breathing or subtle vocal changes—can quietly betray how someone is really feeling inside. The problem is that these micro-movements are easy to miss; reliably detecting them requires exceptionally sensitive sensors. From this perspective, Yue *et al.* [55] introduced a highly stretchable CB/TPU fibrous strain sensor with porous monolith and core-sheath structure for human-motion monitoring, achieving exceptional sensitivity and durability through a one-step coaxial wet-spinning process (Figure 5.19a–d). It was found that the developed sensor enables

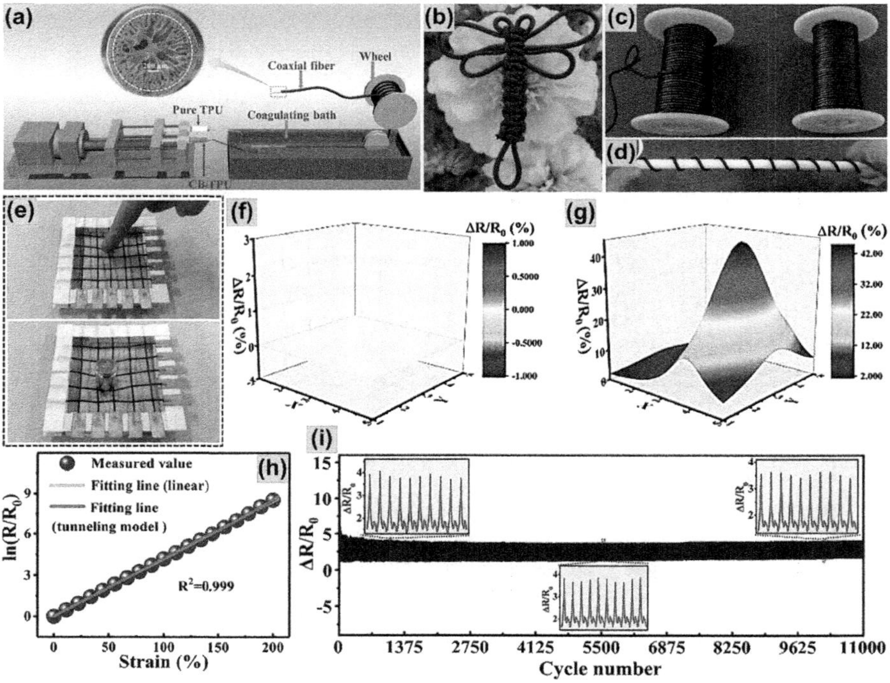

FIGURE 5.19 (a) Preparation procedure of TCTF; (b) a dragonfly woven of TCTF-20 placed on a flower; (c) the as-spun fiber collected on a cylinder with a length of over 5 m; and (d) photographs of the bending of TCTF-20. (e) Digital photographs of a finger (up) and a 25 g weight (down) on the surface of a woven electronic fabric, and corresponding pressure distribution over a 3D area (f, g). (h) Relationship between experimental and theoretical values of $\ln(R/R_0)$ and strain for TCTF-20. (i) $\Delta R/R_0$ of TCTF-20 toward 60% tensile strain for 11,000 cycles.

(This figure was adapted with permission from Ref. [55]. Copyright 2020, Elsevier.)

precise detection of subtle (cheek muscle) and large (knee bending) motions, as well as tactile mapping in a 5 × 5 textile array (Figure 5.19e). Importantly, Figure 5.19f and g presents 3D spatial mappings of relative resistance change across the 5 × 5 textile sample when external stimuli are applied at specific crossing points, showing sharp, localized peaks corresponding precisely to the position of finger pressing (Figure 5.19f) and a 25 g weight (Figure 5.19g), demonstrating excellent spatial resolution and accurate detection of tactile stimuli location. This array-based sensing showed the high sensitivity and durability of the porous core-sheath TCTF-20 sample (TPU-CB@TPU fiber), highlighting the sensor's substantial potential for integration into smart textiles for precise human-motion tracking, pressure mapping, and wearable tactile interfaces. Moreover, this system shows an excellent linear relationship between $\ln(R/R_0)$ and tensile strain up to 200% with $R^2 = 0.999$ (Figure 5.19h), where the experimental data are well fitted by both linear and tunneling model lines, indicating that the resistance response follows tunneling theory and enables predictable, high-accuracy sensitivity over a broad strain range. Notably, Figure 5.19i demonstrates outstanding durability of the porous fiber sensor, exhibiting highly stable and repeatable $\Delta R/R_0$ signals across 11,000 stretching-releasing cycles at 60% strain without noticeable degradation, confirming exceptional long-term reliability and reproducibility for practical wearable motion-monitoring applications. This lightweight (0.28 g cm^{-3}) and weavable TCTF platform advances flexible electronics by overcoming limitations in sensitivity, range, and environmental robustness of prior fiber sensors, with potential for scalable production and robotic applications. Notably, sensor signals indicated that the strain sensor can detect inhalation, exhalation, and other small-range head movements.

In addition to joint movements and body posture, subtle behaviors (such as breathing and vocalizations) may reveal the body's overall state. However, these micro-movements are inherently low in amplitude, making them challenging to capture reliably without the use of highly sensitive sensing technology [8].

To address this shortcoming, MXenes—a relatively new class of two-dimensional transition-metal carbides and nitrides—have recently attracted considerable attention in wearable electronics. Their combination of metallic conductivity, high specific surface area, excellent thermal conductivity, and intrinsically layered structure gives them clear advantages for flexible and textile-integrated devices. In addition, the abundant surface terminations (–O, –OH, –F) introduced during the widely used hydrofluoric acid (HF)-etching process impart hydrophilic character, enabling stable aqueous dispersions. This hydrophilicity greatly simplifies the decoration of textiles through straightforward solution-based techniques such as dip-coating, spray-coating, or printing [56]. Accordingly, Sharma *et al.* [57] developed a polyaniline (PANI)-nanospine-engineered nanofibrous membrane-based piezoresistive sensor for high-performance electronic skins, achieving ultrahigh sensitivity and broad linearity through hybrid hierarchical nanofibers reinforced with MXene and cellulose (Figure 5.20a). For better understanding, Figure 5.20b illustrates the fabrication process of the P-CPCM (PANI-nanospines on the carbonized polyacrylonitrile, cellulose, and MXene) nanofibrous membrane, starting from electrospinning of PAN/cellulose/MXene precursor nanofibers, followed by high-temperature carbonization to form CPCM, and subsequent low-temperature in situ polymerization/

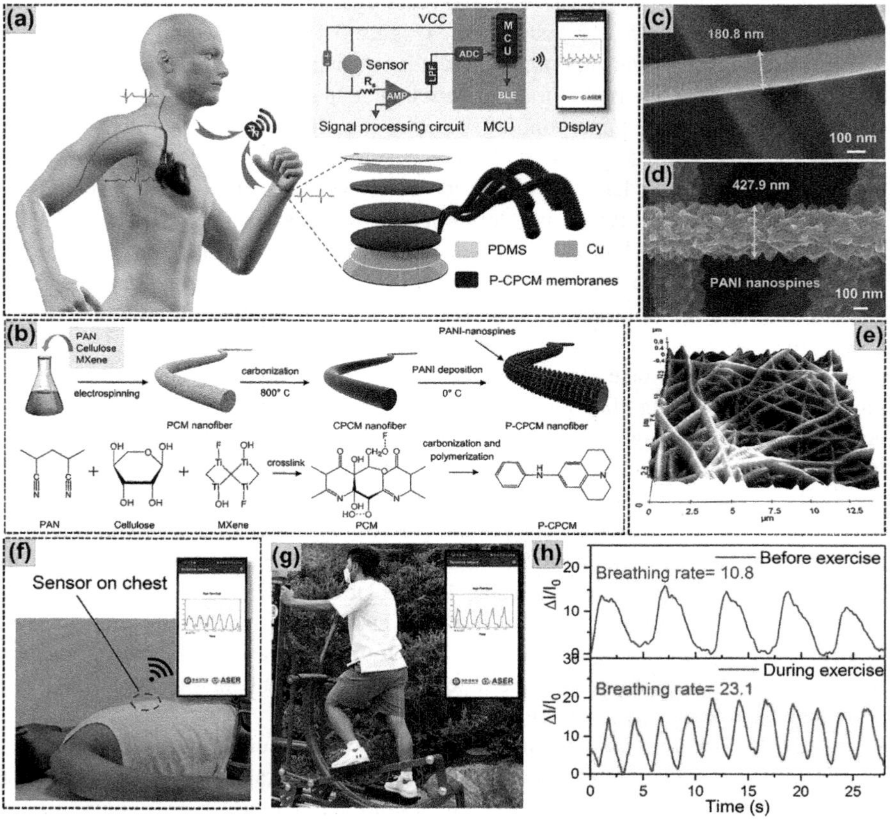

FIGURE 5.20 (a) Schematic showing the electronic skin with the wireless system mounted on the human body for real-time health status monitoring, and a block diagram displaying the complementary signal acquisition system consisting of the low-pass filter, amplifier, controller, Bluetooth device, and mobile unit. (b) Schematic illustrating (up) the fabrication sequence for obtaining the P-CPCM nanofibrous membrane and (down) the chemical process for obtaining the P-CPCM membrane showing the chemical bond between various atoms and molecules in the composites (MXene, cellulose, PAN, and PANI). FESEM images of the (c) CPCM and (d) P-CPCM nanofibers with their high magnification insets showing the diameter of the nanofibers. (e) AFM image of a P-CPCM membrane. (f) Photograph of a volunteer (number 2) with the sensor attached to the chest during deep sleep. The inset shows real-time BR and HR signals monitored on the mobile platform. (g) Photograph of a volunteer (number 1) with the sensor attached to the neck and facemask during a workout. (The inset displays a real-time monitored signal on the mobile platform.) (h) Respiration monitoring before and during exercise.

(This figure was adapted with permission from Ref. [57]. Copyright 2022, Elsevier.)

deposition of polyaniline nanospines, with chemical crosslinking and hydrogen bonding between components enhancing structural stability and conductivity for superior piezoresistive performance. The developed sensor exhibits exceptional sensitivity (179.1 kPa^{-1}), low detection limit (1.2 Pa), wide linear range (0–50 kPa), and durability (>10,000 cycles). In addition, field-emission scanning electron microscopy (FESEM) (Figure 5.20c and d) and atomic force microscopy (AFM) (Figure 5.20e) analyses confirm uniform spine morphology of PANI on the entire surface of the CPCM nanofibrous membrane. Figure 5.20f demonstrates the practical application of the sensor attached to a volunteer's chest during deep sleep, enabling simultaneous, stable, and precise capture of respiratory rate and heart rate signals, highlighting its high sensitivity, low noise, and reliability for noninvasive, continuous physiological monitoring in wearable health devices. Notably, integrated with Bluetooth-enabled wireless systems, the sensor enables real-time monitoring of physiological signals (Figure 5.20g), including radial artery pressure (60.3–68.7 bpm during cycling), jugular venous pulse with three ascent waves (*a*, *c*, and *v*), respiration (10.8–23.1 breaths min^{-1}), and phonation patterns (Figure 5.20h). This P-CPCM platform advances wearable healthcare by mitigating saturation issues in prior microstructures, outperforming in sensitivity, environmental stability (10°C–50°C, 15%–95% RH), and cardiovascular diagnostics.

As another example, Luo *et al.* [56] developed a superhydrophobic, breathable, and multifunctional smart MXene-based textile for wearable sensing electronics, achieving integrated strain and temperature monitoring with personal thermal management through a facile dip-coating method involving polydopamine (PDA) modification of elastic polypropylene fibers, MXene ($Ti_3C_2T_x$) decoration, and PDMS coating to form a hierarchical core–shell structure (Figure 5.21a). As schematically shown in Figure 5.21b–e, the PDA–MXene-PDMS-coated smart textile exhibits excellent waterproofing and breathability, achieved by selective PDMS coating on fiber surfaces that does not block interfiber pores, combined with photothermal conversion capability from the MXene structure. Interestingly, these multifunctional properties of the developed system enable simultaneous monitoring of body motion via piezoresistive strain sensing and temperature via resistance variation, providing a robust, versatile platform for wearable personal health and thermal management applications. Moreover, The PM-3/PDMS-40 textile exhibits exceptional superhydrophobicity (water contact angle (WCA): 151.4°), corrosion resistance (stable WCAs under pH 1–13), breathability (water vapor rate 0.55 kg m^{-2} h^{-1}), and durability against abrasion (50 cycles), UV irradiation (5 h), ultrasonic washing (5 h), and nonlinear relationship between voltage and current due to the material resistance (Figure 5.21f). Outstanding photothermal (saturation ~59°C under 100 mW cm^{-2} for 3600 s) (Figure 5.21g) and electrothermal (89.4°C at 14 V) conversions enable stable Joule heating (> 20 cycles) with UV–Vis absorption > 90% (Figure 5.21h). The textile demonstrates sensitive strain sensing (GF 18 at 50% strain, response time 242 ms, and recovery time of 219 ms with a high stretching speed of 1500 mm min^{-1} for detecting motions like elbow bending (0°–90°), fisting, and walking/running, alongside unique negative temperature coefficient of resistance (TCR, −0.7 to −1.8%/°C at 20°C–50°C) for real-time temperature monitoring (25°C–100°C) with nonlinear response (Figure 5.21i and j).

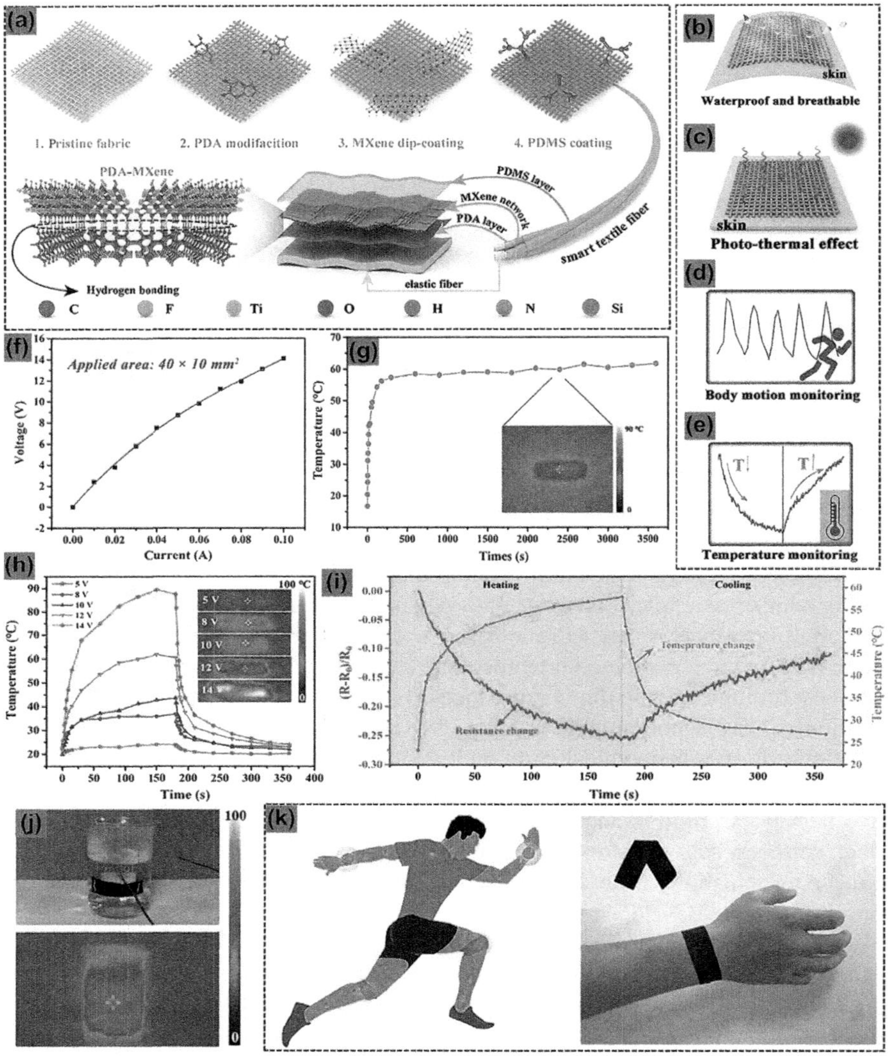

FIGURE 5.21 (a) Schematic illustration of the preparation process and structure of the PM/PDMS textile. (b–e) The PM/PDMS textile attached to human joints monitors human motion and exhibits superior properties and various promising applications. (f) Voltage–current curve of the textile. (g) Photothermal stability of the composite textile under the light irradiation with the power density of 100 mW cm^{-2} for 3600 s; the inset is the IR image of the textile tested for 2400 s. (h) The temperature–time curve of PM-5/PDMS-40 textile under different voltages (the insets show the corresponding IR images of the saturated temperature). (i) The relative resistance and temperature changes of PM-3/PDMS-40 textile with the time under the light-induced heating and natural cooling. (j) The optical photograph of the experimental model (up) and corresponding IR image (down). (k) The smart strain-sensing device worn on the wrist of the volunteer.

(This figure was adapted with permission from Ref. [56]. Copyright 2021, Elsevier.)

This all-in-one PM/PDMS platform advances wearable devices by overcoming MXene oxidation, enhancing environmental robustness, and outperforming prior textiles in multifunctionality, washability, and healthcare applications (Figure 5.21k). Surprisingly, it was found that due to the unique negative thermal coefficient (NTC) property of MXene, the smart textile exhibited the real-time temperature sensing with a high TCR and excellent durability.

Hybrid Signal Sensing

As discussed earlier, wearable sensors are now capable of picking up multiple kinds of signals at once—from body movements and physiological markers (heart rate, skin temperature, etc.) to chemical cues like sweat pH or glucose levels. This provides users with a highly detailed representation of their motion, health status, and even the surrounding environment.

Building on that capability, designers have begun to integrate these sensing modes in more sophisticated ways. Some integrate physical and physiological monitoring into a single fabric platform; others blend physical and biochemical detection. In practice, this usually means integrating different types of sensors into a single textile or strategically placing them at key sites on the body. The result is a hybrid sensing system that delivers multidimensional data without sacrificing comfort, breathability, and skin-friendliness that make textile-based devices so appealing in the first place. It is easy to see why this direction opens up exciting possibilities for the next generation of wearables [10]. Another striking example is a bimodal fabric-based sensor designed to pick up both physical and physiological signals at the same time—using the very same sensing area [58]. In this regard, Zhang *et al.* [58] introduced a high-performance multimodal smart textile integrating a TENG and a thermoresistive micro/nanoporous fiber for simultaneous, interference-free temperature and pulse/touch sensing in wearable health monitoring (Figure 5.22a). This smart textile is fabricated by weaving core–shell structured yarn, where the core consists of six thermoresistive micro/nanoporous fibers (graphene/$Fe_2(MoO_4)_3$/TPU) and an outer nylon monofilament wrapping (Figure 5.22b). The individual thermoresistive porous fiber exhibits a diameter of ~134 μm, excellent sewability (threadable through a needle, Figure 5.22c), and high stretchability up to 50% tensile strain (Figure 5.22d). Photographs in Figure 5.22e and f showcase the resulting smart textile seamlessly integrated into a sweater, highlighting its superior wearability, flexibility, and conformability for practical wearable applications. Fabricated textile achieves fast response/recovery (97/239 ms), and strain-insensitive temperature sensing (0–100% strain), due to the presence of uniform $Fe_2(MoO_4)_3$ structure on graphene, validated by SEM (Figure 5.22g and h) and X-ray diffraction (XRD) (Figure 5.22i). The sensing device exhibits high thermal sensitivity to simultaneously monitor the body temperatures and pulse waves before and after physical exercise (Figure 5.22j) and outstanding durability of the thermal sensor (>5000 cycles at different angles (0°–160°)), enabling precise neck, wrist, and ankle pulse monitoring (Figure 5.22k). This all-in-one platform advances wearable electronics by surpassing prior textile sensors in thermal sensitivity, response speed, and multifunctionality, offering robust solutions for cardiovascular health monitoring and

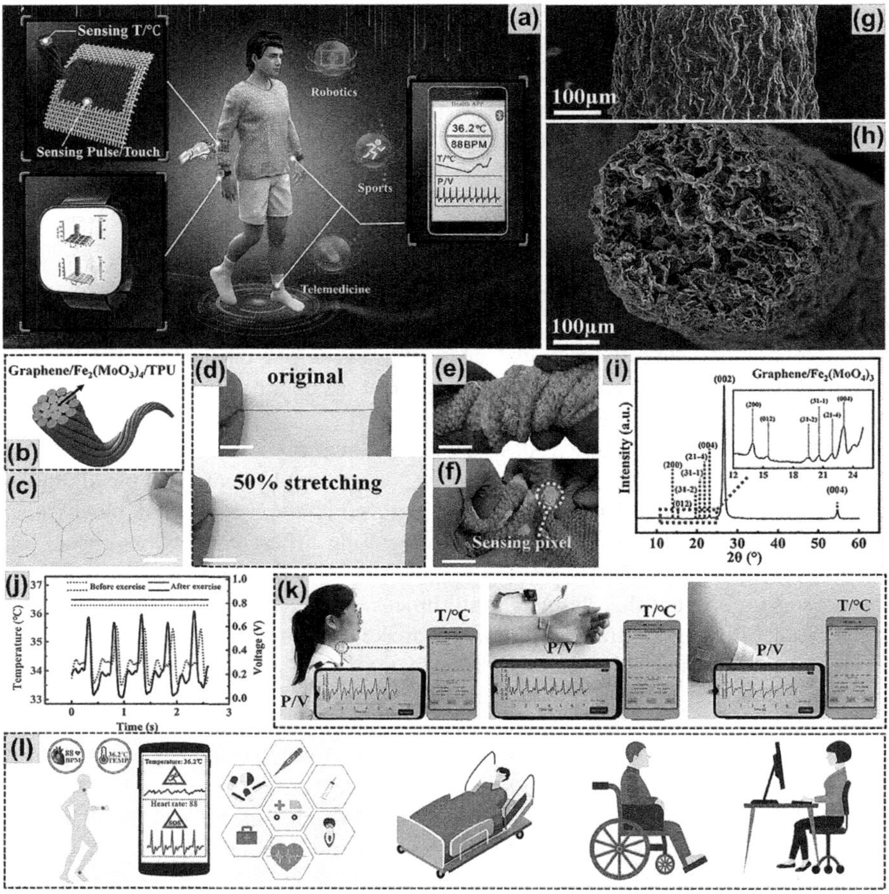

FIGURE 5.22 (a) Schematic illustration of the smart textile worn on the human body to simultaneously monitor temperature and pulse/touch with discriminability. (b) Schematic illustration showing the basic structure of the smart textile. (c) Photograph showing the sewing stitches using the thermoresistive micro/nanoporous fiber. (d) Photographs of the thermoresistive micro/nanoporous fiber in its original state and under stretching. Photographs showing the smart textile (e) under squeezing and (f) under rubbing. (The white textile is the sensing pixel.) SEM images showing the (g) lateral view and (h) cross-sectional view of the micro/nano porous graphene/$Fe_2(MoO_4)_3$/TPU fiber. (i) XRD pattern of the graphene/$Fe_2(MoO_4)_3$ nanocomposite. (The inset shows the enlarged view of the XRD pattern of the $Fe_2(MoO_4)_3$.) (j) The measured body temperatures and pulse waveforms by the smart textile before and after physical exercise. (k) Photographs showing the simultaneous measurement of the body temperature and pulse at the neck, wrist, and ankle. (l) Potential applications of health management systems based on the smart textile.

(This figure was adapted with permission from Ref. [58]. Copyright 2022, Elsevier.)

artificial sensation in diabetes and IoT applications (Figure 5.22l). The two output variables of the sensor were independent. They did not interfere with each other, which could be used for bimodal sensing.

Another example of a physical/physiological hybrid sensing system is a skin-tight electronic shirt that integrates a strain sensor, an ECG sensor, and an EMG sensor. Jin *et al.* [59] developed a highly durable nanofiber-reinforced elastic conductor for skin-tight electronic textiles, integrating silver nanoparticles, PU elastomer, and electrospun PVDF nanofibers to achieve exceptional conductivity and cyclic stability for physiological monitoring (Figure 5.23a and b). This sensing system was fabricated via screen-printing a silver–PU composite with embedded PVDF nanofibers (diameter ~300 nm) and wrinkling to enhance stretchability, and the conductor achieves a high conductivity of 9903 S cm^{-1} (Figure 5.23c) and a low sheet resistance of 0.047 Ω sq^{-1}, with minimal resistance change ($\Delta R/R_0$=0.56 after 5000 cycles at 50% strain) (Figure 5.23d) and stretchability up to 800%, validated by SEM showing a silver-rich top layer due to nanofiber filtering and DMA confirming toughness enhancement (Figure 5.23e). The PVDF nanofibers suppress crack propagation, improving cyclic durability compared to non-reinforced composites (Figure 5.23f). Notably, integrated into a multimodal sensing suit, the conductor enables continuous, high-quality monitoring of electrocardiogram, electromyogram, and motion signals, with stable performance under stretching and sweating (Figure 5.23g and h). This scalable, skin-conformal platform advances wearable smart textiles by overcoming trade-offs in conductivity and durability, outperforming prior nanocomposites in long-term reliability and comfort for healthcare and human–machine interface applications.

In conclusion, to the best of our knowledge, although hybrid signal-sensing textiles have been widely studied, multimodal textiles that combine physical and chemical signals, as well as physical, physiological, and chemical multisensing, are rarely reported; therefore, further research is warranted.

Others

Bridging the communication gap between people who use sign language and those who do not remains a real challenge—sign languages simply are not universal, and most nonsigners have no way to understand them. That is why, in recent years, researchers have increasingly turned to smart textiles as a promising tool. These smart fabrics can detect hand gestures, translate signs into spoken words in real time, and even enable intuitive gesture-based collaboration between humans and machines.

For instance, Liu *et al.* [60] introduced a highly reliable flexible strain sensor based on multi-walled carbon nanotube (MWCNT)-percolated Ecoflex composites (Figure 5.24a), addressing key challenges in wearable electronics by ensuring stable performance under complex microclimates and multidirectional mechanical stimuli. The Ecoflex/carbon nanotube composite fiber (ECCF) sensor was fabricated via a wet-spinning technique (Figure 5.24b), which demonstrates high interfacial strength through a multidirectional separation force analysis between the printed ECCF and the woven fabric substrate. It was observed that the force reaches as high as 0.24, 0.20, 0.18, and 0.19N when the tensile direction is drawn at angles of 45°, 90°, 135°, and laterally 45°, respectively (Figure 5.24c). In addition, the developed sensing

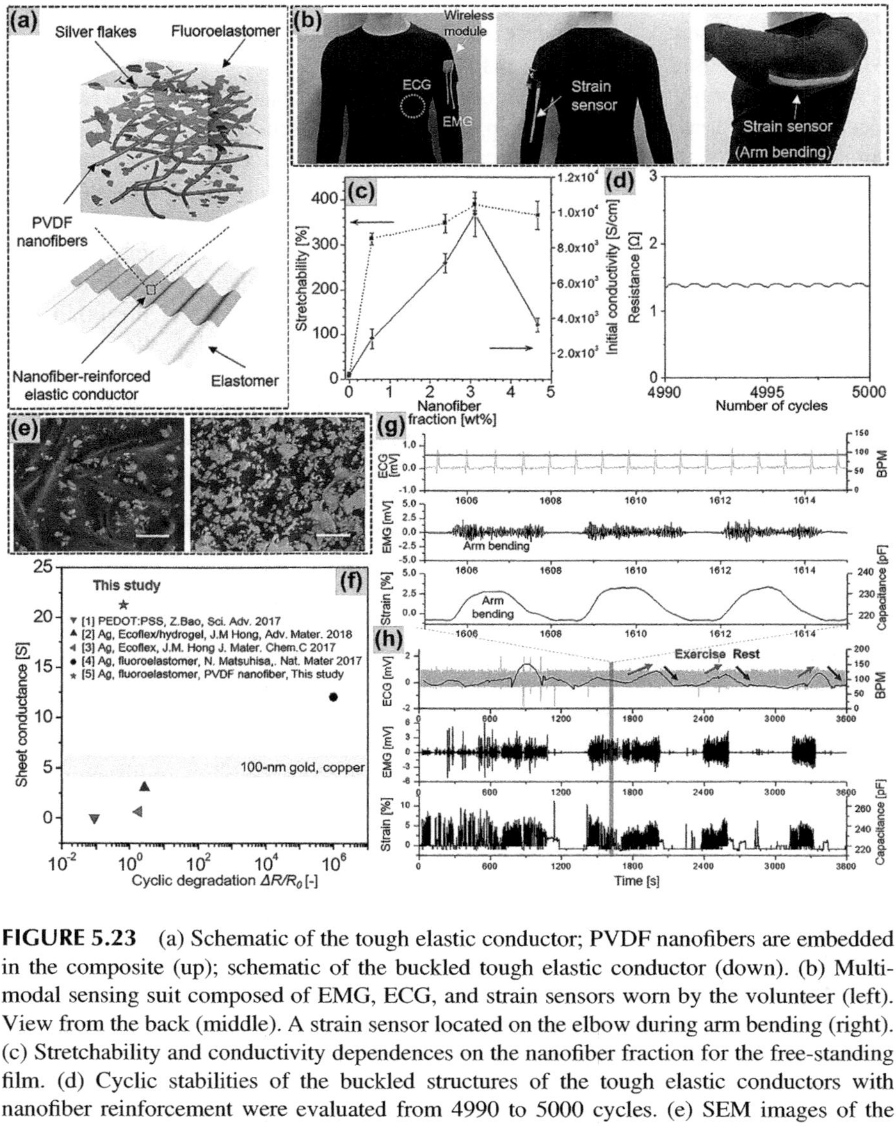

FIGURE 5.23 (a) Schematic of the tough elastic conductor; PVDF nanofibers are embedded in the composite (up); schematic of the buckled tough elastic conductor (down). (b) Multi-modal sensing suit composed of EMG, ECG, and strain sensors worn by the volunteer (left). View from the back (middle). A strain sensor located on the elbow during arm bending (right). (c) Stretchability and conductivity dependences on the nanofiber fraction for the free-standing film. (d) Cyclic stabilities of the buckled structures of the tough elastic conductors with nanofiber reinforcement were evaluated from 4990 to 5000 cycles. (e) SEM images of the frontside (right) and backside (left) of the sample. (f) Comparison of the obtained cyclic durability with those of previously reported stretchable electrodes (strain: 50%, 1000 cycles). (g) EMG, motion, and ECG signals were monitored in real time during the lifting of a 3 kg weight and subsequent release. (h) One hour of continuous monitoring. (BPM obtained from the ECG signal (up), EMG (middle), and capacitance change of the strain sensor (down)).

(This figure was adapted with permission from Ref. [59]. Copyright 2019, American Chemical Society.)

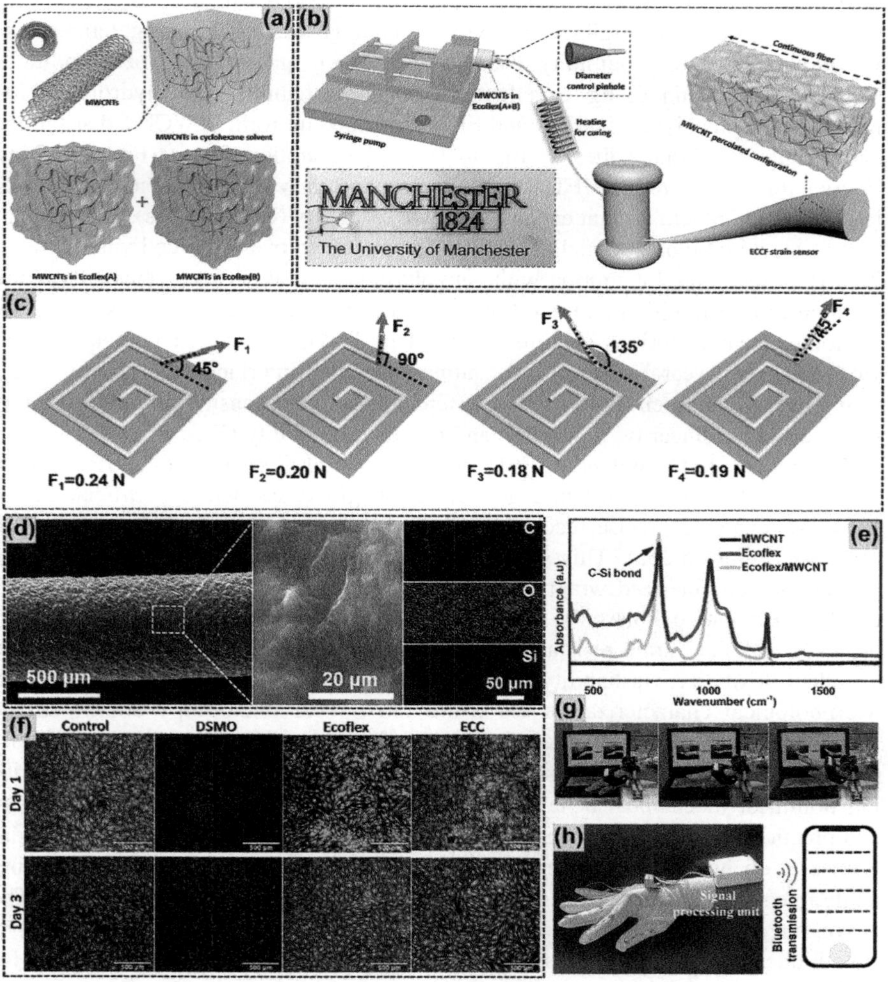

FIGURE 5.24 (a) Illustration of dispersing MWCNTs into cyclohexane solvent through ultrasonic treatment and blending the MWCNTs in cyclohexane with Ecoflex (A) and (B), respectively. (b) Schematic diagram of the fabrication of the ECCF sensor by wet spinning and a 3D-printed pattern (Manchester 1824) acted as an interconnector to light up an LED. (c) Illustration of tensile directions and corresponding separating forces, showing the good interfacial strength between the ECC and woven cotton fabric. (d) Morphology and element distribution on the surface of ECCF16. (e) FTIR spectra of the ECC, pure Ecoflex, and MWCNTs. (f) Investigation of cell viability from live/dead cell assay, showing the good biocompatibility of the ECC. (g) Wrist-bending rehabilitation system for remote robot-assisted exercises, showing the application in man–machine interaction. (h) A smart glove with a fiber sensor sewn into each finger and data visualization via Bluetooth.

(This figure was adapted with permission from Ref. [60]. Copyright 2022, Elsevier.)

device achieves an ultrawide linear sensing range of 860% with a GF of 8.13 and R^2 of 98.92%, showing negligible sensitivity to bending, twisting, pressing, temperature variations (27–45°C), humidity changes (28–95%), and extreme pH conditions (1–14), while maintaining excellent machine washability and hydrophobicity. Morphological analysis via SEM and EDS confirmed uniform MWCNT distribution and strong interfacial adhesion (up to 0.24 N separation force) (Figure 5.24d). Fourier transform infrared (FTIR) and Raman spectra validated chemical bonding (e.g., C–Si at 785 cm^{-1}) that enhances network stability (Figure 5.24e). Biocompatibility tests with human dermal fibroblasts demonstrated over 95% cell viability after 72 h (Figure 5.24f). Integration with convolutional neural networks enabled precise hand gesture recognition and real-time control of a pneumatic wrist exoskeleton for flexion/extension motions (Figure 5.24g). This MWCNT-Ecoflex percolation network advances wearable sensing for human–machine interfaces and motion monitoring by mitigating environmental interference and cross-sensitivity, outperforming prior sensors in linearity, durability, and practical reliability (Figure 5.24h).

In another study, Wu *et al.* [61] produced a full-fiber auxetic-interlaced yarn sensor (AIYS) for a sign-language translation glove assisted by artificial neural networks, enabling precise recognition and real-time translation of the full English alphabet (Figure 5.25a). This sensor was developed through a continuous, mass-producible computerized wrapping spinning technology (Figure 5.25b). The AIYS exhibits a negative Poisson's ratio of −1.5, robust mechanical strength (0.6 cN $dtex^{-1}$), and fast strain-resistance responsiveness (0.025 s) while demonstrating excellent flexibility, washability, weavability, and cyclic durability (>8000 cycles). Morphological characterization via SEM and optical microscopy confirmed the stable interlaced structure of silver-coated polyamide sheath yarns on a PU core (Figure 5.25c), with a viscoelastic constitutive model (Model-II) accurately describing nonlinear mechanical behavior ($R^2 = 0.997$) (Figure 5.25d). Biocompatibility and practical reliability were validated through human-motion detection and translation tests on human skin (after the AIYS is stretched for 8,000 times), showing negligible signal decrease owing to the polymer stress relaxation (Figure 5.25e). The glove, integrated with a 16-channel AIYS array and artificial neural network (ANN) (two hidden layers with 100 nodes each), achieved 99.8% recognition accuracy for 26 letters within 0.25 s, supporting seamless translation of complex sentences and dialogues into text or voice (Figure 5.25f and g). This low-cost AIYS-based system advances assistive wearables for human–machine interfaces and communication by overcoming environmental interference, structural instability, and limited alphabet recognition in prior sensors (Figure 5.25h).

Moreover, Veeramuthu *et al.* [62] developed muscle fiber-inspired electrospun nanostructures reinforced conductive fibers (ERCFs) for smart wearable optoelectronics and energy generators, featuring styrene–butadiene–styrene (SBS) nanofibers with in situ silver nanoparticles (Ag NPs) for strain-insensitive conductivity (Figure 5.26a and b). This sensor was fabricated via electrospinning followed by wet spinning, and the ERCFs achieve exceptional stretchability (711.85% elongation) (Figure 5.26c) and toughness (10.05 MJ m^{-3}), with negligible resistance change under 150% strain and high durability (> 6000 cycles) (Figure 5.26d),

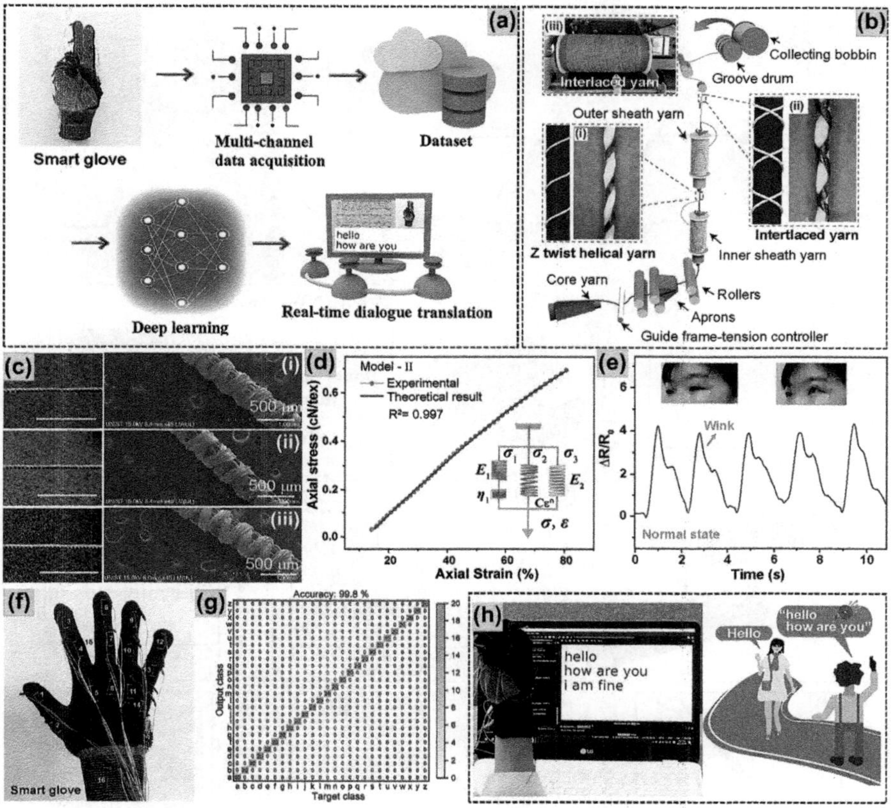

FIGURE 5.25 (a) Diagram of the real-time sign-language translation, showing signal acquisition, data processing, and deep learning paths from the AIYS-array-embedded smart glove to the sign-language translation. (b) Schematic illustration of the continuous yarn-wrapping technology. Insets (i) and (ii) are optical images of the Z-twist helical yarn and interlaced-helical yarn, and (iii) is a picture of collecting bobbins with the AIYS. (c) Optical and SEM images of the AIYS in the states of origin with washing eight times and cycling 8,000 times. (Scale bar of 50 mm.) (d) Comparison of experimental and theoretical viscoelastic mechanical behavior of an AIYS described by Model-II. (e) Detected signals of human winking when volunteers wear the AIYS at the eye corner, the insets are the pictures showing the upper facial expressions around the eyes. (f) Photograph of a smart glove embedded with 16-channel sensors. (g) Confusion matrix for individual recognition of all letters, exhibiting a high overall accuracy of 99.8%. (h) Experiment showing the input sentence in real time from the sign language demonstrated by the volunteer wearing the smart glove. (Illustration shows a signer communicating with a person with the help of the sign-language translation glove.)

(This figure was adapted with permission from Ref. [61]. Copyright 2022, Springer Nature.)

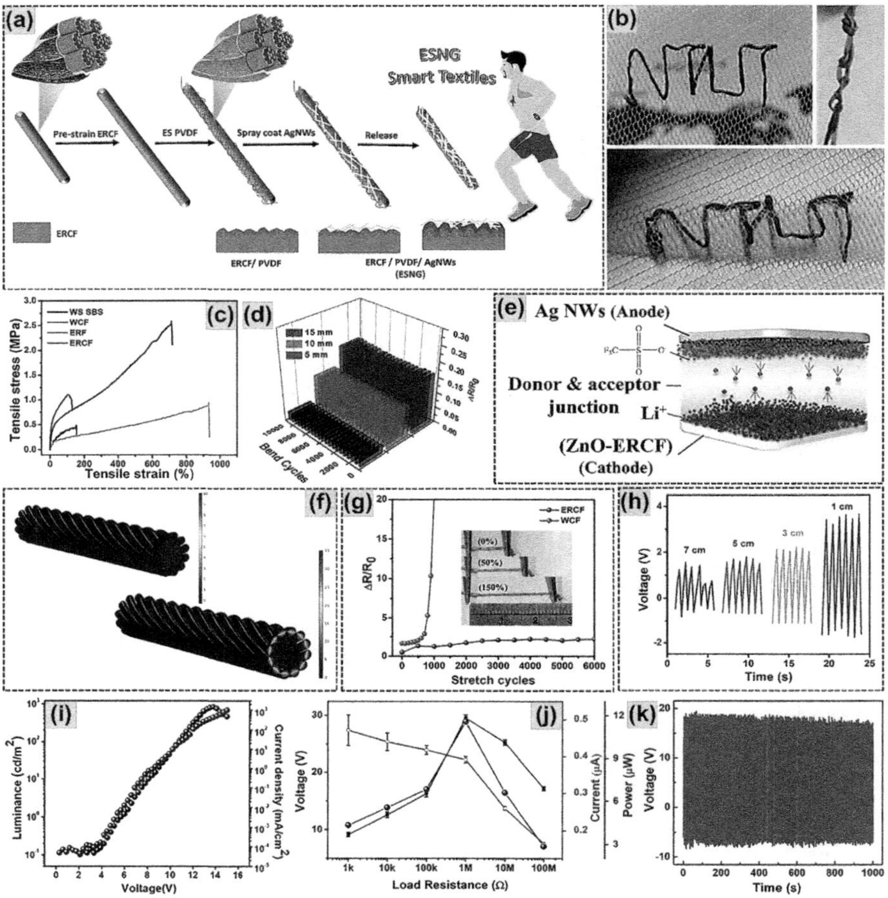

FIGURE 5.26 (a) Schematic representation of the electrospun nanogenerator (ESNG) smart textile fabrication procedure. (b) Sewability, knot, and braid ability of the ERCF electrodes. (c) Stress–strain profile for the substrate and electrodes. (d) ERCF and WCF electrodes on the conductive stability of bending radii. (e) Schematic representation of the ERCF-based LECs. (f) Simulation voltage results for WSNG versus ESNG by COMSOL. (g) Wearable electrodes on stretch release cycles (40 % strain). (h) Fixed wind velocity–distance dependent voltages. (i) Luminance–current density–voltage curve of ERCF LECs. (j) Output voltage and current dependence with respective external load resistances and (k) associated durability measurements.

(This figure was adapted with permission from Ref. [62]. Copyright 2022, Elsevier.)

confirmed by SEM-EDS (Figure 5.26e), FTIR (C-F at 1128 cm^{-1}), XRD (Ag peaks at 38.1°), and thermogravimetric analysis (TGA) analyses. Finite element analysis (FEA) simulations validated superior energy dissipation and stress distribution in ERCFs compared to wet-spun conductive fibers (WCFs) (Figure 5.26f). Integrated into smart gloves, ERCFs enable hysteresis-free gesture recognition for human–machine interfaces via Bluetooth connectivity (Figure 5.26g and h).

For optoelectronic application, ERCF-based light-emitting electrochemical cells (LECs) exhibit yellow emission with a maximum luminance of 830 cd m^{-2} and an efficiency of 1.92 cd A^{-1} at 13.8 V (Figure 5.26i). In the energy-harvesting case, self-poled electrospun twisted polyvinylidene fluoride (PVDF) nanofibers yield piezoelectric outputs of 29.5 V, 0.39 μA, and 11.57 μW, outperforming wet-spun nanogenerators (WSNGs) in breathability and durability up to 30% strain (Figure 5.26j). Interestingly, no obvious degradation in voltage output was observed after the device was subjected to high impact forces over a period of 1000s, demonstrating the durability and robust working characteristics of the device (Figure 5.26k). This versatile ERCF platform advances wearable electronics by addressing mechanical robustness, environmental stability, and integration challenges in prior technologies.

CHALLENGES AND FUTURE PERSPECTIVES

Beyond the established hurdles of durability, power autonomy, and multimodal integration, several emerging challenges are poised to define the next phase of wearable sensors development as the field matures toward real-world deployment. One pressing issue is the management of vast, continuous data streams generated by multimodal sensors, which demands on-device artificial intelligence for real-time analysis, anomaly detection, and privacy-preserving edge computing—without introducing rigid components that undermine textile conformability. Another nascent concern is cybersecurity vulnerability: as smart textiles increasingly interconnect with IoT ecosystems and healthcare networks, protecting against data breaches, signal spoofing, or unauthorized access becomes critical, necessitating lightweight encryption and secure-by-design architectures compatible with flexible substrates. Additionally, the ethical and equitable adoption of these technologies raises new questions, including algorithmic bias in health monitoring (e.g., sensors calibrated predominantly on certain skin tones or body types), accessibility for diverse populations, and the digital divide in low-resource settings where advanced smart textiles may exacerbate healthcare inequalities. Finally, as commercialization accelerates, reconciling performance optimization with circular-economy principles—such as designing for end-of-life disassembly, material recovery, and minimal environmental impact—poses a complex challenge that extends well beyond current sustainability efforts focused on biodegradable conductors alone. Overcoming these evolving obstacles will require not only technological innovation but also close collaboration with ethicists, policymakers, and end users to ensure that smart textiles deliver inclusive, secure, and responsible health benefits.

Despite remarkable progress, several critical challenges must be addressed before textile-based wearable sensors can achieve widespread clinical and consumer adoption. The technical issues that need to be solved in the practical application of smart textiles are listed in the following seven aspects:

Durability and washability: Conductive pathways in smart textiles remain susceptible to repeated mechanical deformation, laundering, and environmental exposure (e.g., humidity, sweat, and UV radiation). Developing robust encapsulation strategies that preserve breathability and flexibility without compromising sensing performance is essential.

Long-term stability and biocompatibility: Aging effects, including oxidation of conductive nanomaterials and delamination of functional coatings, lead to gradual performance degradation. Fully biocompatible, noncytotoxic materials and interfaces that minimize skin irritation during prolonged contact are still needed.

Power supply and energy autonomy: Although flexible batteries and energy-harvesting textiles (triboelectric, piezoelectric, and thermoelectric) have advanced significantly, achieving sufficient power density while maintaining softness, stretchability, and washability remains difficult. Self-powered, maintenance-free systems are a key goal.

Multimodal integration and miniaturization: Combining physical, physiological, and biochemical sensors on a single textile platform without crosstalk or increased bulk is challenging. Seamless integration of signal processing, wireless communication, and data security features into garment-compatible form factors requires further innovation.

Scalability and cost-effective manufacturing: Most high-performance smart textiles reported to date rely on laboratory-scale processes. Transitioning to industry-compatible techniques (e.g., roll-to-roll coating, digital embroidery, or automated knitting/weaving with functional yarns) is crucial for commercial viability.

Standardization and regulatory approval: Lack of standardized testing protocols for mechanical robustness, sensing accuracy under real-world conditions, and biocompatibility hinders clinical translation. Establishing consensus benchmarks and obtaining regulatory clearance for medical-grade devices represent significant hurdles.

Sustainability: The growing smart textile market raises concerns about electronic waste and resource consumption. Future designs should prioritize biodegradable substrates, recyclable conductive materials, and eco-friendly fabrication routes.

CONCLUSION

In conclusion, smart textiles have evolved from conceptual prototypes into sophisticated, garment-integrated platforms capable of continuous, multimodal monitoring of human health and activity. Recent advances in functional materials, textile engineering, and sensor integration have yielded devices with impressive sensitivity, stretchability, and wearer comfort, enabling applications ranging from cardiovascular monitoring and sleep apnea detection to motion recognition and sweat-based metabolite analysis. In addition, hybrid and multimodal systems that simultaneously capture physical, physiological, and biochemical signals are particularly promising, offering richer, more actionable health insights than single-parameter devices.

Nevertheless, substantial challenges—particularly in durability, washability, energy autonomy, scalability, and long-term biocompatibility—must still be overcome to transform laboratory demonstrations into reliable, everyday products. Addressing these barriers through interdisciplinary collaboration among materials scientists, textile engineers, clinicians, and regulatory experts will be essential.

Looking ahead, the convergence of sustainable materials, self-powered operation, artificial-intelligence-driven data analysis, and advanced manufacturing holds tremendous potential to deliver truly intelligent clothing that seamlessly enhances human well-being. As these technologies mature, smart textiles are poised to play a central role in preventive healthcare, remote patient management, and personalized medicine, ultimately contributing to healthier, longer lives in an increasingly connected world.

REFERENCES

[1] M. Dulal, H.R.M. Modha, J. Liu, M.R. Islam, C. Carr, T. Hasan, R.M.S. Thorn, S. Afroj, N. Karim, Sustainable, wearable, and eco-friendly electronic textiles, *Energy & Environmental Materials* 8(3) (2025) e12854. https://doi.org/10.1002/eem2.12854

[2] M. Azeem, M. Shahid, I. Masin, M. Petru, Design and development of textile-based wearable sensors for real-time biomedical monitoring; a review, *The Journal of The Textile Institute* 116(1) (2025) 80–95. https://doi.org/10.1080/00405000.2024.2318500

[3] A. Nag, R.B.V.B. Simorangkir, D.R. Gawade, S. Nuthalapati, J.L. Buckley, B. O'Flynn, M.E. Altinsoy, S.C. Mukhopadhyay, Graphene-based wearable temperature sensors: A review, *Materials & Design* 221 (2022). https://doi.org/10.1016/j.matdes.2022.110971

[4] A. Sankauskaite, R. Pauliukaite, J. Baltusnikaite-Guzaitiene, A. Abraitiene, Smart textile with integrated wearable electrochemical sensors, *Current Opinion in Electrochemistry* 42 (2023). https://doi.org/10.1016/j.coelec.2023.101410

[5] M. Stoppa, A. Chiolerio, Wearable electronics and smart textiles: A critical review, *Sensors*, 14(7) 2014, 11957–11992.

[6] J. Zhao, Y. Fu, Y. Xiao, Y. Dong, X. Wang, L. Lin, A naturally integrated smart textile for wearable electronics applications, *Advanced Materials Technologies* 5(1) (2020) 1900781. https://doi.org/10.1002/admt.201900781

[7] M. Dulal, S. Afroj, J. Ahn, Y. Cho, C. Carr, I.-D. Kim, N. Karim, Toward sustainable wearable electronic textiles, *ACS Nano* 16(12) (2022) 19755–19788. https://doi.org/10.1021/acsnano.2c07723

[8] X. Wei, X. Liang, C. Meng, S. Cao, Q. Shi, J. Wu, *Multimodal electronic textiles for intelligent human-machine interfaces*, OAE Publishing Inc. 2023.

[9] K. Du, R. Lin, L. Yin, J.S. Ho, J. Wang, C.T. Lim, Electronic textiles for energy, sensing, and communication, *iScience* 25(5) (2022). https://doi.org/10.1016/j.isci.2022.104174

[10] W. Xiao, L. Xiaotong, M. Chongguang, C. Shuze, S. Qiongfeng, W. Jun, Multimodal electronic textiles for intelligent human-machine interfaces, *Soft Science* 3(2) (2023) 17. https://doi.org/10.20517/ss.2023.09

[11] A.A. Simegnaw, B. Malengier, G. Rotich, M.G. Tadesse, L. Van Langenhove, Review on the integration of microelectronics for E-textile, *Materials* 14(17) (2021) 5113. https://doi.org/10.3390/ma14175113

[12] *Smart textiles: Fundamentals, Design, and Interaction*, Springer 2017.

[13] H. Luo, B. Gao, Development of smart wearable sensors for life healthcare, *Engineered Regeneration* 2 (2021) 163–170. https://doi.org/10.1016/j.engreg.2021.10.001

[14] J. Shi, S. Liu, L. Zhang, B. Yang, L. Shu, Y. Yang, M. Ren, Y. Wang, J. Chen, W. Chen, Smart textile-integrated microelectronic systems for wearable applications, *Advanced materials* 32(5) (2020) 1901958. https://doi.org/10.1002/adma.201901958

[15] Z. Xu, C. Zhang, F. Wang, J. Yu, G. Yang, R.A. Surmenev, Z. Li, B. Ding, Smart textiles for personalized sports and healthcare, *Nano-Micro Letters* 17(1) (2025) 232. https://doi.org/10.1007/s40820-025-01749-6

[16] B.K. Dejene, M.A. Abtew, A.A. Birlie, M.E. Woldeab, Three-dimensional (3D) knitted spacer textile materials for advanced healthcare solutions: A comprehensive review, *Journal of Industrial Textiles* 54 (2024). https://doi.org/10.1177/15280837241290169
[17] B.K. Dejene, A.D. Gudayu, Exploring the potential of 3D woven and knitted spacer fabrics in technical textiles: A critical review, *Journal of Industrial Textiles* 54 (2024). https://doi.org/10.1177/15280837241253614
[18] S. Wu, T. Dong, Y. Li, M. Sun, Y. Qi, J. Liu, M.A. Kuss, S. Chen, B. Duan, State-of-the-art review of advanced electrospun nanofiber yarn-based textiles for biomedical applications, *Applied Materials Today* 27 (2022). https://doi.org/10.1016/j.apmt.2022.101473
[19] B.K. Dejene, Wearable smart textiles for mood regulation: A critical review of emerging technologies and their psychological impacts, *Journal of Industrial Textiles* 55 (2025) https://doi.org/10.1177/15280837251314190
[20] G.N. Islam, A. Ali, S. Collie, Textile sensors for wearable applications: A comprehensive review, *Cellulose* 27(11) (2020) 6103–6131. https://doi.org/10.1007/s10570-020-03215-5
[21] X. Zhong, H. Hu, H. Fu, Self-cleaning, chemically stable, reshapeable, highly conductive nanocomposites for electrical circuits and flexible electronic devices, *ACS Applied Materials Interfaces* 10(30) (2018) 25697–25705. https://doi.org/10.1021/acsami. 8b07575
[22] K. Singha, A review on coating & lamination in textiles: Processes and applications, *American Journal of Polymer Science* 2(3) (2012) 39–49. https://doi.org/10.5923/j.ajps.20120203.04
[23] J. Cho, J. Moon, K. Jeong, G. Cho, Application of PU-sealing into Cu/Ni electroless plated polyester fabrics for e-textiles, *Fibers Polymers* 8(3) (2007) 330–334. https://doi.org/10.1007/BF02877279
[24] Y.J. Yun, W.G. Hong, D.Y. Kim, H.J. Kim, Y. Jun, H.-K. Lee, E-textile gas sensors composed of molybdenum disulfide and reduced graphene oxide for high response and reliability, *Sensors and Actuators B: Chemical* 248 (2017) 829–835. https://doi.org/10.1016/j.snb.2016.12.028
[25] Y. Liu, K.-N. Zhang, Y. Zhang, L.-Q. Tao, Y.-X. Li, D.-Y. Wang, Y. Yang, T.-L. Ren, Flexible, wearable, and functional graphene-textile composites, *Applied Physics Letters* 110(26) (2017). https://doi.org/10.1063/1.4990530
[26] W.G. Whittow, A. Chauraya, J. Vardaxoglou, Y. Li, R. Torah, K. Yang, S. Beeby, J. Tudor, Inkjet-printed microstrip patch antennas realized on textile for wearable applications, *IEEE Antennas Wireless Propagation Letters* 13 (2014) 71–74. https://doi.org/10.1109/LAWP.2013.2295942
[27] P. Schäl, I. Juhász Junger, N. Grimmelsmann, A. Ehrmann, Development of graphite-based conductive textile coatings, *Journal of Coatings Technology and Research* 15(4) (2018) 875–883. https://doi.org/10.1007/s11998-017-0024-5
[28] T. Yan, Z. Wang, Z.-J. Pan, A highly sensitive strain sensor based on a carbonized polyacrylonitrile nanofiber woven fabric, *Journal of Materials Science* 53(16) (2018) 11917–11931. https://doi.org/10.1007/s10853-018-2432-z
[29] S. Li, H. Li, Y. Lu, M. Zhou, S. Jiang, X. Du, C. Guo, Advanced textile-based wearable biosensors for healthcare monitoring, *Biosensors* 13(10) (2023) 909. https://doi.org/10.3390/bios13100909
[30] L. Van Langenhove, *Smart Textiles for Medicine and Healthcare: Materials, Systems and Applications*, Elsevier 2007.
[31] P. Gründler, *Chemical Sensors: An Introduction for Scientists and Engineers*, Springer 2007.
[32] X. Zhang, H. Ju, J. Wang, *Electrochemical Sensors, Biosensors and Their Biomedical Applications*, Academic Press 2011.

[33] L. Jabbour, C. Gerbaldi, D. Chaussy, E. Zeno, S. Bodoardo, D. Beneventi, Microfibrillated cellulose–graphite nanocomposites for highly flexible paper-like Li-ion battery electrodes, *Journal of Materials Chemistry* 20(35) (2010) 7344–7347. https://doi.org/10.1039/C0JM01219J

[34] L. Hu, H. Wu, F. La Mantia, Y. Yang, Y. Cui, Thin, flexible secondary Li-ion paper batteries, *ACS Nano* 4(10) (2010) 5843–5848. https://doi.org/10.1021/nn1018158

[35] A.M. Gaikwad, D.A. Steingart, T. Nga Ng, D.E. Schwartz, G.L. Whiting, A flexible high potential printed battery for powering printed electronics, *Applied Physics Letters* 102(23) (2013). https://doi.org/10.1063/1.4810974

[36] W. Fan, Q. He, K. Meng, X. Tan, Z. Zhou, G. Zhang, J. Yang, Z.L. Wang, Machine-knitted washable sensor array textile for precise epidermal physiological signal monitoring, *Science Advances* 6(11) (2020) eaay2840. https://doi.org/10.1126/sciadv.aay2840

[37] J. Dong, D. Wang, Y. Peng, C. Zhang, F. Lai, G. He, P. Ma, W. Dong, Y. Huang, I.P. Parkin, T. Liu, Ultra-stretchable and superhydrophobic textile-based bioelectrodes for robust self-cleaning and personal health monitoring, *Nano Energy* 97 (2022). https://doi.org/10.1016/j.nanoen.2022.107160

[38] Z. Zhao, Q. Li, Y. Dong, J. Gong, Z. Li, J. Zhang, Core-shell structured gold nanorods on thread-embroidered fabric-based microfluidic device for Ex Situ detection of glucose and lactate in sweat, *Sensors and Actuators B: Chemical* 353 (2022). https://doi.org/10.1016/j.snb.2021.131154

[39] S. Takamatsu, T. Lonjaret, D. Crisp, J.-M. Badier, G.G. Malliaras, E. Ismailova, Direct patterning of organic conductors on knitted textiles for long-term electrocardiography, *Scientific Reports* 5(1) (2015) 15003. https://doi.org/10.1038/srep15003

[40] R. Lin, H.-J. Kim, S. Achavananthadith, Z. Xiong, J.K. Lee, Y.L. Kong, J.S. Ho, Digitally-embroidered liquid metal electronic textiles for wearable wireless systems, *Nature Communications* 13(1) (2022) 2190. https://doi.org/10.1038/s41467-022-29859-4

[41] T. He, Q. Shi, H. Wang, F. Wen, T. Chen, J. Ouyang, C. Lee, Beyond energy harvesting - multi-functional triboelectric nanosensors on a textile, *Nano Energy* 57 (2019) 338–352. https://doi.org/10.1016/j.nanoen.2018.12.032

[42] K. Meng, S. Zhao, Y. Zhou, Y. Wu, S. Zhang, Q. He, X. Wang, Z. Zhou, W. Fan, X. Tan, A wireless textile-based sensor system for self-powered personalized health care, *Matter* 2(4) (2020) 896–907. https://doi.org/10.1016/j.matt.2019.12.025

[43] X. Zhao, Y. Zhou, J. Xu, G. Chen, Y. Fang, T. Tat, X. Xiao, Y. Song, S. Li, J. Chen, Soft fibers with magnetoelasticity for wearable electronics, *Nature Communications* 12(1) (2021) 6755. https://doi.org/10.1038/s41467-021-27066-1

[44] H. Ha, T. Suryaprabha, C. Choi, Z.A. Chandio, B. Kim, S. Lim, J.Y. Cheong, B. Hwang, Recent research trends in textile-based temperature sensors: A mini review, *Nanotechnology* 34(42) (2023). https://doi.org/10.1088/1361-6528/ace913

[45] S. Afroj, N. Karim, Z. Wang, S. Tan, P. He, M. Holwill, D. Ghazaryan, A. Fernando, K.S. Novoselov, Engineering graphene flakes for wearable textile sensors via highly scalable and ultrafast yarn dyeing technique, *ACS Nano* 13(4) (2019) 3847–3857. https://doi.org/10.1021/acsnano.9b00319

[46] S. Rahmani Charvadeh, J. Ghalibafan, Investigation of textile permittivity under the influence of fever for designing a wearable temperature sensing antenna, *Biosensors and Bioelectronics: X* 12 (2022). https://doi.org/10.1016/j.biosx.2022.100264

[47] J. Landsiedel, W. Root, N. Aguiló-Aguayo, H. Duelli, T. Bechtold, T. Pham, Multi-point flexible temperature sensor array and thermoelectric generator made from copper-coated textiles, *Sensors* 21(11) (2021) 3742. https://doi.org/10.3390/s21113742

[48] Y. Wu, X. Dai, Z. Sun, S. Zhu, L. Xiong, Q. Liang, M.-C. Wong, L.-B. Huang, Q. Qin, J. Hao, Highly integrated, scalable manufacturing and stretchable conductive core/shell fibers for strain sensing and self-powered smart textiles, *Nano Energy* 98 (2022). https://doi.org/10.1016/j.nanoen.2022.107240

[49] Y. Pan, C. Yang, Z. Du, J. Xue, G. Jin, H. Mou, Highly stretchable and sensitive sensing composite yarn based on multidimensional structure modulation for wearable E-textiles, *Applied Materials Today* 42 (2025). https://doi.org/10.1016/j.apmt.2025.102591

[50] X. Wang, Z. Liu, T. Zhang, Flexible sensing electronics for wearable/attachable health monitoring, *Small* 13(25) (2017) 1602790. https://doi.org/10.1002/smll.201602790

[51] X. Cao, Y. Jie, N. Wang, Z.L. Wang, Triboelectric nanogenerators driven self-powered electrochemical processes for energy and environmental science, *Advanced Energy Materials* 6(23) (2016). https://doi.org/10.1002/aenm.201600665

[52] Y. Jie, H. Zhu, X. Cao, Y. Zhang, N. Wang, L. Zhang, Z.L. Wang, One-piece triboelectric nanosensor for self-triggered alarm system and latent fingerprint detection, *ACS Nano* 10(11) (2016) 10366–10372. https://doi.org/10.1021/acsnano.6b06100

[53] S. Ahn, Y. Cho, S. Park, J. Kim, J. Sun, D. Ahn, M. Lee, D. Kim, T. Kim, H. Shin, J.-J. Park, Wearable multimode sensors with amplified piezoelectricity due to the multi local strain using 3D textile structure for detecting human body signals, *Nano Energy* 74 (2020). https://doi.org/10.1016/j.nanoen.2020.104932

[54] Y. Li, X. Miao, J.Y. Chen, G. Jiang, Q. Liu, Sensing performance of knitted strain sensor on two-dimensional and three-dimensional surfaces, *Materials & Design* 197 (2021). https://doi.org/10.1016/j.matdes.2020.109273

[55] X. Yue, Y. Jia, X. Wang, K. Zhou, W. Zhai, G. Zheng, K. Dai, L. Mi, C. Liu, C. Shen, Highly stretchable and durable fiber-shaped strain sensor with porous core-sheath structure for human motion monitoring, *Composites Science and Technology* 189 (2020). https://doi.org/10.1016/j.compscitech.2020.108038

[56] J. Luo, S. Gao, H. Luo, L. Wang, X. Huang, Z. Guo, X. Lai, L. Lin, R.K.Y. Li, J. Gao, Superhydrophobic and breathable smart MXene-based textile for multifunctional wearable sensing electronics, *Chemical Engineering Journal* 406 (2021). https://doi.org/10.1016/j.cej.2020.126898

[57] S. Sharma, A. Chhetry, P. Maharjan, S. Zhang, K. Shrestha, M. Sharifuzzaman, T. Bhatta, Y. Shin, D. Kim, S. Lee, J.Y. Park, Polyaniline-nanospines engineered nanofibrous membrane based piezoresistive sensor for high-performance electronic skins, *Nano Energy* 95 (2022). https://doi.org/10.1016/j.nanoen.2022.106970

[58] X. Zhang, S. Tang, R. Ma, Z. Chen, J. Zhuo, L. Cao, J. Yang, G. Yang, F. Yi, High-performance multimodal smart textile for artificial sensation and health monitoring, *Nano Energy* 103 (2022). https://doi.org/10.1016/j.nanoen.2022.107778

[59] H. Jin, M.O.G. Nayeem, S. Lee, N. Matsuhisa, D. Inoue, T. Yokota, D. Hashizume, T. Someya, Highly durable nanofiber-reinforced elastic conductors for skin-tight electronic textiles, *ACS Nano* 13(7) (2019) 7905–7912. https://doi.org/10.1021/acsnano.9b02297

[60] Z. Liu, Z. Li, Y. Yi, L. Li, H. Zhai, Z. Lu, L. Jin, J.R. Lu, S.Q. Xie, Z. Zheng, Y. Li, J. Li, Flexible strain sensing percolation networks towards complicated wearable microclimate and multi-direction mechanical inputs, *Nano Energy* 99 (2022). https://doi.org/10.1016/j.nanoen.2022.107444

[61] R. Wu, S. Seo, L. Ma, J. Bae, T. Kim, Full-fiber auxetic-interlaced yarn sensor for sign-language translation glove assisted by artificial neural network, *Nano-Micro Letters* 14 (1) (2022) 139. https://doi.org/10.1007/s40820-022-00887-5

[62] L. Veeramuthu, C.-J. Cho, M. Venkatesan, R. Kumar. G. H.-Y. Hsu, B.-X. Zhuo, L.-J. Kau, M.-A. Chung, W.-Y. Lee, C.-C. Kuo, Muscle fibers inspired electrospun nanostructures reinforced conductive fibers for smart wearable optoelectronics and energy generators, *Nano Energy* 101 (2022) 107592. https://doi.org/10.1016/j.nanoen.2022.107592

6 Fluorescent and Chromatic Eco-Friendly Wearables

Lea Sophie Kämper, Lea Schulz, Anicka Siegert, and Boris Mahltig

INTRODUCTION

In recent years, the megatrends of sustainability and safety have become increasingly relevant in various aspects of life. Awareness of these issues has been influenced by various international developments [1, 2]. Eco-friendly products cover these aspects, especially under consideration of environmental aspects, but social aspects are less considered. The textile industry is taking up these megatrends within individual corporate cultures, as well as along the entire value chain, and is striving for optimisation there [1, 3]. These areas can be addressed directly on the textile using various finishing and finishing processes [4, 5]. One simple way to introduce a safety issue to a textile product is to use fluorescent and luminescent textiles due to the increased visibility gained from fluorescent effects [6].

Textile materials are themselves not fluorescent; so to achieve this property, a textile functionalisation has to be performed by adding components which are fluorescent. Functional components can be applied on textiles using different strategies, such as wet-chemical processing, coating, printing or spin-doping [7–11]. In case of fluorescent modification, quite common methods are the use of fluorescent disperse dyes for synthetic fibres or the application of fluorescent pigments in a printing application [12–14].

As the current market offering shows, such finishes are used in the field of functional and safety clothing, in addition to decorative elements in everyday products [15–18]. The extent to which these options address and implement the topic of eco-friendly properties will be discussed in this chapter, especially with regard to commercially available products in this field. In fact, a textile product is a quite complex product combining different materials and production steps. For this, the development of an eco-friendly fluorescent textile should be done according to a holistic approach covering not only the fluorescent dye as a functional component but also the textile fabric material, the source and production processes.

The main parameters influencing the eco-friendly performance of a fluorescent textile are summarised schematically in Figure 6.1. Considering these parameters,

DOI: 10.1201/9781003507208-6

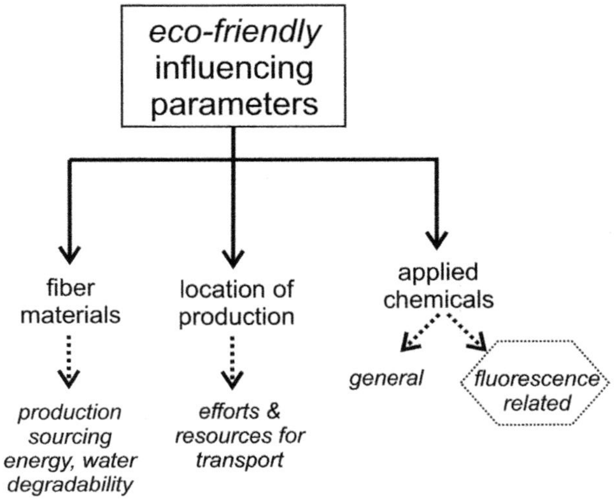

FIGURE 6.1 Schematic overview of parameters influencing the eco-friendly performance of textile products with fluorescent properties.

a textile product can be ranked to value its eco-friendly performance. It should be clear that a product cannot have an absolute eco-friendly performance, but by considering most parameters, a high eco-friendly rank might be achieved. These parameters can be categorised into three directions. The first category is related to the fibres used to build up the textile product. This category is valid for all types of textiles and, of course, not only for fluorescent textiles. The fibre materials can be ranked for their eco-friendly performance according to production processes, sourcing, materials, water and energy consumption [19–21]. The second category is related to the location where the textile material is produced. A production nearer to the place of final use and consumption is more eco-friendly, because fewer resources are needed for transporting [22]. The third category is related to chemicals applied to the fibres and fabrics for several reasons [23]. Here, it can be distinguished between chemicals which are used in general without specific relation to the fluorescent properties of a textile and textiles which are dedicated to fluorescent properties.

The chemicals related to the fluorescence properties can be fluorescent dyes or pigments. Also, binders or crosslinking components needed for fixation and washing stability of fluorescent pigment belong to this category. The eco-friendly performance of those fluorescent-related chemicals can be ranked due to different factors. An overview of these factors is schematically depicted in Figure 6.2. The first parameter is the toxicity of the chemical itself. Other parameters are related to production processes and the biodegradability after use. The fastness properties count especially if long-term use of the textile is aimed. Finally, application processes have to be considered and the possible use of further components as a binder or cross-linker.

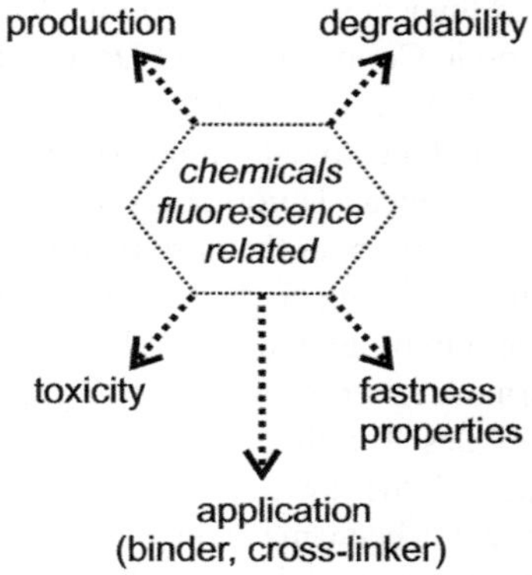

FIGURE 6.2 Schematic overview of parameters influencing the eco-friendly performance of chemicals added for fluorescent applications.

FLUORESCENT AND LUMINOUS TEXTILES

The physical phenomenon of fluorescence describes the behaviour of materials that, when irradiated with ultraviolet (UV) light, send back part of the light with a longer wavelength [24]. This causes the materials to glow, but unlike phosphorescence, there is no noticeable afterglow once the UV light source has dried up [24, 25]. In textiles, this effect is achieved when they are treated with optical brighteners or other fluorescent dyes or pigments are applied or incorporated into the fibres [26, 27]. For most applications, optical brighteners and fluorescent dyes are excited with UV light, mainly in the wavelength range from 320 to 390 nm, related to the UV light present in the sunlight [28]. For optical brightener, the emission of blue light is aimed [29]. In contrast, fluorescent dyes in general can exhibit emission maxima in the complete range of visible light, depending on the desired colour effect. Fluorescent dyes are used on textiles for several reasons as fashion aspects or to integrate signal colours in highly visible safety clothing. Together with phosphorescent textiles, fluorescent textiles can be accounted for in the group of passive luminescent textiles.

Consequently, there is a group of active luminous textiles. Here, no pigments are used to create a luminous effect, but the textile substrates are equipped with various conductors. First, there are electroluminescent textiles (EL textiles) which have a conductive electrode structure, textiles with integrated optical fibres and finally luminous textiles with integrated light-emitting diodes (LEDs) [30, 31].

ECO-FRIENDLY PERFORMANCE AND TEXTILES

The textile industry is expected to produce 25% of global CO_2 emissions by 2025. These attention-grabbing figures have led to the issue of eco-friendly products becoming more and more important [32]. Since the production of a conventional garment puts a strain on the environment in many different ways, there are a few points where action can be taken [33, 34]. To achieve this, different efforts are made, for example, the Textile Alliance was launched in 2014 on the initiative of the Federal Ministry for Economic Cooperation and Development (BMZ). This alliance consists of 130 members from business, politics and civil society. The aim is to achieve social, economic and ecological improvements along the entire textile supply chain. Further, the European Union (EU) presented its new textile strategy in March 2022. At its core, this is about the introduction of eco-design specifications for textiles. According to this, clothing on the European market might be in the future subject to specifications on durability and repairability [35].

An eco-friendly performance is not only given by a pollutant-free and human-ecological production of textiles. Textile-ecological criteria cover the entire life cycle of textile products, starting with the raw materials and ending with disposal [36, 37]. Significant improvements have been made in the areas of occupational, environmental and consumer protection in recent years. One major issue in this regard is, for example, the legal regulation regarding human-toxicological and ecologically critical textile auxiliaries and dyes. The azo dyes that fall into this category, which are produced based on carcinogenic amines, are banned in the German and European markets, for example [38]. However, it should be remarked that several azo dyes are still allowed for their use as food colourants [39].

According to the Textile Labeling Act (German law), all textiles must be labelled with the raw material. This provides information about the fibres contained in the textile but not about the textile finishing of a product [40, 41]. From the consumer protection organisation's point of view, this can be seen as an informative gap, since dyes and finishing agents remain in and on the fibres and are of relevance due to eco-friendly performance. Textile finishing is 'now possible at a very high ecological quality level, but this is still far from being standard in all production countries' [38]. Through skin contact that the consumer has with the garment, he also comes into contact with chemicals that were not washed out during the manufacturing process. Here, it must be determined whether absorption into the body through direct contact could lead to harmful effects. The assessment of chemicals for clothing is very complex due to the different variables. However, years of discussion on this topic have led to many efforts being made to completely avoid the effects on people and the environment or to keep them as minimal as possible [38].

Special attention is paid to the functional and outdoor clothing sector when it comes to eco-friendly textile products. A large proportion of these products are made of synthetic fibres, which are obtained from mostly non-renewable resources. Synthetic fibre materials decompose very slowly after emission into the environment. For this, the establishment of a raw material cycle is attractive. Initial approaches in this direction are, for example, the introduction of a raw material cycle for pure polyester clothing, which was started in 1994 through a corporate merger involving the mountain sports equipment supplier Vaude, or the use of polyester fibres from recycled polyethylene terephthalate (PET) beverage bottles, which the American outdoor equipment supplier Patagonia first committed to in 1993 [38].

When considering the sustainability of textile products that are treated with fluorescent pigments, one must examine the human and environmental compatibility of the treatment, as well as the manufacturing resources, conditions and impacts on recyclability.

INNOVATIVE FLUORESCENT MATERIALS

Fluorescent dyes are broadly used in textile applications, and many types of fluorescent dyes are commercially available [42]. Nevertheless, there are still efforts in research and development of new fluorescent materials with better eco-friendly performance. These research efforts often follow strategies such as using bio-based non-toxic chemicals for the preparation of fluorescent components or the use of innovative, more eco-friendly application techniques. Some impressive developments are described as follows. Instead of synthetic dyes, bio-based natural dyes can be used to implement fluorescence properties on textile materials [43]. However, during the dyeing process, the originally present fluorescent properties can also be erased [44]. Natural dyes can also be applied using new application techniques such as microwave-assisted dyeing done with Curcumin on acrylic fibres [45]. Fluorescent components can be realised by the treatment of natural compounds such as amino acids in a hydrothermal microwave-supported treatment [46]. The used amino acids in this production are bio-based and non-toxic, so an eco-friendly starting point for the production of fluorescent components is given by this microwave process. Also, synthetic fluorescent disperse dyes can be realised by cleaner production using supercritical CO_2 as a solvent [47].

SOL–GEL APPLICATIONS

Using sol–gel applications, inorganic coatings can be applied onto textiles [48]. Components like fluorescent dyes can be embedded into these coatings or attached onto the coating surface [49, 50]. By this, fluorescent active textiles can be produced [51–53]. However, many sol–gel recipes contain high ratio of organic solvent, such as ethanol. Compared to recipes with water as main solvent, recipes containing organic solvents are less advantageous related to eco-friendly performance. The use of organic solvents is also disadvantageous due to cost reasons and work safety. For these reasons, several efforts are done to decrease the ratio of organic solvent in sol–gel recipes [54, 55].

TEXTILE PRODUCT EXAMPLES

Various selected products and manufacturers that sell luminous textiles for different areas of application are presented in the following. Selected textile objects are described as examples, and the respective manufacturer's approach to the topic of eco-friendly performance is discussed.

LUMINOUS TEXTILES FOR EVERYDAY USE

Luminous textiles, which absorb and re-emit light, are an important safety feature in various everyday objects and increase visibility in dark or confusing situations [56]. In recent years, they have become an integral part of daily lives of many consumers. The luminosity of these textiles is based mainly on the phenomena of fluorescence and phosphorescence [57]. Areas of application range from leisure to sportswear, with them playing a particularly important role in protective clothing for road traffic. The various areas of application for luminous textiles are examined in more detail below using product examples.

SPORT CLOTHES

In the area of sportswear, fluorescent elements are integrated to increase the visibility of people during outdoor activities. The intense colours are intended to prevent accidents, especially in poor lighting conditions. One product example in this area is a hat produced and supplied by the company Trigema. This hat is equipped with a reflective logo and is made of an elastic and roughened microfibre material with fluorescent colouring. Trigema states that the product is made in Germany and that climate-neutral shipping is offered. In addition, Trigema has the ‘Green Button’ seal, which verifies state-certified environmental and social standards [58]. This company claims that resource-saving production is a central aspect, which is an aspect of eco-friendly performance. Resource-saving production is claimed to be one central aspect of the German textile company Trigema. It is stated that production in Germany, according to strict requirements and guidelines, makes the products more eco-friendly than clothing from the Far East, as larger amounts of CO_2 are saved through shorter transport routes [59].

A related product example is offered by the company iQ-UV with an orange hat promoted mainly for use as a UV protective textile. However, this hat is also equipped with fluorescent properties. The light conversion of UV light to visible light by optical brighteners and fluorescent dyes is well described for the application of UV protective textiles [60, 61].

This orange hat from iQ-UV is made of fibre from 85% Nylon and 15% Elastane. The fluorescent emission exhibits a strong maximum at around 620 nm related to orange colouration (compare Figure 6.3). Two smaller emission maxima are observed at 427 and 490 nm. According to company information, three aspects of eco-friendly performance can be set for this product, and related labels are mentioned. First, the Oeko-Tex standard refers to safety regarding chemicals. Second, ‘100% made in Europe’ is related to the production in Poland specifically for the European market. Finally, the label ‘Ocean Plastic’ is related to the use of recycled polymers sourced from waste in the sea, mainly Nylon from old fishing nets [62].

A typical product example from the field of sports clothes is a cycling jacket (Vaude Virt Softshell II cycling jacket), offered by the company Vaude. This jacket contains a fluorescent outer material made of recycled polyester. According to the manufacturer, the jacket bears the ‘Green Shape Label’, which stands for products that are made from eco-friendly materials and produced in an environmentally friendly way. The jacket is also marked with the ‘Green Button’ [63]. Further, the company claims climate-neutrally compensated production. ‘Climate-neutrally compensated’ means that all greenhouse gases caused by products, services or processes are offset by climate protection measures. However, this does not fundamentally reduce emissions. Rather, the CO_2 emitted is calculated, and the

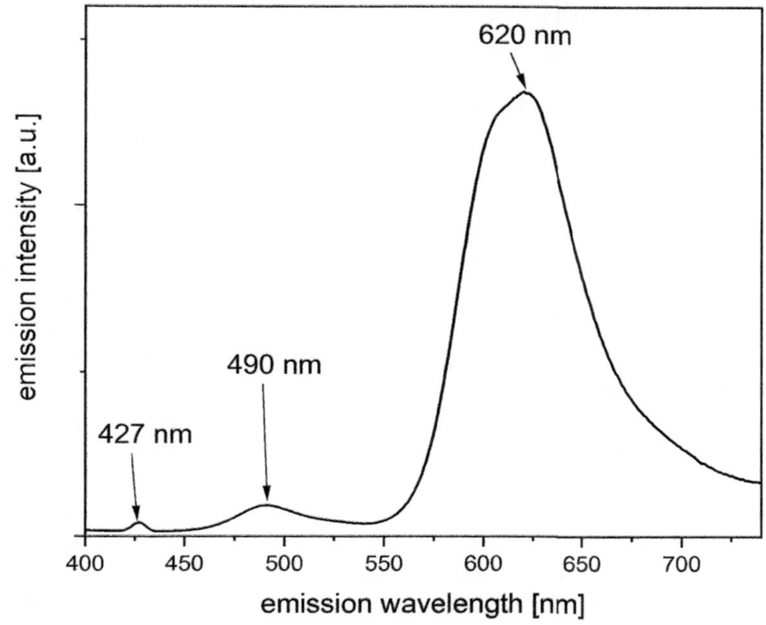

FIGURE 6.3 Fluorescence emission spectrum of a commercially available textile product—orange hat from iQ-UV. The emission spectrum is recorded for excitation with UV-A light at 380 nm.

emissions are offset through financial support of recognised climate protection projects [64]. A key element of Vaude's eco-friendly strategy is the Green Shape label. Another important aspect is recycling. A large proportion of Vaude products are made from recycled materials. Vaude has also set itself the goal of switching all water-repellent membranes and equipment to non-fluorinated alternatives by 2025 [63, 65].

TEXTILES WITH AFTER-GLOWING EFFECT

Bed linen and blankets are typical textile products with the after-glowing effect offered in the section of home textiles. A product example in this field is luminescent bed linen for children supplied by the company Herding [66]. This bed linen is made of 100% cotton, and the luminescent effect is achieved through phosphorescent dyes. In this application, the luminescent effect does not serve the purpose of any safety application—it is merely for entertainment. Similar colour effects are also used in children's nightwear. The luminous effect only occurs in the dark and lasts for a certain time, depending on the dye used. There are numerous products on the market with similar properties. According to the manufacturer, the bed linen considered here has special sustainability features. According to the manufacturer, the product is certified according to Oeko-Tex. In addition, the product is washable at 60°C and suitable for tumble drying, which can contribute to the longevity of the product [66]. Another product example for after-glowing textiles is a hoodie with a luminescent skeleton motif from Boden. The luminous effect is intended to entertain the wearer. The manufacturer provides information about the materials used [67]. Boden has its own sustainability goals for the company on its website [68].

A product example for children's clothes is a pair of children's pyjamas with after-glowing prints offered in 2024 by a German discount supermarket. The afterglow print exhibits yellow/green colouration. However, only small areas of this textile product are printed with the phosphorescent material, such as the small star motif presented in Figure 6.4. The fluorescent emission exhibits a maximum at 518 nm according to the yellow/green colour impression (Figure 6.5). According to the accompanying product information, this pyjama is made from 100% cotton, and Oeko-Tex certification is given. Further, the label 'investing in better cotton' is attached, referring to the approach of better and more eco-friendly production of cotton fibres [69]. For this, mainly two aspects of eco-friendly performance are considered—the general chemical safety of the product and the improved sourcing of the cotton fibres.

Unfortunately, no further information on the type of phosphorescent agent in this product is given alongside. By investigation of the print using scanning electron microscopy (SEM), the applied phosphorescent pigments can be clearly identified in the printed area (Figure 6.6). As stated in an earlier report, these pigments are based on zinc sulphide (ZnS). In SEM images, areas containing chemical elements of higher atomic mass appear brighter, giving a kind of material contrast [70]. By this, the phosphorescent zinc sulphate can be distinguished from the surrounding binder and other probable organic pigments which are also parts of the printing recipe.

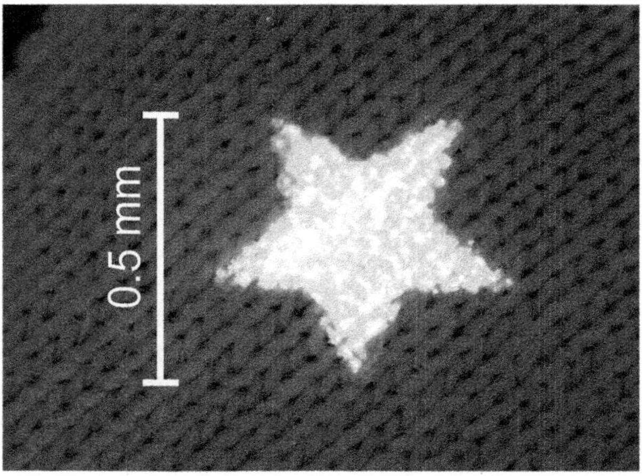

FIGURE 6.4 Photograph taken from a printed area of a commercially available textile product with afterglow effect—children's pyjamas with phosphorescent print.

The photograph is taken using a digital microscope working with UV-A light.

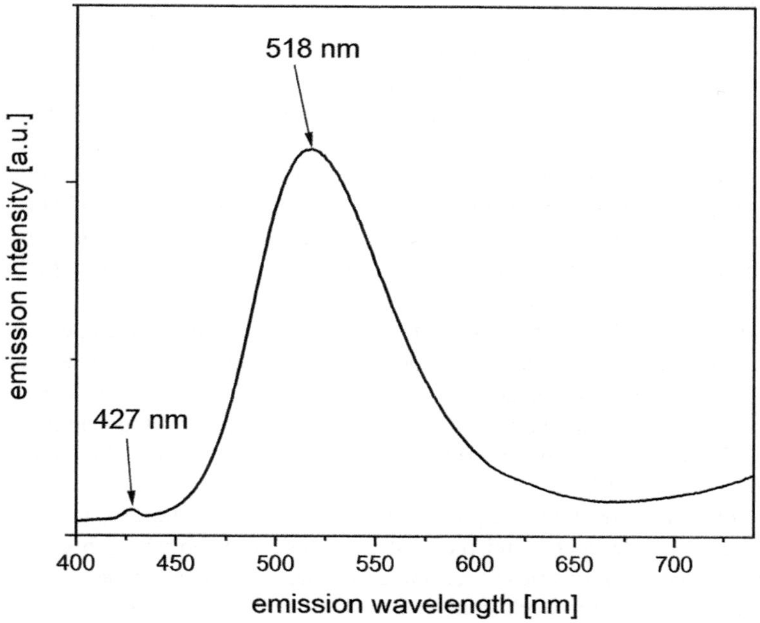

FIGURE 6.5 Fluorescence emission spectrum of a commercially available textile product with afterglow effect—children's pyjamas with phosphorescent print. The emission spectrum is recorded for excitation with UV-A light at 380 nm.

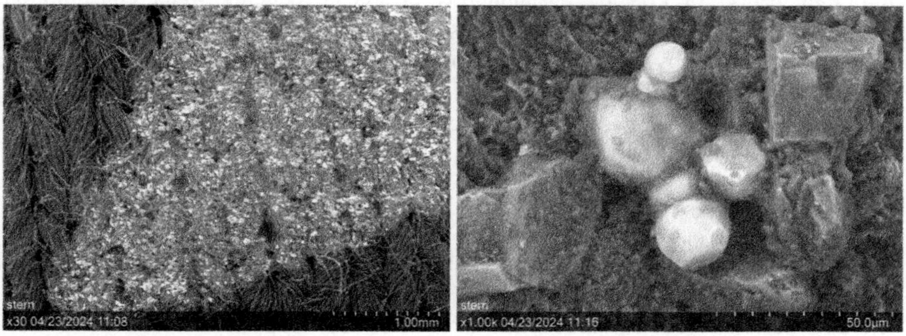

FIGURE 6.6 SEM images taken from a printed area of a commercially available textile product with afterglow effect—children's pyjamas with phosphorescent print. Left image: an overview with low magnification; right image: a detailed view of the applied pigments with high magnification.

FUNCTIONAL AND PROTECTIVE CLOTHES

The purpose of warning and protective clothing is to ensure high visibility in dangerous situations. The clothing must be visible in daylight as well as at dusk and in the dark. It is divided into three classes based on the minimum areas of the fluorescent background material and the retroreflective material [71].

WORKWEAR

One product example in the area of workwear with fluorescent properties is work trousers offered by the company Mascot.

These work trousers are a material combination of a background material dyed with fluorescent orange and knees trimmed with black fabric. Further, there are two strips of reflective material at the level of the lower leg. According to the manufacturer, the product has been manufactured in accordance with the REACH regulation and is certified according to Oeko-Tex. In addition, the materials of the product are dyed by a Bluesign-certified supplier [72]. The addressed eco-friendly performance is based on chemical safety.

A winter overall is another product that has been specially developed for worker protection. The overall is made of 100% polyester and has been dyed fluorescent orange-red. The legs are made of a contrasting fabric in black. The sleeves, trouser legs and torso are equipped with reflective strips. According to the manufacturer, the product is free of polyvinyl chloride, phthalates, perfluorochemical (PFC) and other harmful chemicals [73].

The Fristads work shorts are part of the Fristads Green Collection and are made of lightweight organic cotton and recycled polyester. The fabric has been dyed with fluorescent yellow. Two strips of reflective material are attached to the legs. According to the manufacturer, the work shorts are made of 100% sustainable materials and are Oeko-Tex certified [74]. The entire production chain of the 'green collection' is designed to keep the impact on the environment as low as possible. All products are certified with an Environmental Product Declaration (EPD). This EPD contains detailed information on the overall environmental impact of the product. Another important aspect of Fristads' sustainability efforts is reducing water consumption. The Green Collection products are dyed using a special dyeing process, which reduces water consumption by 75% compared to conventional dyeing processes. [74]. The eco-friendly performance of this product is related to three aspects—chemical safety, used fibre materials and decreased water consumption.

LUMINESCENT PRINTS

The above-presented and discussed examples of commercialised fluorescent and phosphorescent textile products are all addressed to at least one aspect of eco-friendly performance. However, the producers and suppliers address this eco-friendly performance mainly on the origin of the used fibre material and the general chemical safety. The type of fluorescent material used is not disclosed or part of the discussion of eco-friendly performance. Mainly, there is no information on how the fluorescent and phosphorescent materials are considered with respect to eco-friendly performance.

To close this gap, the following luminescent prints on textile fabrics are reported, which are especially executed for this chapter. For this experimental evaluation, several commercially available fluorescent and phosphorescent pigments are considered, which are dedicated to environmentally friendly properties. Suitable application recipes and application techniques are given. Light emission performance and rubbing stability are part of the discussion, showing performance but also the limits of luminescent prints on textile fabrics.

EVALUATED PIGMENTS

Altogether six different luminescent pigments are evaluated which are supplied by three different companies—Textilchemie Dr. Petry GmbH (Reutlingen, Germany), CHT Germany GmbH (Tübingen, Germany) and Kremer Pigmente GmbH (Aichstetten, Germany) (compare Table 6.1). The four pigments from Textilchemie Dr. Petry and CHT are fluorescent and are supplied as environmentally friendly pigments. These pigments are distributed as a dispersion, and one main issue is the formaldehyde-free recipe [75–78].

The two pigments supplied by Kremer are phosphorescent afterglow pigments. They are supplied as dry powders [79, 80]. The phosphorescent pigment green exhibits a yellow-green afterglow in the dark. It consists of ZnS, whose phosphorescent activity is stimulated by small additions of copper chloride [79]. The phosphorescent pigment red exhibits a red-orange afterglow. It is built up by a combination of the phosphorescent pigment green and a not further disclosed red daylight luminescent colourant [80].

COATING PREPARATION

For application onto textile substrates, the different colour pigments have to be combined with a binding agent to support sufficient adhesion to the textile surface. A total of five different combinations of binding agent and the respective pigment are prepared: 1%, 3%, 5%, 10% and 15%. As a binding agent, the printing paste Printperfekt 226 FF supplied by the company CHT (Tübingen, Germany) is used. This ready-made printing paste is transparent, formaldehyde-free and suitable for colour printing on light textiles and nonwovens. In addition, this formaldehyde-free screen printing paste is characterised by a purely aqueous base, suitable for environmentally friendly colour printing [81]. For each current print application, an amount of 50 g is prepared and adjusted to the respective percentage combination (see Table 6.2).

TABLE 6.1
Overview of Evaluated Luminescent Pigments

No.	Pigment Name	Colour	Supplier	References
P1	Perilen Magenta	Pink	Dr. Petry	[76]
P2	Perilen Yellow	Yellow	Dr. Petry	[76]
CHT1	Bezafluor Rosa FF	Pink	CHT	[77]
CHT2	Bezafluor Gelb FF	Yellow	CHT	[78]
K1	Phosphorescent Pigment Green	Yellow, green afterglow	Kremer	[79]
K2	Phosphorescent Pigment Red	Red, orange	Kremer	[80]

TABLE 6.2
Overview of the Composition of Used Coating Agents

Pigment Content [%]	Addition of Binder [g]	Addition of Colour Pigment [%]
1	49.5	0.5
3	48.5	1.5
5	47.5	2.5
10	45.0	5.0
15	42.5	7.5

FIGURE 6.7 Printing table arrangement—before and after printing applications.

After preparation of the printing paste recipes, screen printing is performed for application onto cotton fabrics. This is done on a printing table from Johannes Zimmer, as shown in Figure 6.7. The colour recipes are printed on a cotton fabric measuring 30 × 40 cm. First, the cotton fabric is placed on the surface and aligned using adhesive strips. Then, the selected printing stencil is placed on top. It is important to note that this must cover the fabric well to prevent the colour from getting onto the printing table. The machine is then switched on, and a metal rod is brought into position using a magnet. This is placed so that the desired area can be printed on. The colour is positioned over a large area in front of the metal rod, and then the rod is moved over the stencil in two passes (back and forth) using magnet technology (see Figure 6.7). The finished fabric samples are hung on the laboratory high-temperature steamer, the MATHIS Laboratory HAT Steamer Type GD, to dry. These fabrics are attached to the row of needles. The drying process is done at 120°C for 5 minutes. The machine is then heated up to 160°C, and the print on the fabric sample is finally sealed for 3 minutes.

ANALYTICAL METHODS

For visualisation of fluorescent and phosphorescent effects, a UV-A lamp is used for illumination of coated textile samples. This lamp is a type PL-023.2UV supplied by Kunzer (Willy Kunzer GmbH, Forstinning, Germany). In order to examine the samples with phosphorescent pigments, the dyed fabric samples are illuminated for 30 seconds and then tested to see how long and how intense the colours remained.

The rubbing stability of applied fluorescent prints is tested by using a crockmeter (James Heal). Here, the rubbing stability is tested by rubbing a cotton test fabric over the printed fabrics. Two test arrangements are done, either with dry or wet cotton test fabric. After rubbing, the colour change of the cotton test fabrics is evaluated with daylight and also under UV light. To determine the rub fastness, grades between 1 and 5 were awarded. 5 means that no abrasion has taken place, whereas 1 means that clear abrasion residues are visible on the test sample. This evaluation scheme comes from the grey scale for evaluating bleeding according to ISO 105-AO3.

For recording fluorescence emission spectra of treated textiles, a spectrofluorophotometer RF 6000 (Shimadzu, Japan) is used. The excitation wavelength is set for all measurements to 380 nm. Electron microscopy is performed by using a scanning electron microscope (SEM), TM3000 Tabletop (Hitachi, Japan). The textile samples are investigated using a microscope as prepared, and no further treatment of the samples is done before microscopic investigation.

PRINTING RESULTS AND SAMPLE PERFORMANCE

Colour Impression with Visible Light

The different colour pigments from different manufacturers are easily combined with the binder and then distributed for printing. The higher the proportion of colour pigment in relation to the binding agent, the brighter the colour is. However, it is foreseeable that the differentiation in the luminous intensity becomes smaller as the proportion of colour pigment increases. For example, there is a large difference in intensity between 1% and 5% recipe, whereas the difference between 10% and 15% is significantly smaller. The constellation with 1% colour pigment appears significantly paler compared to the highest pigment proportion of 15%. The colour pigments from Textilchemie Dr. Petry and CHT, in comparison, both show an increase in intensity with higher colour pigment proportions. From the first visual impression, the colour pigments from the CHT group, the Bezafluor FF pigments, radiate a brighter colour intensity. Due to the fact that the pigments from Kremer are so-called afterglow pigments, they are not compared but assessed separately. Under normal lighting conditions, the afterglow colours are sometimes barely visible, depending on the constellation composition between the colour pigments and the binder. In the 1% recipe in particular, it is difficult to see the print at all with the naked eye without UV light. Here, too, it is clear that the higher the colour pigment content, the more visible the print is.

Colour Impression with UV Light

In relation to UV intensity, it can be stated that the greater the proportion of colour pigments, the more radiant and intense the intensity is. What is striking about both the pigments from Textilchemie Dr. Petry and those from the CHT group is that the difference in intensity between the 10% and 15% constellation is no longer particularly strong.

The afterglow phosphorescent pigments from Kremer, green and red, provide similar findings to the colour pigments from Textilchemie Dr. Petry. An overview of printed samples containing the phosphorescent pigments K1 and K2 in increasing concentration is given in Figure 6.8. It is clear that the more intense the constellation between colour pigment and binder, the significantly higher the intensity. In addition to the general intensity test, the UV lamp from Kunzer is also used to test how long the colours afterglow. The afterglow effect is clearly visible in the dark 5 minutes after illumination with UV light.

FIGURE 6.8 Prints containing the pigments K1 and K2 in increasing concentration (left to right).

Photographs under illumination with UV-light.

RUBBING STABILITY

The rubbing fastness is tested for all prepared printed cotton fabrics. Four different test set-ups for rubbing fastness are performed for each sample—rubbing under dry and wet conditions; evaluation of the rubbed test fabric under illumination with visible light or UV light. The results on rubbing fastness are shown as a function of increasing pigment concentration exemplarily for samples with coatings containing the two pigments, fluorescent pigment P1 and phosphorescent pigment K1 (Figure 6.9). The performance of these pigments is quite representative of the recipe application with the other currently investigated pigments. When it comes to rub fastness under dry conditions, it is immediately apparent that there is hardly any abrasion or bleeding detected for evaluation with visible light. Overall, it can be seen that the abrasion is significantly more intense, especially when wet. When dry, abrasion could only be seen in the higher percentage constellations. The samples are also examined with illumination under UV light. For the phosphorescent, the increasing abrasion with increasing pigment concentration under the wet rubbing test, is only visible if the evaluation is done under UV light, probably because the colour intensity of these samples under visible light is low.

FLUORESCENCE PROPERTIES

The fluorescence properties of all prepared cotton fabrics are determined by measuring the fluorescence emission spectra (Figures 6.10–6.12). The emission spectra are recorded for excitation with UV-A light at 380 nm. The fluorescence emission spectra of the two Perilen pigments P1 and P2 are compared in Figure 6.10. Depending on their colouration, one strong emission maximum is observed. For samples with pigment P1, a second, smaller emission maximum at around 430 nm is detected. For the strong maximum, an increase is determined with increasing

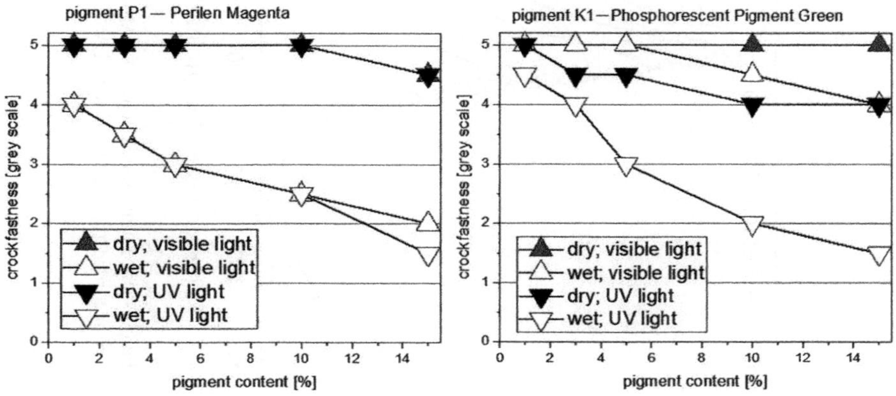

FIGURE 6.9 Rubbing stability of prints containing the pigments Perilen Magenta and Phosphorescent Pigment Green as a function of increasing pigment content. The rubbing fastness is determined with a dry and wet rubbing specimen and evaluated under visible light and UV light.

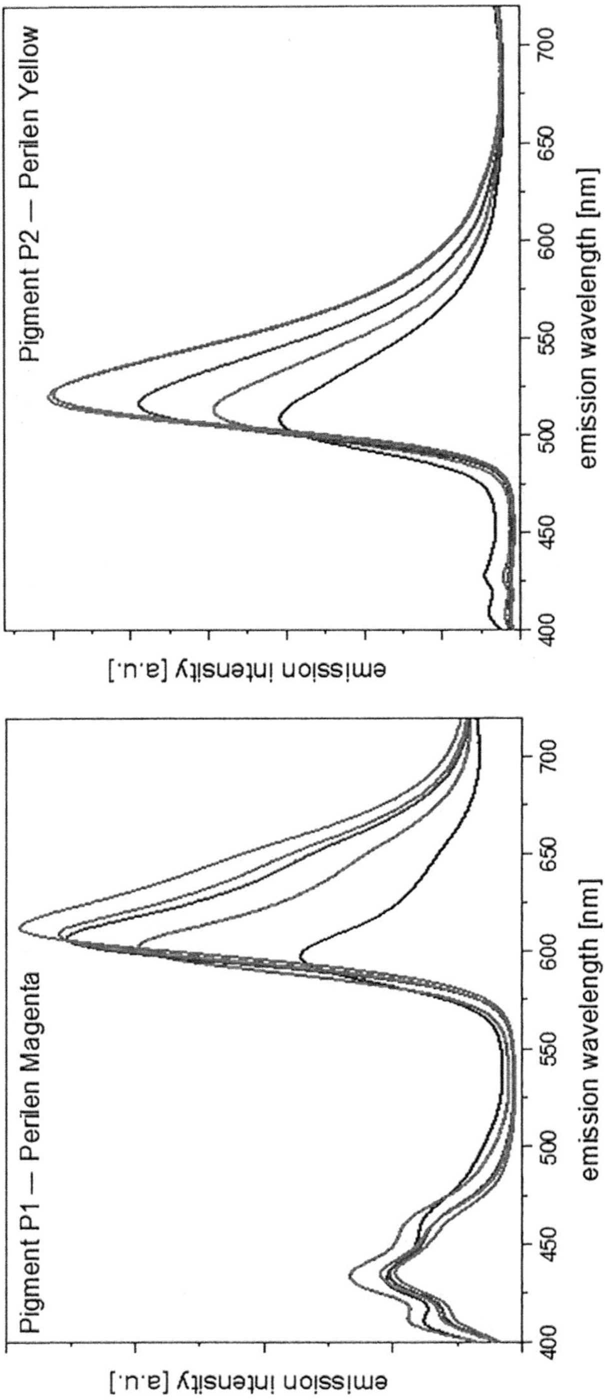

FIGURE 6.10 Fluorescence emission spectra of textiles coated with different Perilen pigments of increasing concentration in the printing paste. The emission spectra are recorded for excitation with UV-A light at 380 nm.

pigment concentration. However, this increase is especially strong for the increase of pigment concentration from 1% to 5%. In contrast, the increase in pigment concentration from 5% to 15% shows less effect of emission intensity. Here, a kind of saturation effect could be estimated. The fluorescent emission spectra for samples with both pigments C1 and C2 are presented in Figure 6.11 and show nearly similar performance.

It is also remarkable that with increasing pigment concentration, not only did the emission intensity increase, but the emission maxima also shifted to higher wavelengths. Both the saturation effect and the shift of the emission maximum with increasing dye concentration are phenomena which are well described for organic fluorescence dyes in solutions [82, 83]. By this, it is clear that for a good fluorescence effect, recipes with medium fluorescence pigment concentrations are advantageous compared to recipes with low or higher concentrations. The use of medium concentration of fluorescence pigments is also advantageous for economic and environmental reasons and enhances the eco-friendly performance of such an application as textile print. Additionally advantageous is the better abrasion stability of the pigment-containing prints with medium fluorescent pigment concentration.

The fluorescence emission spectra of cotton fabrics coated with phosphorescent pigments K1 and K2 are presented for increasing pigment concentration and shown in Figure 6.12. These coatings show a significant difference in emission behaviour compared to the fluorescence pigments described above and also to each other. Coated samples containing phosphorescent pigment K1 exhibit a clear emission maximum at around 530 nm, which is continuously increasing with pigment concentration, so no saturation effect is observed. Also, no shift in emission maximum is observed. This is probably the main difference between the investigated organic and inorganic luminescent components. The pigment K2 is a combined product of an organic and an inorganic luminescent part. The emission maximum at around 520 nm is related to the inorganic component and exhibits no shift at higher concentrations of pigment. The emission maximum at around 580 nm is related to the organic component and shows a shift in the emission maximum.

Microscopic Investigations

Microscopic investigations on the coated fabrics are done to illustrate the regular pigment distribution on the textile surface after printing application. In Figure 6.13, an image of the uncoated cotton fabric is compared to a fabric containing a print with the pigment P2 applied in medium pigment concentrations. The cotton fibre structure is clearly visible, as well as parts of the binder and regularly distributed pigments. The size of the pigment particles is quite small in comparison to the cotton fibre diameter. Due to the fact that the pigment P2 is organic, the material contrast to the organic fibre material is low and the pigment can only be detected by a different surface topography.

Figure 6.14 presents images of a cotton fabric coated with a recipe containing the phosphorescent pigment K1. These images are recorded in different magnifications to support a general overview of the fabric surface and a further detailed view. The surface topography differs significantly from that of the cotton fabric coated with the

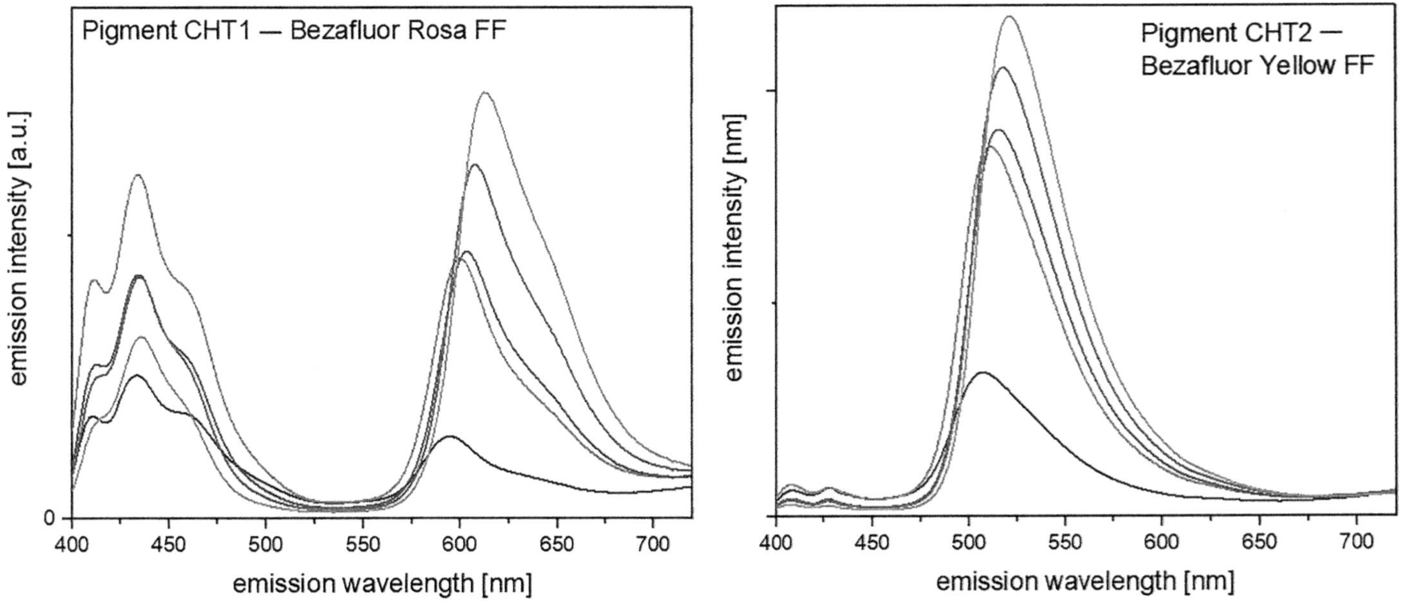

FIGURE 6.11 Fluorescence emission spectra of textiles coated with different Bezafluor pigments of increasing concentration in the printing paste. The emission spectra are recorded for excitation with UV-A light at 380 nm.

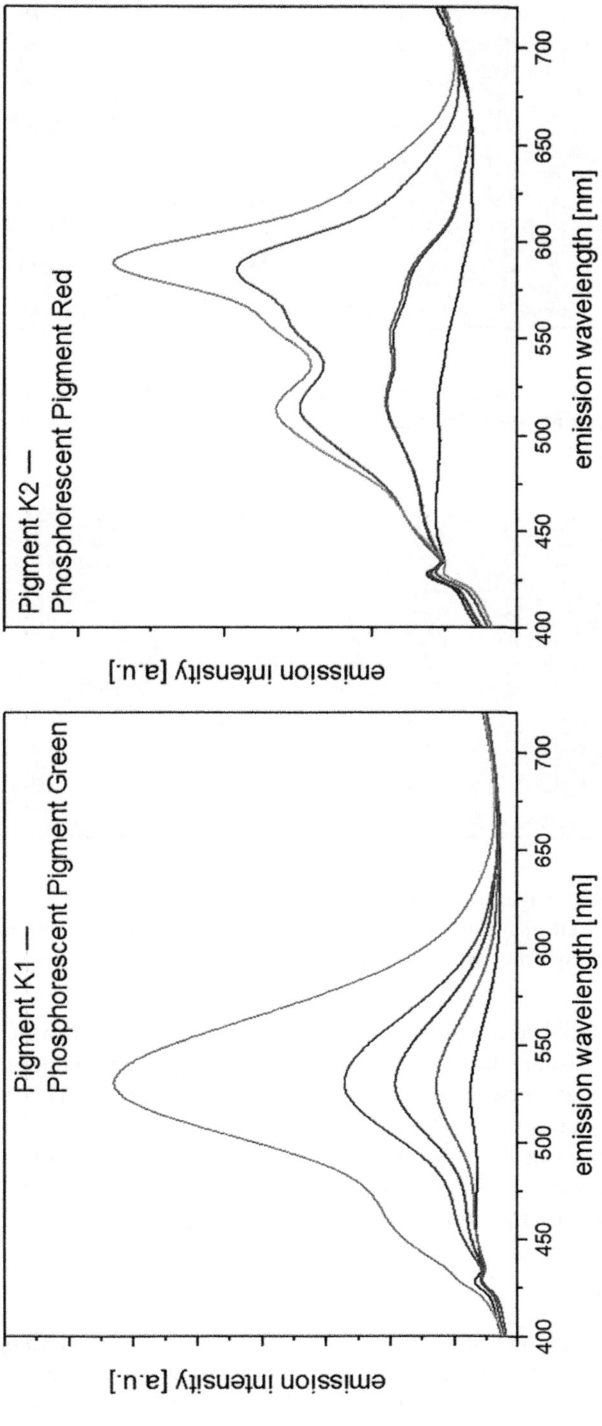

FIGURE 6.12 Fluorescence emission spectra of textiles coated with different phosphorescent pigments of increasing concentration in the printing paste. The emission spectra are recorded for excitation with UV-A light at 380 nm.

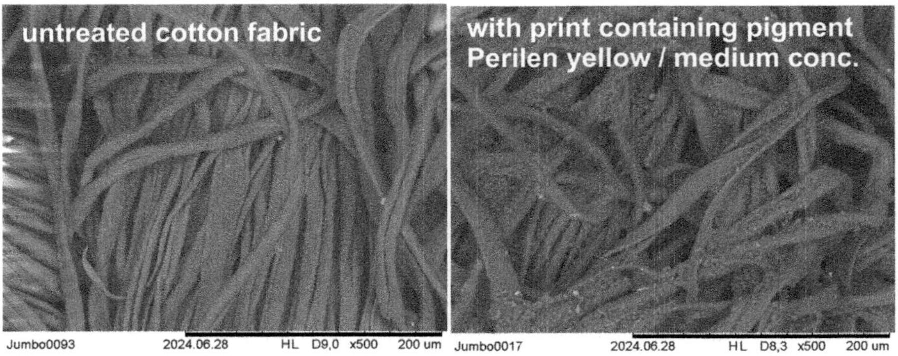

FIGURE 6.13 SEM images taken from cotton fabric—left: uncoated cotton fabric; right: coated fabric with print containing the fluorescent pigment Perilen yellow/Pigment P2 in medium concentrations.

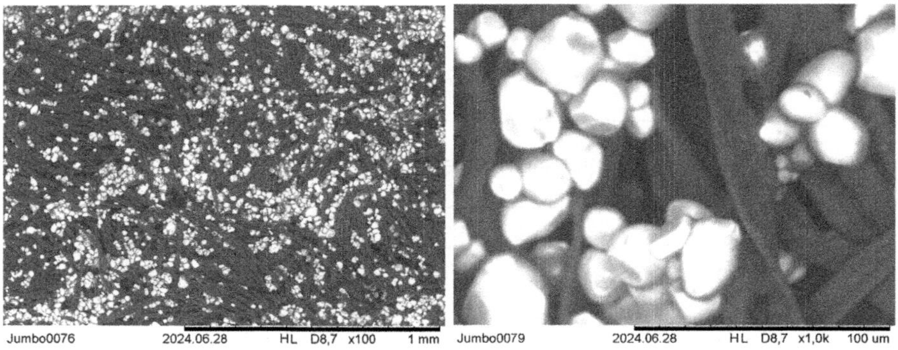

FIGURE 6.14 SEM images taken from cotton fabric printed with phosphorescent pigment green/pigment K1. Left image as an overview with low magnification; right image detailed view on the applied pigments with high magnification.

organic pigment P2. The inorganic pigments are similar in size or even larger than the cotton fibres. Also, a clear material contrast is given due to the inorganic nature of the pigments, mainly based on ZnS. The use of larger pigment particles is often related to lower abrasion stability or a higher ratio of applied binder. For this, inorganic pigments can be disadvantageous compared to the use of organic pigments. However, for a phosphorescent application, these inorganic pigments are required to gain the desired luminescent functional property.

CONCLUSION

Luminescent pigments are already used in various areas of the clothing industry. When it comes to the eco-friendly aspect, it is found that most companies are very secretive about the use of their fluorescence dye and colour pigments. This is due, among other things, to the Textile Labeling Act. This only obliges manufacturers to provide information about the fibres contained in the textile but not about other materials used, for example, for haberdashery and equipment. This also eliminates the obligation to provide information about the processing processes used. In the case discussed here, this concerns the luminous pigments.

In conclusion, it can be said that the topics of eco-friendly production and products are integral parts of the textile and clothing industry. Due to the fact that there is no clear information on the colour pigments used, a specific statement on eco-friendly performance in this direction is difficult to make. However, by different labels and certifications, a statement on the general chemical safety of textile products is given.

Nevertheless, many companies and their product examples provide high efforts to different aspects of eco-friendly products—as fibre materials used, chemical safety, place of production and delivery routes. These areas are as well important part of eco-friendly luminescent textile products.

For the estimation of a complete eco-friendly performance of a luminescent textile product, a holistic approach considering different aspects of materials and production has to be used. Further, it has to be excepted that there is no fundamental eco-friendly product.

REFERENCES

[1] Youn, C., & Jung, H. J. (2021). Semantic network analysis to explore the concept of sustainability in the apparel and textile industry. *Sustainability*, 13(7), 3813. DOI: 10.3390/su13073813

[2] Zukunftsinstitut GmbH (Frankfurt am Main, Germany), web page on megatrends for safety, https://www.zukunftsinstitut.de/blog-megatrend-sicherheit (accessed 2024-20-08).

[3] Luján-Ornelas, C., Güereca, L. P., Franco-García, M. L., & Heldeweg, M. (2020). A life cycle thinking approach to analyse sustainability in the textile industry: A literature review. *Sustainability*, 12(23), 10193. DOI: 10.3390/su122310193

[4] Kumar, P. S., & Joshiba, G. J. (2020). Sustainability in dyeing and finishing. *Sustainability in the Textile and Apparel Industries: Production Process Sustainability*, 165–177. DOI: 10.1007/978-3-030-38545-3_7

[5] Plakantonaki, S., Kiskira, K., Zacharopoulos, N., Chronis, I., Coelho, F., Togiani, A., Kalkanis, K., & Priniotakis, G. (2023). A review of sustainability standards and ecolabeling in the textile industry. *Sustainability*, 15(15), 11589. DOI: 10.3390/su151511589

[6] Santos, G., Marques, R., Silva, S., Oliveira, J., Castro, P., Pereira, C., & Pinheiro, M. (2021). Innovative high-visibility protective clothing development. *Textiles*, 1(3), 405–418. DOI: 10.3390/textiles1030021

[7] Mahltig, B., Rabe, M., & Muth, M. (2019). Textiles, dyeing, and finishing. *Kirk-Othmer Encyclopedia of Chemical Technology*, 1–35. DOI: 10.1002/0471238961.0609140903011201.a01.pub2

[8] Verbič, A., Gorjanc, M., & Simončič, B. (2019). Zinc oxide for functional textile coatings: Recent advances. *Coatings*, 9(9), 550. DOI: 10.3390/coatings9090550

[9] Sbai, S. J., Boukhriss, A., Majid, S., & Gmouh, S. (2020). The recent advances in nanotechnologies for textile functionalization. *Advances in Functional and Protective Textiles*, 531–568. DOI: 10.1016/B978-0-12-820257-9.00020-5

[10] Radetić, M. (2013). Functionalization of textile materials with silver nanoparticles. *Journal of Materials Science*, 48, 95–107. DOI: 10.1007/s10853-012-6677-7

[11] Mahltig, B., Günther, K., Askani, A., Bohnet, F., Brinkert, N., Kyosev, Y., Weide, T., Krieg, M., & Leisegang, T. (2017). X-ray-protective organic/inorganic fiber–along the textile chain from fiber production to clothing application. *The Journal of the Textile Institute*, 108(11), 1975–1984. DOI: 10.1080/00405000.2017.1303122

[12] Youm, K., Kumar, S., & Koh, J. (2022). Synthesis and spectral Properties of fluorescent Phthalimidylhydrazone disperse dyes and their application to poly (ethylene terephthalate) dyeing. *Fibers and Polymers*, 23(9), 2667–2678. DOI: 10.1007/s12221-022-0150-2

[13] Li, M., Zhang, L., Peng, H., & Fu, S. (2018). Preparation of fluorescent pigment latex and its application on binder-free printing of cotton fabrics. *Journal of Applied Polymer Science*, 135(6), 45826. DOI: 10.1002/app.45826

[14] Baatout, K., Saad, F., Baffoun, A., Mahltig, B., Kreher, D., Jaballah, N., & Majdoub, M. (2019). Luminescent cotton fibers coated with fluorescein dye for anti-counterfeiting applications. *Materials Chemistry and Physics*, 234, 304–310. DOI: 10.1016/j.matchemphys.2019.06.007

[15] Park, S. (2019). An evaluation of the suitability of fluorescent fabrics and retroreflective materials for road traffic warning clothing in compliance with international standards. *Fashion and Textiles*, 6(1), 34. DOI: 10.1186/s40691-019-0190-4

[16] Park, S. J., & Kim, S. W. (2020). Designing and fabricating of the high-visibility smart safety clothing. *Science of Emotion and Sensibility*, 23(4), 105–116. DOI: 10.14695/KJSOS.2020.23.4.105

[17] Mahltig, B., Ernst, V., & Schröder, L. (2023). Exemplarily view on selected fluorescence textile products. *Communications in Development and Assembling of Textile Products*, 4 (1), 61–69. DOI: 10.25367/cdatp.2023.4.p61-69

[18] Mahltig, B., Heil, C., Kaub, S., & Kapadiya, J. N. (2024). The use of phosphorescence micromaterials for commercial textile products. *Communications in Development and Assembling of Textile Products*, 5(1), 1–10. DOI: 10.25367/cdatp.2024.5.p1-10

[19] Periyasamy, A. P., & Militky, J. (2020). Sustainability in regenerated textile fibers. *Sustainability in the Textile and Apparel Industries: Sourcing Synthetic and Novel Alternative Raw Materials*, 63–95. DOI: 10.1007/978-3-030-38013-7_4

[20] Patti, A., Cicala, G., & Acierno, D. (2020). Eco-sustainability of the textile production: Waste recovery and current recycling in the composites world. *Polymers*, 13(1), 134. DOI: 10.3390/polym13010134

[21] Raja, A. S. M., Arputharaj, A., Saxena, S., & Patil, P. G. (2019). Water requirement and sustainability of textile processing industries. *Water in textiles and fashion*, 155–173. DOI: 10.1016/B978-0-08-102633-5.00009-9

[22] Raian, S., Ali, S. M., Sarker, M. R., Sankaranarayanan, B., Kabir, G., Paul, S. K., & Chakrabortty, R. K. (2022). Assessing sustainability risks in the supply chain of the textile industry under uncertainty. *Resources, Conservation and Recycling*, 177, 105975. DOI: 10.1016/j.resconrec.2021.105975

[23] Luo, Y., Song, K., Ding, X., & Wu, X. (2021). Environmental sustainability of textiles and apparel: A review of evaluation methods. *Environmental Impact Assessment Review*, 86, 106497. DOI: 10.1016/j.eiar.2020.106497

[24] Hesse, M., Meier, H., & Zeeh, B. (2008). *Spectroscopic Methods in Organic Chemistry*. Georg Thieme Verlag, Stuttgart.

[25] Vik, M., Vikova, M., & Kasparova, M. (2014). Decay of phosphorescent warning design on textile substrates. *Applied Mechanics and Materials*, 440, 112–117. DOI: 10.4028/www.scientific.net/AMM.440.112

[26] Dekanić, T., Čorak, I., Tarbuk, A., Hofmann, J., & Marković, N. (2020). The influence of the optical brighteners concentration on the degree of whiteness and UV protection of cotton fabric. *Tekstil*, 69(4-6), 80–91.

[27] Wang, H., & Sun, J. (2022). Wool fabrics colored with fluorescent pigment emulsion: Their color performance and fluorescent properties. *Journal of Natural Fibers*, 19(3), 1069–1083. DOI: 10.1080/15440478.2020.1788685
[28] Egambaram, O. P., Kesavan Pillai, S., & Ray, S. S. (2020). Materials science challenges in skin UV protection: A review. *Photochemistry and Photobiology*, 96(4), 779–797. DOI: 10.1111/php.13208
[29] Chathoth, A. M., Subba Rao, A. N., Nair, S., Nagarajappa, G. B., & Pandey, K. K. (2023). Luminescent transparent wood from a woody cellulosic template treated with an optical brightener. *Journal of Applied Polymer Science*, 140(28), e54028. DOI: 10.1002/app.54028
[30] Hu, B., Li, D., Ala, O., Manandhar, P., Fan, Q., Kasilingam, D., & Calvert, P. D. (2011). Textile-based flexible electroluminescent devices. *Advanced Functional Materials*, 21 (2), 305–311. DOI: 10.1002/adfm.201001110
[31] Graßmann, C., Lempa, E., Rabe, M., Kitzig, A., Naroska, E., & Neukirch, B. (2017). Electroluminescent textile for therapeutic applications. *Advances in Science and Technology*, 100, 73–78. DOI: 10.4028/www.scientific.net/AST.100.73
[32] Schmidt P., Lead innovation (Vienna, Austria), webpage on circular economy in the textile industry, https://www.lead-innovation.com/insights/blog/kreislaufwirtschaft-in-der-textilindustrie#:~:text=Nachhaltigkeit%20wird%20auch%20in%20der (accessed 2024-20-08).
[33] Moazzem, S., Crossin, E., Daver, F., & Wang, L. (2021). Environmental impact of apparel supply chain and textile products. *Environment, Development and Sustainability*, 1–19. DOI: 10.1007/s10668-021-01873-4
[34] Toprak, T., & Anis, P. (2017). Textile industry's environmental effects and approaching cleaner production and sustainability, an overview. *Journal of Textile Engineering & Fashion Technology*, 2(4), 429–442. DOI: 10.15406/jteft.2017.02.00066
[35] Umweltbundesamt (Dessau, Germany), webpage on sustainability and clothing, https://www.umweltbundesamt.de/umwelttipps-fuer-den-alltag/haushalt-wohnen/bekleidung#so-bringen-sie-mehr-nachhaltigkeit-in-ihren-kleiderschrank (accessed 2024-20-08).
[36] Chen, H. L., & Burns, L. D. (2006). Environmental analysis of textile products. *Clothing and Textiles Research Journal*, 24(3), 248–261. DOI: 10.1177/0887302X06293065
[37] Sandin, G., & Peters, G. M. (2018). Environmental impact of textile reuse and recycling: A review. *Journal of Cleaner Production*, 184, 353–365. DOI: 10.1016/j.jclepro.2018.02.266
[38] Grundmeier, A.-M. (2019). *Bekleidung und Gesundheit*, Schneider Verlag Hohengehren GmbH, Baltmannsweiler, Germany.
[39] Grimm, H.-U., & Ubbenhorst, B. (2023). *Chemie im Essen*, Droemer Knaur GmbH, Munich, Germany.
[40] Hahn, C. (2020). Textilkennzeichnung. Moderecht: Rechtsgrundlagen für Designer, Hersteller und Händler. 47–48. DOI: 10.1007/978-3-658-29514-1_7
[41] IHK München und Oberbayern (Munich, Germany), online brochure for textile labeling, https://www.bayern-innovativ.de/de/unternehmen/ihk-muenchen-und-oberbayern (accessed 2024-26-08).
[42] Christie, R. M. (2011). Fluorescent dyes. *Handbook of Textile and Industrial Dyeing*, 562–587. DOI: 10.1533/9780857093974.2.562
[43] Rather, L. J., Azam, M., Shabbir, M., Bukhari, M. N., Shahid, M., Khan, M. A., Haque, Q.M.R. & Mohammad, F. (2016). Antimicrobial and fluorescence finishing of woollen yarn with Terminalia arjuna natural dye as an ecofriendly substitute to synthetic antibacterial agents. *RSC Advances*, 6(45), 39080–39094. DOI: 10.1039/C6RA02717B

[44] Lee, H., Mahltig, B. (2025). Dyeing with extracts from Amaranth plant: an investigation in respect to mordanting procedures and fluorescence properties. *Communications in Development and Assembling of Textile Products*, 6(1), 1–12. https://doi.org/10.25367/cdatp.2025.6.p1-12

[45] Popescu, V., Astanei, D. G., Burlica, R., Popescu, A., Munteanu, C., Ciolacu, F., Ursache, M., Ciobanu, L. & Cocean, A. (2019). Sustainable and cleaner microwave-assisted dyeing process for obtaining eco-friendly and fluorescent acrylic knitted fabrics. *Journal of Cleaner Production*, 232, 451–461. DOI: 10.1016/j.jclepro.2019.05.281

[46] Mahltig, B., Greiler, L. C., & Haase, H. (2018). Microwave assisted conversion of an amino acid into a fluorescent solution. *Acta Chimica Slovenica*, 65(4), 865–874. DOI: 10.17344/acsi.2018.4513

[47] Hou, J., Xiong, X., Jiao, C., Huang, X., Fu, D., Zhao, H., & Li, Y. (2022). Cleaner production of disperse florescent dyes in supercritical CO_2 and their applications in dyeing polyester fabric. *Dyes and Pigments*, 202, 110250. DOI: 10.1016/j.dyepig.2022.110250

[48] Mahltig, B., & Textor, T. (2008). *Nanosols and Textiles*. World Scientific, Singapore.

[49] Avnir, D., Kaufman, V. R., & Reisfeld, R. (1985). Organic fluorescent dyes trapped in silica and silica-titania thin films by the sol-gel method: Photophysical, film and cage properties. *Journal of Non-Crystalline Solids*, 74(2–3), 395–406. DOI: 10.1016/0022-3093(85)90081-X

[50] Takahashi, Y., Maeda, A., Kojima, K., & Uchida, K. (2000). Luminescence of dyes doped in a sol–gel coating film. *Journal of Luminescence*, 87, 767–769. DOI: 10.1016/S0022-2313(99)00393-2

[51] Yuan, W., Grethe, T., & Mahltig, B. (2023). Sol-gel coatings with the fluorescence dye Rhodamine B for optical modification of cotton. *Communications in Development and Assembling of Textile Products*, 4(1), 1–17. DOI: 10.25367/cdatp.2023.4.p1-17

[52] Boukhriss, A., Roblin, J. P., Aaboub, T., Boyer, D., & Gmouh, S. (2020). Luminescent hybrid coatings prepared by a sol–gel process for a textile-based pH sensor. *Materials Advances*, 1(4), 918–925. DOI: 10.1039/D0MA00211A

[53] Raduly, F. M., Rădițoiu, V., Rădițoiu, A., Frone, A. N., Nicolae, C. A., Răut, I., Constantin, M. & Grapin, M. (2023). Multifunctional finishing of cotton fabric with curcumin derivatives coatings obtained by Sol–Gel method. *Gels*, 9(5), 369. DOI: 10.3390/gels9050369

[54] Xing, Y., Yang, X., & Dai, J. (2007). Antimicrobial finishing of cotton textile based on water glass by sol–gel method. *The Journal of Sol-Gel Science and Technology*, 43, 187–192. DOI: 10.1007/s10971-007-1575-1

[55] Bakar, N. H. A., Yusop, H. M., Ismail, W. N. W., & Zulkifli, N. F. (2023). Sol-gel finishing for protective fabrics. *Biointerface Research in Applied Chemistry*, 13, 283. DOI: 10.33263/BRIAC133.283

[56] Shih, J. Y., Lai, S. L., & Cheng, H. T. (2013). Characteristics and applications of luminous fibers. *Applied Mechanics and Materials*, 368, 702–707. DOI: 10.4028/www.scientific.net/AMM.368-370.702

[57] Sharma, R., & Bairagi, N. (2018). The role of photoluminescent pigments in textiles. *Trends in Textile Engineering & Fashion Technology*, 2(2), 1–7. DOI: 10.31031/TTEFT.2018.02.000533

[58] Trigema W. Grupp KG (Burladingen, Germany), company webpage on sustainability, https://www.trigema.de/unternehmen/nachhaltigkeit/ (accessed 2024-20-08).

[59] Trigema W. Grupp KG (Burladingen, Germany), company webpage on high visibility hat with orange coloration, https://www.trigema.de/herren/fitness-und-laufbekleidung/Muetze-mit-reflektierendem-Logo-leuchtorange-0.html (accessed 2024-20-08).

[60] Grancaric, A.M., Tarbuk, A., & Botteri, L. (2014). Light conversion and scattering in UV protective textiles. *AUTEX Research Journal*, 14(4), 247–258. DOI: 10.2478/aut-2014-0025
[61] Mahltig, B., Leuchtges, G., & Holstein, P. (2022). T-shirts: An overview and comments on price range, functional materials and European production. *Tekstilna industrija*, 70(4), 4–13. DOI: 10.5937/tekstind2204004M
[62] IQ-Company AG, iQ-UV (Walluf, Germany), product webpage on orange hat for UV protection, https://iq-arbeitsschutz.de/uv-beanie-skull-mit-reflektierendem-logo/2115007260-3-S-M (accessed 2024-21-08).
[63] VAUDE Sport GmbH & Co.KG (Tettnang, Germany), company webpage with sustainability report, https://nachhaltigkeitsbericht.vaude.com/ (accessed 2024-20-08).
[64] Utopia GmbH (München, Germany), webpage on climate neutrality and compensation, https://utopia.de/sponsored-content/das-steckt-hinter-den-kompensationsarten_222604/ (accessed 2024-20-08).
[65] Hübner, K. (2023). Dem Regen trotzen: 200 Jahre wetterfeste Chemie. *Chemie in unserer Zeit*, 57(6), 360–367. DOI: 10.1002/ciuz.202300013
[66] Klaus Herding GmbH (Rhede, Germany), company webpage on phosphorescent textile product, https://herding-shop.de/paw-patrol-bettwaesche-glow-in-the-dark/https://www.fristads.com/de-de/produkt-info/sortimente/fristads-green
[67] JP Boden B.V. (Rijsenhout, Netherlands), company webpage on phosphorescent product, https://www.bodendirect.de/de-de/leuchtender-skelett-hoodie-fuer-halloween-rauch grau-leuchtendes-skelett/sty-b1991-gry (accessed 2024-20-08).
[68] JP Boden B.V. (Rijsenhout, Netherlands), company webpage on sustainability aspects, https://www.bodendirect.de/de-de/nachhaltigkeit?intid=D09_About_Us_CB5_CTA (accessed 2024-20-08).
[69] Better Cotton Initiative (London, UK), organization webpage, https://bettercotton.org/who-we-are/ (accessed 2024-21-08).
[70] Mahltig, B., & Grethe, T. (2022). High-performance and functional fiber materials: A review of properties, scanning electron microscopy SEM and electron dispersive spectroscopy EDS. *Textiles*, 2(2), 209–251. DOI: 10.3390/textiles2020012
[71] Seidu, R. K., Sun, L., & Jiang, S. (2024). A systematic review on retro-reflective clothing for night-time visibility and safety. *The Journal of the Textile Institute*, 115(7), 1138–1150. DOI: 10.1080/00405000.2023.2212194
[72] Mascot International GmbH (Flensburg, Germany), company webpage on fluorescent textile product with high visibility, https://www.mascot.de/de/mascot-accelerate-safe-19078-511/#14010 (accessed 2024-20-08).
[73] Workwear Group ApS (Spjald, Denmark), company webpage on workwear with high visibility, https://guenstige-arbeitskleidung.de/Helly-Hansen-UC-ME-Winteroverall%2C-Hi-Vis-Rot%2FEbony/138723.html (accessed 2024-20-08).
[74] Fristads GmbH (Norderstedt, Germany), company webpage on workwear and sustainability aspects, https://www.fristads.com/de-de/produkt-info/sortimente/fristads-green (accessed 2024-20-08).
[75] Textilchemie Dr. Petry GmbH (Reutlingen, Germany), company web page on printing techniques and chemical products, https://www.drpetry.de/produkte/produktuebersicht/druckerei/ (accessed 2024-20-08).
[76] Textilchemie Dr. Petry GmbH (Reutlingen, Germany), product web page for Perilen pigments, https://www.drpetry.de/en/textile-news/new-fluorescent-perilen-fl-pigments/ (accessed 2024-20-08).
[77] CHT Germany GmbH (Tübingen, Germany), product webpage for Bezafluor Rosa FF https://solutions.cht.com/cht/web.nsf/id/pa_de_productdetail.html?Open&pID=PDE-TUEB-BC59PC-DE (accessed 2024-20-08).

[78] CHT Germany GmbH (Tübingen, Germany), product webpage for Bezafluor Gelb FF https://solutions.cht.com/cht/web.nsf/id/pa_de_productdetail.html?Open&pID=PDE-TUEB-BC59P9-DE&shkey=01040811&prange=WW (accessed 2024-20-08).

[79] Kremer Pigmente GmbH (Aichstetten, Germany), product web page for Phosphorescent Pigment Green https://www.kremer-pigmente.com/en/shop/pigments/56500-phosphorescent-pigment-green.html (accessed 2024-20-08).

[80] Kremer Pigmente GmbH (Aichstetten, Germany), product web page for Phosphorescent Pigment Red https://www.kremer-pigmente.com/en/shop/pigments/pearlescent-effect-pigments/phosphorescent-pigments/56650-phosphorescent-pigment-red.html (accessed 2024-20-08).

[81] CHT Germany GmbH (Tübingen, Germany), product webpage for printing paset Printperfekt 226FF, https://solutions.cht.com/cht/web.nsf/id/pa_de_productdetail.html?Open&pID=PDE-TUEB-AHXEMS-DE&prange=WW (accessed 2024-20-08).

[82] Barzan, M., & Hajiesmaeilbaigi, F. (2018). Investigation the concentration effect on the absorption and fluorescence properties of Rhodamine 6G dye. *Optik*, 159, 157–161. DOI: 10.1016/j.ijleo.2018.01.075

[83] Fonin, A. V., Sulatskaya, A. I., Kuznetsova, I. M., & Turoverov, K. K. (2014). Fluorescence of dyes in solutions with high absorbance. Inner filter effect correction. *PloS One*, 9(7), e103878. DOI: 10.1007/s10895-020-02622-y

7 Eco-Friendly Photonic Wearables

Md. Nazif Hasan Chowdhury and
Md. Reazuddin Repon

INTRODUCTION

Electronic textiles are gradually becoming part of our everyday lives. The three most common functions of electronic textiles are the generation, storage and utilization of electricity. Different methods can be adapted to produce smart textiles; photoelectric cells can be used to incorporate solar energy to produce electricity for certain functions [1]. This conversion of electricity from light can be made possible using the photovoltaic effect, where the valence band of a photoelectric component absorbs light and excitedly escapes after jumping to the conduction band. Then, the free excited electrons reach a junction from where they are transported to built-in textile components to produce an electric stimulus, to be precise, an electric current [2]. This is the very common mechanism of photovoltaic textiles, which can be regarded as photonic wearables too. The necessity of these photonic wearables is increasing drastically in all sectors, especially in healthcare and biomedical applications where light energy is incorporated with sensors to observe patient vitals remotely, without even bothering them. This has been termed as electronic point of care system [3]. Except medical sector, these wearable electronics are used in protective apparels, sports, military gears, safety equipment, energy storage and harvesting and a lot more [4]. These applications will be discussed vividly in the later part of this chapter.

Although electronic textiles are gaining popularity heavily, there are some drawbacks in the manufacturing processes and final products. In most of the cases, synthetic fibres and other non-biodegradable plastic materials are used for making electronic textiles, especially for photoelectric cells. These components have a deleterious effect on the earth when dumped after use. They produce e-waste which is mostly non-perishable and enhances the carbon footprint. Moreover, they are mostly non-recyclable, have a short lifespan and the microplastic pollution from these products not only impacts the earth's surface but also the marine ecosystem. Here comes the significance of eco-friendly photonic wearables. Usually, conventional materials which are used for photonic wearables are plastics, heavy metals and synthetic polymers which are mostly non-biodegradable and cause a threat to nature. For this reason, biodegradable biomaterials are now being used for wearable electronics as they are reusable and after being worn out, they can be decomposed by nature without harming the natural ecosystem. These materials can be derived from

DOI: 10.1201/9781003507208-7

multifarious sources such as plants, proteins and several biocomposites [5]. In this chapter, the materials for green textiles, factors affecting the efficacy, applications, challenges and prospects of green wearables will be discussed in detail.

MATERIALS FOR GREEN WEARABLES

As discussed earlier, the most commonly used conventional materials for photoelectronics are harmful to nature. So, they cannot be endorsed for commercial applications any further, and they need to be replaced with sustainable green alternatives eventually. Using biodegradable materials in photonic electronics introduces several advances. First, they do not cause harm to nature as they do not leave carbon footprints, and they decompose with the earth smoothly after being used. Second, they are biocompatible and cause minimal harm to the human body after being attached to it for a prolonged period. Biodegradable materials such as PLA, PCL and PHA are being used for manufacturing photonic wearables recently, due to the above-mentioned benefits [6]. Besides, natural fibres are supplanting synthetic products in making of green wearables. Although green wearables can endorse a plethora of materials in manufacturing, some of them are shortly highlighted here (Table 7.1).

ELECTROLUMINESCENT FIBRES AND FABRICS

Electroluminescent fibres are flexible light-emitting fibres that produce light in the presence of an alternating electrical field. They have a phosphor layer which is usually made from metal and coated around a thin conductive core. When AC voltage is applied, the electrons in the phosphor get excited and emit visible light, the textile component thus glows in darkness. These fabrics not only demonstrate aesthetic appeal, but also they can be used for functional purposes such as medical applications, traffic light signal and many more. Ultra-stretchable woven fibres can be used in an electroluminescent product to show pixel-based arbitrary patterns controlled by a mobile application [18]. These electroluminescent materials are replacing traditional rigid light probes due to their flexibility and convenience. They are being used in the military and police as they are portable adaptive camouflage [19]. These materials can be used actively or passively, which means by photoluminescence or triboluminescence [20]. There have been very few but insightful successful initiatives to produce electroluminescent materials. Hong et al. used a solid-state electrolyte ion gel derived from a polymer and ionic liquid and then dyed the composite with a luminophore dye to produce a light-emitting fabric [21]. Song et al. worked on developing a soft textile-embedded PDMS+ZnS composite to impart uniform lucidity compared to typical coplanar displays [22]. Hu et al. worked on developing a textile display by PAN fibre at the glass transition temperature, interweaving with a transparent fibre electrode which demonstrated a quite organized light emission at a remarkably low cost [23]. Xie et al. developed a silk-fibroin-based electroluminescent fibre with triboelectric sensing to be used for smart wearables

TABLE 7.1
Materials Used for Green Wearables

Material	Application Area	Advantages	References
Polylactic acid (PLA)	Biodegradable sensors, wearable photonics	Recyclable, non-toxic, corn-based material, so decomposable in nature	[7]
Polyhydroxyalkanoates (PHB/PHA)	Bioplastic as a replacement for conventional plastic in electronic textiles, medical corsetry	Intercellular polyester, biocompatible, cost-efficient	[8]
Silk and paper	Energy harvesting, light emission	Biobased and biodegradable materials, satisfy SDG-12 and 13	[9]
Peptides and DNA	Biomedical applications such as tissue engineering, biosensors for photonic wearables	Abundant in nature, biocompatible, excellent electrochemical properties	[10]
Poly(3,4-ethylenedioxythiophene): *p*-toluene sulfonic acid (PEDOT: PTS) and polyurethane (PU) composite	Electronic skin (E-skin) for human	Superior mechanical flexibility, soft material with biocompatibility	[11]
Carbon nanotube	Energy storage for supercapacitors and batteries, bio signal monitoring system	Excellent electrical conductivity, mechanical strength and flexibility	[12, 13]
Wood-based cellulose nanofibrils (CNF)	Radio frequency (RF) circuits, digital logic circuits	Bio-resorbable, biocompatible	[14, 15]
Natural fibres (cotton, jute, flax etc.)	Wearable electronics, optical textiles, photonic materials for aesthetic and functional applications	Availability, excellent moisture properties, biodegradability, sustainability	[16, 17]

such as sports gear [24]. There are more textile materials which are being integrated and experimented everyday to provide affordable but best-performing electroluminescent textiles. Among them, different synthetic polymers, fibre mesh, an wide array of textile-based biocomposites are notable, and these photonic wearables are being used within a huge bracket including commercial sports goods, biomedical corsetry, military gears and many more and new application scopes are being introduced with a skyrocketing pace [25, 26].

FACTORS AFFECTING PERFORMANCE OF ECO-FRIENDLY PHOTONIC WEARABLES

From the discussion till now, it is quite obvious that eco-friendly photonic wearables have equipped a permanent space in the global technical textile market. Although, some factors are to be considered as they can affect the performance of these photonic smart wearables. Some of the most important factors are being delineated here.

Energy Efficiency

Energy efficiency is a crucial determinant in the performance of eco-friendly photonic wearables. If the energy efficiency is in upper spectrum, then it provides longer battery life as well as longer operation time. Also, it generates less heat if the energy efficiency is good. Overall, it ensures the stability of the product and opens room for a sleek and convenient design. If efficient energy is ensured, it guarantees the implementation of eco-friendly power sources in different arenas such as triboelectric nanogenerators and solar cells. These efficient energy sources are creating a boom in the field of Internet of Things (IoT). Personal wearable items such as smartwatches, smart jewelleries and electronic clothing has launched a new field of research, called Internet of Wearable Things (IoWT), where these wearables are connected to mobile applications in personal devices and gaining high popularity gradually [27].

Lighting Technology

Lighting technology also plays a vital role in the performance of photonic wearables. Lighting technology refers to the method and material for producing light sources in photonic wearables to control and operate the lighting for aesthetic and functional applications. Lighting technology in photonic wearables has been used for human clothing and interior decoration such as interior decorative lighting on wall panel [28, 29], as well as functional applications such as monitoring patients in healthcare [30]. Robust lighting technology ensures better visibility of light functions in garments, enhances safety for specialized products, improves optical properties and enhances fashion.

TYPE OF MATERIALS

The performance of photonic wearables largely depends on the types of materials used here. Different light-emitting diodes (LED), optical fibres, cellulose nanocrystals, eco-friendly materials such as natural fibres, biopolymers etc. are used in photonic wearables. If they are integrated perfectly based on the required functionality goal, then they can ensure excellent performance properties. There are some light-responsive materials which absorb light and can show optical properties such as fluorescence, phosphorescence and plasmonic and photothermal effects. These materials are quantum dots, different nanoparticles and semiconductors [31]. Moreover, there are some light-delivering materials which focus on light transmission and emissions such as LED. Also, sometimes stretchable electronic materials are used. Different types of inorganic and natural fibres as well as biopolymers are also used in photonic wearables for dedicated functionality [32–34]. These materials serve as photodetectors [35], wearable health solutions [36], as well as other flexible photonic materials [37]. These particular types of materials can directly impact the distinguishing factors of photonic wearable such as optical performance, flexibility, durability, comfort and safety. Newer materials are being endorsed with this list, and the performance of photonic electronics is ameliorating the standard day by day.

SENSORS AND DATA ACCURACY

The types of sensor those are used in a photonic textile can impact the safety and functionality for consumers. Also, the data from these inputs need to be meticulous, or chances of discrepancies arise. Lack of accuracy and unauthorized or inappropriate sensors might cause risk to the wearer. Also, these data are used for analysing possible flaws and developing next models. So, this is a highly important feature to assess the performance of photonic textiles.

CONNECTIVITY AND COMMUNICATION

A common characteristic of electronic textile is that it must have the ability to connect to other devices. Wearable wireless networks are used to create a connection between human body and any device using a wireless network such as Bluetooth or Wi-Fi [38]. However, there might be some challenges which are possible to overcome. Smart connectivity makes photonic wearables safe and smart for consumers, giving them the ability to remotely control and customize devices as per their convenience. However, increased connectivity might also spike the demand for energy, green photonic wearables mostly use reusable and natural sources to produce electricity, such as solar energy or mechanical energy from body, therefore, this is not a big deal.

USER INTERFACE AND INTERACTION

User interface and interaction depend on the bond of connectivity as discussed previously. If the user interface is up to date, it becomes cozy for users. Especially, the young generation looks for a smart interface and added features. So, the major wearable electronics manufacturers are investing in this feature obviously. Mostly, the recent textronics are becoming popular due to their compact size, flexibility, lightweight but robust structure, immunity to electromagnetic interference and parallel sensing through optical multiplexing [39].

ENVIRONMENTAL CONDITIONS

Environmental conditions such as temperature, humidity and moisture, ultraviolet (UV) and light exposure, mechanical and chemical stress etc. can have an impact on the performance of photonic wearables. For instance, overheating can degrade and distort the material. Humidity can make these things prone to short circuits.

Mechanical stress and chemical exposure beyond the normal limit can distort the morphology of these photonic materials, hence denigrating the performance simultaneously. Hence, it is obvious that electronic wearable or photonic wearable performance solely depends on the factors associated with their manufacturing environment; so it is mandatory to maintain the parameters and conditions correctly [40].

SIZE AND COMFORT

Size and comfort properties are associated with the features of photonic wearables. Usually, size determines the energy consumption level. So, if the size of the electronic textile is smaller, it will be cooler. Photonic thermal wears are now used in cold regions. As it has the remote control of heat management, it ensures comfort for the wearer [41]. Stretchable electronic materials are being used in manufacturing wearable electronics which are regarded as next-generation comfortable textronics, suitable for monitoring in medical applications. It is regarded as 'comfort to wear' if it does not disturb the daily life of the wearer. This level of comfort is being engineered in the modern electronic textiles [42]. Comfort will impact the wearable electronics market largely as the market is expected to expand 15.09% between 2022 and 2027 [43]. Size and comfort affect the user acceptance, ease of mobility, thermal comfort and aesthetic appeal.

APPLICATIONS OF ECO-FRIENDLY PHOTONIC WEARABLES

Photonic wearables are gaining popularity day by day in almost every sphere. Major application areas cover safety and aesthetics of wearer, healthcare sector, entertainment and environmental monitoring. Major application areas are being delineated here.

Safety and Visibility

Eco-friendly photonic wearables are used for safety and comfort applications, also for monitoring people in need of extra attention. Photonic wearable textiles are used to monitor kids and juveniles as they demonstrate illuminated visibility. It can also be used for monitoring critical patients admitted into hospitals. Sensors can be programmed, thus transforming colour according to the situation, which enhances the safety and monitoring of the wearer more practical and convenient. Not only that, already in this chapter, it has been discussed how this technology helps in interior décor apart from cloth tech; the implementation is now not only confined to aesthetic interior for visual satisfaction but also been used for aerospace and submarine applications [44]. Optical fibres and laser technology are being integrated with these photonic sensors for clear visibility and safety [45]. However, there are a huge number of applications of photonic textiles or photonic wearables in healthcare which will be discussed later, and some successful installations in building interiors, transportation textiles and other fields; it is still very little explored in the domain of safety apparel; although there are huge scopes to focus on this field. Photonic wearables can be used for uniforms of industry workers, firefighters, rescue workers, military, traffic police etc., for better visibility in dark, hazy conditions as well as remote monitoring. It will definitely add an extra layer of safety for them. After the emergence of the fourth industrial revolution, industrial workers' safety has become a prime concern for industrialists and policymakers and photonic wearables can be a robust option to make it happen [46]. However, there is a risk that in contact with water or any liquid, these photonic sensors can cause short circuit and any unexpected occurrence, also the lights can be damaged due to pressure, contact or thrust in hazardous condition. So, a more robust and sustainable design of photonic wearables is required to use them in appropriate conditions for safety and visibility.

Health and Wellness

Most of the photonic wearable applications are available in healthcare, and this particular area is developing rapidly. From routine health checkups to dealing with critical conditions, these photonic wearables are used everywhere [30]. Photonic nanomaterials are being integrated with wearable electronics due to their excellent physical and mechanical features [47]. Nowadays, modern hospitals and healthcare organizations are introducing photonic wearables for patient uniforms, so that they do not have to go and manually check them frequently, but they can remotely observe the patients without even disturbing them. Photonic sensors can track the movement of inner organs such as heartbeat, respiration and oxygen level in blood. Light-based patches can monitor skin conditions, these sensors can be used in bandages so that the condition of ulcer can be assessed without opening the bandage. Skin-like flexible electronics, which are termed as electronic skin or e-skin, can be used to measure the real-time vitals of the human body [48, 49]. These photonic skins can track breathing, cardiac condition, mental state, blood flow, etc., which makes it easy for health professionals to monitor patients [50–54]. Photonic sensors are used not only for monitoring by visualization, but they also have applications in generating new medicine, drug delivery and medical diagnosis methods [55–59]. Apart from this, being well is a prime concern for everyone today. People try to maintain fitness so they do not have to visit hospitals. People work out, perform indoor and outdoor training, sports for their wellbeing. These photonic sensor-based electronic textiles help them with monitoring body conditions. Wearable electronics are fitness trackers and aesthetic textiles at the same time, so they are gaining popularity among athletes and the mass community, as sportswear and gym outfits [60, 61].

Entertainment and Events

Photonic electronics are popular in the fashion industry, stage decoration, home furnishing and different events. Wearable technology has drastically blended fashion, technology, art and science in the 21st century, but the ultimate goal of wearable technology is to transform fashion from ancient to modern and make it more diversified [62]. And this has happened by two core functions, sensor data receiving and exchanging with other devices, wearable technology focuses on 'design in

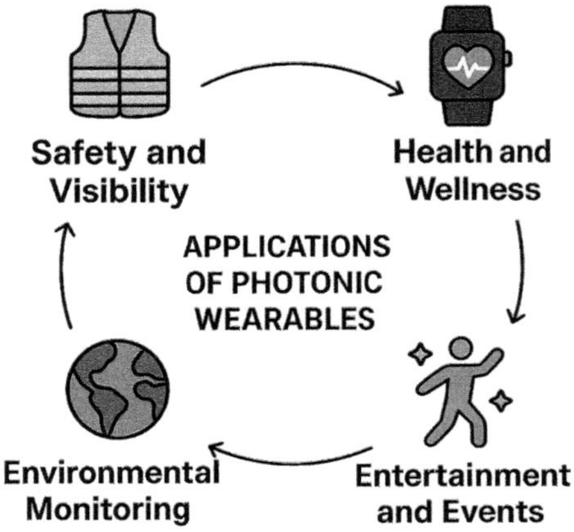

FIGURE 7.1 Applications of eco-friendly photonic wearables.

motion' to make the fashion industry more vibrant and digital [63]. Polymeric optical fibre integrated textiles or photonic textiles are becoming popular in the fashion industry due to its ambience and lucrative attire, beside other technological advantages [64, 65]. Besides fashion industry, photonic wearable technology is being endorsed with entertainment industry such as concerts, movie studio etc. for light effects and interior decoration.

Environmental Monitoring

Photonic wearables consist of light-responsive materials which can detect UV radiations, pollution in nature, factors like temperature and humidity. Meteorologists, ecologists and environmental scientists can use photonic wearable technology to get an update on the environment. Textiles dyed with luminophore dyes can react to external stimuli such as chemical or environment index. These can be smart outdoor clothing for tourists and backpackers, as well as any adventure travellers. Although research in this particular application domain is negligible, there are enough scopes to integrate photonic wearables for environmental monitoring applications (Figure 7.1).

CURRENT CHALLENGES AND FUTURE PROSPECTS

There is a saying that 'nothing is unmixed blessing on earth'. Which means there will be some advantages and disadvantages to everything. Photonic wearables are not exceptional from them; there are some current challenges in the manufacturing of these photonic wearables, but there are some bright future prospects as well. The main challenges included are sustainability of material, energy consumption, manufacturing cost, durability, wearer comfort, etc. In the previous discussion of this chapter, the core advantages of photonic wearables have been delineated. Now, some current challenges and future prospects will be discussed together for a better and clear understanding of readers.

Material Sustainability versus Performance

Different types of materials are used in the manufacturing of eco-friendly photonic wearables, which have already been discussed in this chapter [31–34]. One problem is that truly biodegradable and environmentally friendly materials have weak strength and can be degraded in the presence of heat, moisture and other environmental factors. As a result, there may be an impact on performance such as poor illumination and weak visibility. However, for this, we cannot allow toxic materials to take over again, only because they are robust and cheap. Considering the long-term vision, and for the sake of the environment as well as the ecosystem, it is a must to deal with these challenges and make eco-friendly options more contemporary and advanced. A matter of hope is that researchers are working dedicatedly to substitute deleterious materials with green alternatives and eco-friendly photonic wearables are improving drastically.

Energy Efficiency and Power Sources

Green materials often lack the ability to consume energy as much as compared to other commercial materials due to their weak polymer orientation, resulting in poor lighting and shabby visibility. Also, the natural components cannot provide longer backup compared to metal nanoparticles. To solve this challenge, scientists are now delving into integrated photonics, where several sensor elements can be assembled onto a single microchip, reducing the size and cost simultaneously [66]. Besides, they are trying to explore novel materials to enhance sensor efficacy so that the power consumption can be calibrated and unified. By tackling these issues, eco-friendly wearable photonic or biosensors can bring a revolution in the power industry. With the help of the buzzing trend of nanotechnology and AI, these smart biosensors will be much more accurate to be installed in industrial inspection as they will be able to catch the slightest discrepancy in the formation of the product [67, 68]. Research in developing energy efficiency is going on, and in lieu of metal ion batteries, renewable energy is being introduced everywhere as a power source, which will sort out this challenge smartly.

Durability and Environmental Stability

As discussed earlier, eco-friendly materials usually have a shorter lifespan as they easily get degraded under natural conditions such as heat and moisture. On the contrary, commercial photonic sensors can provide better durability and performance, but they can cause harm to the human body. We have no option to stick with eco-friendly alternatives for the sake of earth and mankind. To protect the eco-friendly components from deterioration, eco-friendly coating and encapsulation can be used which will deter the heat and moisture from interacting with biosensors; as a result, the longevity of these wearables will increase and better performance shall be derived. Though there are not many advanced research in eco-coating and encapsulation and as these are newer fields, new researchers can explore this future prospect of green photonic wearables.

Manufacturing Scalability and Accuracy

Although several studies have been done on developing eco-friendly photonic wearables and biosensors, there is not yet any sustainable plan or layout for large-scale commercial production. Usually, the conventional manufacturing objectives for electronic textiles such as roll-to-roll printing, low-temperature deposition and natural fibre integration can be time-consuming and expensive in the long run. Hence, more innovation in green technology, 3D printing, low-temperature deposition is required to solve this challenge. Another challenge in sensors is extracting accurate data, especially for medical tools and health monitoring. Now, biosensors are used for monitoring vitals of the human body such as heart rate, blood flow, respiration, etc., which have already been discussed before [50–54]. The problem is sometimes these sensors cannot extract data and show accurate results due to some hidden constraints. A study has shown that the epidermis of human skin can be an impediment for accurate information to the sensor and thus it can alter the physical, mechanical, chemical or optical modality of the sensor, for which extracting exact readings gets tougher [69]. Using greener alternatives can solve this issue as they are biocompatible and tend to show better accuracy.

Functional Integration and User Comfort

It can be difficult to embed photonic sensors in a wearable electronic which is soft and breathable and at the same time to keep it lightweight. Yes, it may be challenging but not impossible. Stretchable ultra-thin flexible electronics can contribute to advancements in nanofibre photonics, and seamless integration of fabric can solve the challenges to a great extent. It has to be kept in mind that aesthetics and comfort go hand in hand from a customer's perspective. So, at the same time, it has to be perfectly functional, aesthetic and cozy for the wearer (Figure 7.2).

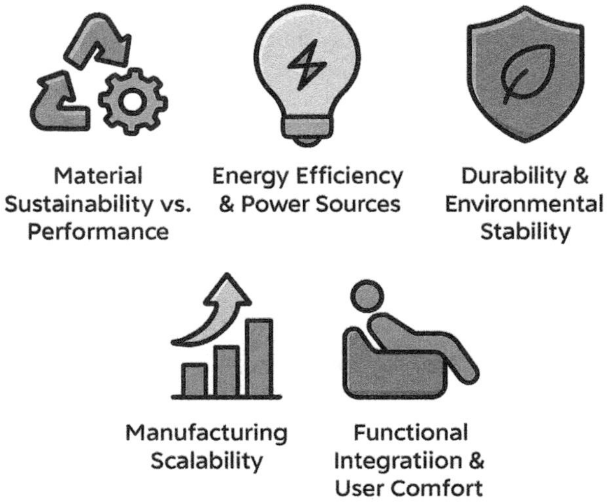

FIGURE 7.2 Current challenges and future prospects of eco-friendly photonic wearables.

CONCLUSION

In this chapter, a brief overview of eco-friendly photonic wearable is provided along with the materials used, factors affecting the performance of photonic wearables, electroluminescent fibres and fabrics. Also, the applications of eco-friendly photonic wearables have been discussed, from where it is quite obvious how these green photonic wearables have become a crucial part for safety, monitoring, medical monitoring, as well as in fashion industry and entertainment. Lastly, some shortcomings have been addressed for eco-friendly photonic wearables along with their possible solution and future prospects. Finally, this chapter delineates the necessity of eco-friendly photonic wearables in our everyday life along with their pros and cons and overcomes the possible challenges with practical insights to utilize the potential of this more efficiently, helping readers of this chapter to have a clean overview of eco-friendly photonic wearables.

REFERENCES

[1] B. Maurizio, and R. Moro. "Low-cost fabrication, eco-friendly materials, and easy integration: the new technological paradigm for the future wireless sensor networks." *2013 European Microwave Conference*. IEEE, 2013. https://ieeexplore.ieee.org/document/6686792

[2] B. A. Gregg, "Excitonic Solar Cells," *J. Phys. Chem. B*, vol. 107, no. 20, pp. 4688–4698, May 2003, doi: 10.1021/jp022507x

[3] G. Chen, X. Xiao, X. Zhao, T. Tat, M. Bick, and J. Chen, "Electronic Textiles for Wearable Point-of-Care Systems," *Chem. Rev.*, vol. 122, no. 3, pp. 3259–3291, Feb. 2022, doi: 10.1021/acs.chemrev.1c00502

[4] G. Ehrmann, and A. Ehrmann, "Electronic Textiles," *Encyclopedia*, vol. 1, no. 1, pp. 115–130, Jan. 2021, doi: 10.3390/encyclopedia1010013

[5] T. Kim, "Research Article ECO - Friendly Biodegradable Materials for Wearable Device Applications," vol. 06, no. 03, pp. 9507–9510, 2025.

[6] L. Wu, X. Shi, and Z. Wu, "Recent Advancements and Perspectives of Biodegradable Polymers for Supercapacitors," *Adv. Funct. Mater.*, vol. 33, no. 16, Apr. 2023, doi: 10.1002/adfm.202211454

[7] S. Jha, B. Akula, H. Enyioma, M. Novak, V. Amin, and H. Liang, "Biodegradable Biobased Polymers: A Review of the State of the Art, Challenges, and Future Directions," *Polymers*, vol. 16, no. 16, p. 2262, Aug. 2024, doi: 10.3390/polym16162262

[8] B. E. DiGregorio, "Biobased Performance Bioplastic: Mirel," *Chem. Biol.*, vol. 16, no. 1, pp. 1–2, Jan. 2009, doi: 10.1016/j.chembiol.2009.01.001

[9] C. Lee, S. Kim, and Y. Cho, "Silk and Paper: Progress and Prospects in Green Photonics and Electronics," *Adv. Sustain. Syst.*, vol. 6, no. 2, Feb. 2022, doi: 10.1002/adsu.202000216

[10] G. T. Tihan, M. Mindroiu, I. Rau, L. M. Nóbrega de Assis, A. Pawlicka, and R. G. Zgarian, "The electrochromic device performance with DNA based electrolyte," *Mater. Chem. Phys.*, vol. 241, p. 122349, Feb. 2020, doi: 10.1016/j.matchemphys.2019.122349

[11] H. Kai, W. Suda, Y. Ogawa, K. Nagamine, and M. Nishizawa, "Intrinsically Stretchable Electrochromic Display by a Composite Film of Poly(3,4-ethylenedioxythiophene) and Polyurethane," *ACS Appl. Mater. Interfaces*, vol. 9, no. 23, pp. 19513–19518, Jun. 2017, doi: 10.1021/acsami.7b03124

[12] K. I. Bolotin, K. J. Sikes, J. Hone, H. L. Stormer, and P. Kim, "Temperature-Dependent Transport in Suspended Graphene," *Phys. Rev. Lett.*, vol. 101, no. 9, p. 096802, Aug. 2008, doi: 10.1103/PhysRevLett.101.096802

[13] E. K. Lee et al., "Fractal Web Design of a Hemispherical Photodetector Array with Organic-Dye-Sensitized Graphene Hybrid Composites," *Adv. Mater.*, vol. 32, no. 46, Nov. 2020, doi: 10.1002/adma.202004456
[14] Y. H. Jung et al., "High-Performance Green Flexible Electronics Based on Biodegradable Cellulose Nanofibril Paper," *Nat. Commun.*, vol. 6, 2015, doi: 10.1038/ncomms8170
[15] H. Rahmani et al., "Next-Generation IoT Devices: Sustainable Eco-Friendly Manufacturing, Energy Harvesting, and Wireless Connectivity," *IEEE J. Microwaves*, vol. 3, no. 1, pp. 237–255, Jan. 2023, doi: 10.1109/JMW.2022.3228683
[16] N. Laras Agustina, "No Tit'ליle," ペインクリ ック学会治療指針2. pp. 1–9, 2019, doi: 10.23919/EuMC.2013.6686792
[17] M. Dulal et al., "Toward Sustainable Wearable Electronic Textiles," *ACS Nano*, vol. 16, no. 12, pp. 19755–19788, Dec. 2022, doi: 10.1021/acsnano.2c07723
[18] H. Mi et al., "Electroluminescent Fabric Woven by Ultrastretchable Fibers for Arbitrarily Controllable Pattern Display," *ACS Appl. Mater. Interfaces*, vol. 13, no. 9, pp. 11260–11267, Mar. 2021, doi: 10.1021/acsami.0c19743
[19] S. Xu, X. Fan, S. Chen, F. Manshaii, J. Yin, and J. Chen, "Advances in Luminescent Fibers for Interactive Smart Textiles," *cMat*, vol. 1, no. 3, Dec. 2024, doi: 10.1002/cmt2.33
[20] Y. Jiang et al., "Solid-State Intramolecular Motions in Continuous Fibers Driven by Ambient Humidity for Fluorescent Sensors," *Natl. Sci. Rev.*, vol. 8, no. 4, Apr. 2021, doi: 10.1093/nsr/nwaa135
[21] K. Hong, K. G. Cho, D. C. Lim, J. Y. Lee, and K. H. Lee, "Light Emitting Fabrics Based on Luminophore Dye-Doped Ion Gel Electrolyte Microfibers," *Dye. Pigment.*, vol. 154, pp. 188–193, Jul. 2018, doi: 10.1016/j.dyepig.2018.02.052
[22] S. Song, B. Song, C.-H. Cho, S. K. Lim, and S. M. Jeong, "Textile-Fiber-Embedded Multiluminescent Devices: A New Approach to Soft Display Systems," *Mater. Today*, vol. 32, pp. 46–58, Jan. 2020, doi: 10.1016/j.mattod.2019.08.004
[23] X. Hu et al., "Non-noble Metal Electroluminescent Fibers for Visual Monitoring and Interaction," *Adv. Fiber Mater.*, vol. 7, no. 1, pp. 227–239, Feb. 2025, doi: 10.1007/s42765-024-00480-y
[24] Q. Xie et al., "Flexible Silk-Fibroin-Based Electroluminescent Fiber with External-Field-Driven Touch Response and Triboelectric Sensing for Smart Wearables," *Adv. Funct. Mater.*, Aug. 2025, doi: 10.1002/adfm.202514650
[25] Q. Xie et al., "Design Principles of Electroluminescent Devices Based on Different Electrodes and Recent Advances Toward Their Application in Textiles," *J. Mater. Chem. C*, vol. 13, no. 18, pp. 8934–8957, 2025, doi: 10.1039/d5tc00209e
[26] B. Hu et al., "Textile-Based Flexible Electroluminescent Devices," *Adv. Funct. Mater.*, vol. 21, no. 2, pp. 305–311, Jan. 2011, doi: 10.1002/adfm.201001110
[27] W. B. I. N. Qaim et al., "Towards Energy Efficiency in the Internet of Wearable Things: A Systematic Review," vol. 8, 2020, doi: 10.1109/ACCESS.2020.3025270
[28] K. M. Gerhardsson, and T. Laike, "User Acceptance of a Personalised Home Lighting System Based on Wearable Technology," *Appl. Ergon.*, vol. 96, p. 103480, Oct. 2021, doi: 10.1016/j.apergo.2021.103480
[29] Z. Yang, Z. Wang, J. Zhang, C. Huang, and Q. Zhang, "Wearables Can Afford," in *Proceedings of the 13th Annual International Conference on Mobile Systems, Applications, and Services*, New York, NY, USA: ACM, May 2015, pp. 317–330. doi: 10.1145/2742647.2742648
[30] R. Song, S. Cho, S. Khan, I. Park, and W. Gao, "Lighting the Path to Precision Healthcare: Advances and Applications of Wearable Photonic Sensors," *Adv. Mater.*, Jan. 2025, doi: 10.1002/adma.202419161
[31] G.-H. Lee et al., "Multifunctional materials for implantable and wearable photonic healthcare devices," *Nat. Rev. Mater.*, vol. 5, no. 2, pp. 149–165, Jan. 2020, doi: 10.1038/s41578-019-0167-3

[32] S. T. Parker et al., "Biocompatible Silk Printed Optical Waveguides," *Adv. Mater.*, vol. 21, no. 23, pp. 2411–2415, Jun. 2009, doi: 10.1002/adma.200801580
[33] A. Dupuis et al., "Prospective for biodegradable microstructured optical fibers," *Opt. Lett.*, vol. 32, no. 2, p. 109, Jan. 2007, doi: 10.1364/OL.32.000109
[34] D. R. Sparta, A. M. Stamatakis, J. L. Phillips, N. Hovelsø, R. van Zessen, and G. D. Stuber, "Construction of implantable optical fibers for long-term optogenetic manipulation of neural circuits," *Nat. Protoc.*, vol. 7, no. 1, pp. 12–23, Jan. 2012, doi: 10.1038/nprot.2011.413
[35] S. Cai, X. Xu, W. Yang, J. Chen, and X. Fang, "Materials and Designs for Wearable Photodetectors," *Adv. Mater.*, vol. 31, no. 18, May 2019, doi: 10.1002/adma.201808138
[36] T. Park, J. W. Leem, Y. L. Kim, and C. H. Lee, "Photonic Nanomaterials for Wearable Health Solutions," *Adv. Mater.*, Feb. 2025, doi: 10.1002/adma.202418705
[37] C. Zhang and Y. S. Zhao, "Flexible Photonic Materials and Devices: Synthetic Strategies, Sensing Properties, and Wearable Applications," *Adv. Mater.*, Dec. 2024, doi: 10.1002/adma.202415856
[38] D. Arbia, M. Alam, Y. Moullec, and E. Hamida, "Communication Challenges in on-Body and Body-to-Body Wearable Wireless Networks—A Connectivity Perspective," *Technologies*, vol. 5, no. 3, p. 43, Jul. 2017, doi: 10.3390/technologies5030043
[39] Y. Wu et al., "Photonic Interfaces: an Innovative Wearable Sensing Solution for Continuous Monitoring of Human Motion and Physiological Signals," *Small Methods*, Jun. 2025, doi: 10.1002/smtd.202500727
[40] X. Tao, "Wearable electronics and photonics," *Wearable Electron. Photonics*, pp. 1–250, 2005, doi: 10.1533/9781845690441
[41] J. K. Tong, X. Huang, S. V. Boriskina, J. Loomis, Y. Xu, and G. Chen, "Infrared-Transparent Visible-Opaque Fabrics for Wearable Personal Thermal Management," *ACS Photonics*, vol. 2, no. 6, pp. 769–778, Jun. 2015, doi: 10.1021/acsphotonics.5b00140
[42] T. Shimura, S. Sato, P. Zalar, and N. Matsuhisa, "Engineering the Comfort-of-Wear for Next Generation Wearables," *Adv. Electron. Mater.*, vol. 9, no. 9, Sep. 2023, doi: 10.1002/aelm.202200512
[43] R. Zou et al., "Self-Powered and Self-Sensing Wearable Devices from a Comfort Perspective," *Device*, vol. 2, no. 11, p. 100466, Nov. 2024, doi: 10.1016/j.device.2024.100466
[44] A. Cutolo et al., "Innovative Photonic Sensors for Safety and Security, Part II: Aerospace and Submarine Applications," *Sensors*, vol. 23, no. 5, p. 2417, Feb. 2023, doi: 10.3390/s23052417
[45] M. Alessandra Cutolo and G. Breglio, "(INVITED)Interferometric Fabry-Perot sensors for ultrasound detection on the tip of an optical fiber," *Results Opt.*, vol. 6, p. 100209, Jan. 2022, doi: 10.1016/j.rio.2021.100209
[46] E. Svertoka et al., "Wearables for Industrial Work Safety: A Survey," *Sensors*, vol. 21, no. 11, p. 3844, Jun. 2021, doi: 10.3390/s21113844
[47] J. Han, D. Kim, J. Kim, G. Kim, P. Fischer, and H. Jeong, "Plasmonic Nanostructure Engineering with Shadow Growth," *Adv. Mater.*, vol. 35, no. 34, Aug. 2023, doi: 10.1002/adma.202107917
[48] C. Zhao, J. Park, S. E. Root, and Z. Bao, "Skin-Inspired Soft Bioelectronic Materials, Devices and Systems," *Nat. Rev. Bioeng.*, vol. 2, no. 8, pp. 671–690, Jun. 2024, doi: 10.1038/s44222-024-00194-1
[49] H. Li et al., "Photonic Skin for Photonic-Integration-Based Wearable Sensors," *Optica*, vol. 12, no. 2, p. 190, Feb. 2025, doi: 10.1364/OPTICA.546458
[50] H. Jeong et al., "Differential Cardiopulmonary Monitoring System for Artifact-Canceled Physiological Tracking of Athletes, Workers, and COVID-19 Patients," *Sci. Adv.*, vol. 7, no. 20, May 2021, doi: 10.1126/sciadv.abg3092

[51] H. U. Chung et al., "Binodal, Wireless Epidermal Electronic Systems with In-Sensor Analytics for Neonatal Intensive Care," *Science*, vol. 363, no. 6430, Mar. 2019, doi: 10.1126/science.aau0780
[52] H. Wang et al., "Carbon-Based Flexible Devices for Comprehensive Health Monitoring," *Small Methods*, vol. 7, no. 2, Feb. 2023, doi: 10.1002/smtd.202201340
[53] H. Hu et al., "A wearable cardiac ultrasound imager," *Nature*, vol. 613, no. 7945, pp. 667–675, Jan. 2023, doi: 10.1038/s41586-022-05498-z
[54] Y. Zhao et al., "Ultra-conformal skin electrodes with synergistically enhanced conductivity for long-time and low-motion artifact epidermal electrophysiology," *Nat. Commun.*, vol. 12, no. 1, p. 4880, Aug. 2021, doi: 10.1038/s41467-021-25152-y
[55] A. Cutolo et al., "Ultrasound Waves in Tumors Via Needle Irradiation for Precise Medicine," *Sci. Rep.*, vol. 12, no. 1, p. 6513, Apr. 2022, doi: 10.1038/s41598-022-10407-5
[56] M. A. Cutolo, A. Migliaccio, L. Altucci, A. Cutolo, and A. Cusano, "An Innovative High Frequency Hyperthermia Approach Against SARS-CoV-2 and Related Virus: Feasibility Analysis," Feb. 24, 2021. doi: 10.21203/rs.3.rs-231129/v1
[57] B. Carotenuto et al., "Optical Fiber Technology Enables Smart Needles for Epidurals: An In-Vivo Swine Study," *Biomed. Opt. Express*, vol. 10, no. 3, p. 1351, Mar. 2019, doi: 10.1364/BOE.10.001351
[58] S. Zuppolini et al., "Label-Free Fiber Optic Optrode for the Detection of Class C β-lactamases Expressed by Drug Resistant Bacteria," *Biomed. Opt. Express*, vol. 8, no. 11, p. 5191, Nov. 2017, doi: 10.1364/BOE.8.005191
[59] G. Quero et al., "Long Period Fiber Grating Nano-Optrode for Cancer Biomarker Detection," *Biosens. Bioelectron.*, vol. 80, pp. 590–600, Jun. 2016, doi: 10.1016/j.bios.2016.02.021
[60] S. Lee et al., "Nanomesh Pressure Sensor for Monitoring Finger Manipulation Without Sensory Interference," *Science*, vol. 370, no. 6519, pp. 966–970, Nov. 2020, doi: 10.1126/science.abc9735
[61] S. Yoo et al., "Responsive Materials and Mechanisms as Thermal Safety Systems for Skin-Interfaced Electronic Devices," *Nat. Commun.*, vol. 14, no. 1, p. 1024, Feb. 2023, doi: 10.1038/s41467-023-36690-y
[62] I. Hrga, "Wearable technologies: Between fashion, art, performance, and science (fiction)," *Tekstilec*, vol. 62, no. 2, pp. 124–136, 2019, doi: 10.14502/Tekstilec2019.62.124-136
[63] J. Birringer and M. Danjoux, "Wearable Technology for the Performing Arts," in *Smart Clothes and Wearable Technology*, Elsevier, 2023, pp. 529–571. doi: 10.1016/B978-0-12-819526-0.00005-9
[64] S. Koo and Y. Chae, "Wearable Technology in Fashion," in *Leading Edge Technologies in Fashion Innovation*, Cham: Springer International Publishing, 2022, pp. 35–57. doi: 10.1007/978-3-030-91135-5_3
[65] S. Jabari, A. Shaikh, Ç. Genç, O. Buruk, J. Virkki, and J. Hamari, "A Systematic Literature Review on Computational Fashion Wearables," *Int. J. Human–Computer Interact.*, vol. 40, no. 22, pp. 6913–6940, Nov. 2024, doi: 10.1080/10447318.2023.2269007
[66] M. A. Butt, X. Mateos, and R. Piramidowicz, "Photonics Sensors: A Perspective on Current Advancements, Emerging Challenges, and Potential Solutions (Invited)," *Phys. Lett. A*, vol. 516, p. 129633, Aug. 2024, doi: 10.1016/j.physleta.2024.129633
[67] G. Alagappan et al., "Leveraging AI in Photonics and Beyond," *Photonics*, vol. 9, no. 2, p. 75, Jan. 2022, doi: 10.3390/photonics9020075
[68] J. Z. Hassan, A. Raza, Z. U. Din Babar, U. Qumar, N. T. Kaner, and A. Cassinese, "2D Material-Based Sensing Devices: An Update," *J. Mater. Chem. A*, vol. 11, no. 12, pp. 6016–6063, 2023, doi: 10.1039/D2TA07653E
[69] J. Heikenfeld et al., "Wearable Sensors: Modalities, Challenges, and Prospects," *Lab Chip*, vol. 18, no. 2, pp. 217–248, 2018, doi: 10.1039/C7LC00914C

8 Synthetic Polymer Nanofibre Films with Smart and Active Agents for Food Packaging

Steven J. McNeil

INTRODUCTION

Food packaging is essential for ensuring the safety, quality, and shelf life of food. It serves to protect food from external factors such as microorganisms, moisture, oxygen and dust while also providing mechanical support and containment. Given the critical role of food packaging in human health and the substantial market it represents for packaging manufacturers, significant research efforts have been directed towards enhancing the performance and sustainability of food packaging. A key area of focus in this research is the development of active and smart packaging technologies, see Figure 8.1.

Active packaging can be broadly categorised into two types: release systems and absorption/scavenging systems. Release systems include active agents such as antimicrobials, which are designed to inhibit the growth of harmful microorganisms on food. Absorption or scavenging systems, on the other hand, are intended to remove undesirable substances from the packaged food, such as ethylene and moisture. Smart packaging also falls into two main categories: indicators and responsive systems. Indicator systems provide information about a specific property of the food, such as a visual indication of pH. Responsive systems, in contrast, are designed to react to certain conditions by performing a specific action, such as releasing an antimicrobial agent when the microbial load on the food exceeds a predetermined threshold. Multiple agents can be incorporated into the same nanofibre film, and some compounds are both smart and active. For instance, anthocyanins are antimicrobial and halochromic, that is, they can change colour in response to changes in pH.

Nanofibre films, also known as nanofibre membranes, are at the forefront of advancements in food packaging technology. This prominence arises from their unique amalgamation of properties, including an exceptionally high specific surface area, the ability to incorporate smart and active agents, low density, mechanical flexibility and their adaptability for customisation. For instance, parameters such as

DOI: 10.1201/9781003507208-8

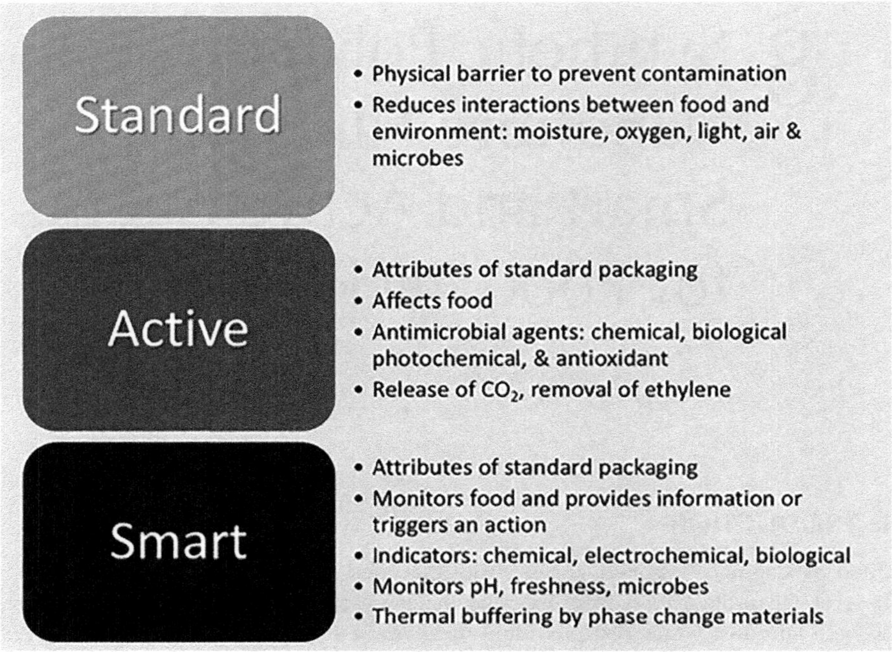

FIGURE 8.1 Comparison of the features of standard, active and smart packaging.

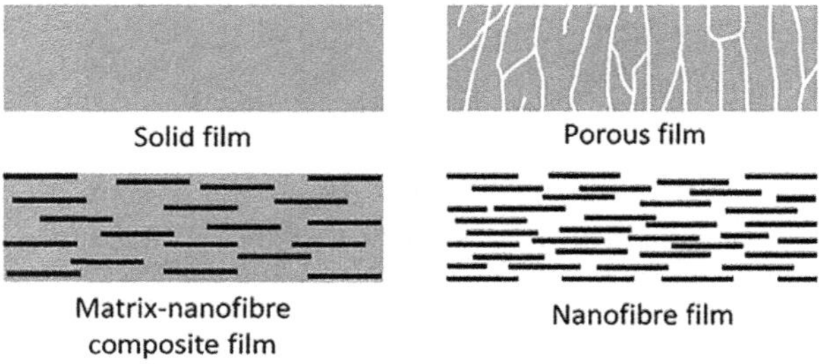

FIGURE 8.2 Types of packaging films, cross-sections.

permeability, fibre diameter, wettability and the distribution of agents in the fibre cross-section can be precisely tailored to meet specific requirements. The high surface area of nanofibrous packaging facilitates significant interaction with the food, thereby enhancing their smart and active functionalities.

Nanofibres are strictly defined as fibres having diameters less than 100 nm; however, they are generally considered to encompass fibres with diameters ranging from 100 to 1,000 nm. Films can be fabricated solely from nanofibres or as composites comprising both nanofibres and non-fibrous polymers, see Figure 8.2. The structure of films made entirely of nanofibres makes them highly flexible compared to solid films, thus enabling them to conform to the shape of packaged food [1, 2].

The development of nanofibrous smart active food packaging is progressing concurrently with advancements in other fields where nanofibrous films are employed. These fields include packaging for non-food items, medicine—particularly in drug delivery and tissue engineering, filtration, environmental remediation and advanced textiles [3–9].

This chapter reviews smart and active nanofibre food packaging derived from synthetic polymers. It examines the four predominant methods utilised for preparing films, the 18 specific synthetic polymers employed, as well as the smart and active agents incorporated into these films, and their consequent effects on the preservation and quality of packaged food. The polymer polylactic acid (PLA) is included in this review as it is synthesised industrially, albeit from biologically derived monomers. Biopolymers such as polysaccharides and proteins, are discussed in a separate chapter of this book, which also covers their applications in blends and composites with synthetic polymers [10].

To maintain the conciseness of this review, it does not endeavour to cover every method for the preparation of nanofibres and nanofibrous films. Instead, it concentrates on techniques that have exhibited potential in the realm of food packaging. Notably excluded are some innovative methods such as plasma-induced synthesis [11]. Additionally, methods that are well established for the production of microfibres but currently appear inefficient for nanofibre fabrication, such as melt blowing, melt extrusion, wet spinning, islands-in-the-sea synthesis, template synthesis and phase separation, are also omitted [12, 13].

PREPARATION OF NANOFIBRES AND NANOFIBRE FILMS

This section discusses the methods of preparing nanofibres and nanofibre films that can be applied to a diverse range of synthetic polymers. The methods are electrospinning (ES), centrifugal spinning (CF), solution blow spinning (SBS) and microfluidic spinning (MFS). In these processes, the spun fibres are typically collected in the form of films as they are produced. The properties of these films can be modified by post-spinning processes (see Section Post-spinning Processing). As well as being used in nanofibrous films, nanofibres can also be combined with non-fibrous films to make composite films with specific properties.

ELECTROSPINNING

Since it was first documented in 1887, ES has evolved to become the predominant method for making nanofibre films [14, 15]. In the most straightforward and widely used type of ES, a polymer is passed through a needle (spinneret), and an electric field stretches a droplet at the end of the needle to form what is known as a Taylor Cone. When a sufficiently high voltage is applied, the electrostatic repulsion between molecules in the droplet is strong enough to overcome the surface tension, and a jet of material leaves the droplet, traveling towards a collecting electrode (collector), which is typically earthed, or otherwise maintained at a different voltage than the spinneret. Charges accumulate on the surfaces of the streaming jets during flight as they dry/solidify into fibres. Small irregularities in the shapes of the streaming jets cause an uneven accumulation of charges and the resulting electrostatic repulsion causes bending and stretching, a phenomenon known as 'whipping', which produces uniform fibres with nanoscale diameters, which accumulate on the collector, see Figure 8.3. The fibres are charged when they arrive at the collector and this charge must be removed to prevent electrostatic repulsion of subsequently arriving fibres; hence, collectors are electrically conductive [16]. The fibres deposited on the collector build up over time, forming an increasingly thicker film.

The polymer used for spinning may be either in solution or melt form, with the corresponding solidification of the jet occurring due to either solvent evaporation or cooling. The flow rate of the polymer solution or melt through a single spinneret typically ranges from 0.5 to 2 mL/h, which is sufficient to replenish the droplet by replacing the jetted material. The potential difference between the spinneret and the collector is generally 5–30 kV, with a typical separation distance of 10–20 cm [14, 16]. The preceding discussion underscores the significance of viscosity, surface tension, electrical conductivity and the rate of solidification of the polymer in ES.

Multiple spinnerets can be utilised to increase the production rate of nanofibre films. High productivity can also be attained through needleless systems, wherein multiple droplets of polymer are employed concurrently, see Figure 8.3 [17]. The uniformity of

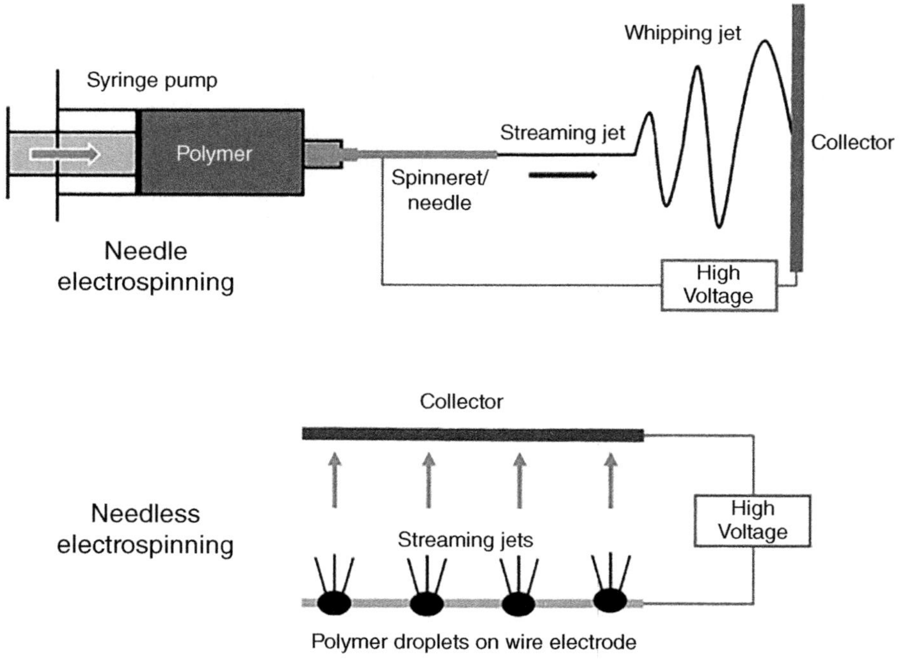

FIGURE 8.3 Electrospinning.

the resultant nanofibre films can be enhanced by incorporating auxiliary electrodes to modulate the electric field acting on the polymer jets [18].

A significant strength of the electrospinning process is its adaptability to specific requirements, which can be achieved by modifying the polymer solution or melt, the spinneret, the collector and the voltage. For instance, to create a film from incompatible polymers that cannot be co-dissolved in the same solvent, multiple spinnerets can be employed, each supplied with a different polymer, to produce a film composed of different types of fibre. Additionally, coaxial fibres, featuring distinct polymers in the core and sheath, can be fabricated using coaxial needles [19]. Active and smart compounds can be incorporated into the polymer prior to the spinning process, and some illustrative recent studies are discussed below and listed in Tables 8.1 and 8.2.

The most basic collector configuration is a stationary flat conductor, while more complex designs are employed for specific applications. For example, by rotating a flat collector at high speeds (ranging from 500 to 4,500 rpm), fibres can be deposited with high alignment, thereby enhancing certain mechanical properties [114]. In gap electrospinning, the collector is two parallel wires with incoming fibres spanning these wires in parallel [115]. In high-throughput systems, the spun fibres are directed onto moving collectors, such as rotating cylinders or continuous belts [116].

To assist the electrospinning of high-viscosity polymers, a rotating spinneret has been developed in which this spinneret applies a high shear force to the polymer [117]. Most studies on nanofibre ES have utilised direct current. Although alternating current ES has been explored for some time [118, 119], it has not been extensively investigated in the context of packaging. In alternating current ES, the fibres carry minimal charge during their flight, resulting in reduced repulsion and the formation of films with highly aligned fibres. Notably, for food packaging applications, the collector does not need to be conductive, allowing fibres to be deposited onto insulating surfaces, which could potentially include food.

Solution Blow Spinning

SBS, also referred to as energy spinning and blow spinning, involves introducing polymer solution into a stream of pressurised gas, typically air. This technique is closely related to airbrushing, and commercial airbrushes have been utilised. SBS was first documented in 2009 [120] and has become the second most widely investigated method for preparing nanofibrous food packaging films and has been the subject of several comprehensive reviews [121–124]. Nanofibres produced via this method have found applications in diverse fields, including food packaging, textiles, medicine and membrane separation.

In this process, a polymer solution is injected into a coaxial, high-speed airflow, see Figure 8.4. The airflow generates a shear force on the surface of the polymer, which exceeds the polymer's surface tension, thereby forming a jet of polymer. The airflow stretches and dries the polymer jets into fibres as they travel to a collector. The collector must permit air to pass through it, to avoid distorting the airflow.

TABLE 8.1
Synthetic Polymer Nanofibre Films with Active Agents

Polymers	Preparation	Agents	Activities[a]	Food	References
Ethylene vinyl alcohol	ES	Cuminaldehyde	ABac	Meat	[20]
PCL	ES	Silver nanoparticles on hemp powder	ABac, AFun	Grapes	[21]
PCL	ES	Ethylene, titanium dioxide	Release and removal of ethylene[b]	Bananas	[22]
PCL	ES	Laurel and clove essential oils	ABac, AOx	Fish	[23]
PCL	ES	Rutin, chitosan	ABac, AOx	Fish	[24]
PCL	ES	Sage extract	ABac, AOx	–	[25]
PCL	ES	gamma-Decalactone	AFun	Oranges	[26]
PCL	ES	Zeolite, silica	Absorption of histamine	–	[27]
PCL	ES	*Ganoderma lucidum* extract	AOx	Fish	[28]
PCL, PVP, PCL/PVP	SBS	Chlorogenic acid	ABac	–	[29]
PEO	CS	Silver nanoparticles	ABac	–	[30]
PEO	ES	Bacteriophage	ABac	–	[31]
PEO	ES	Cinnamon essential oil	ABac	Meat	[32]
PEO	ES	Tea tree oil	ABac	Meat	[33]
PEO	ES	Nisin	ABac	Cheese	[34]
PEO	ES	Aloe vera extract	AOx	–	[35]
PLA	SBS	Potassium permanganate, proprietary humidity absorbent	Removal of ethylene and water vapour	Tomatoes	[36]
PLA	SBS	1-Methylcyclopropene	Retardation of ripening	Bananas	[37]
PLA	SBS	Essential oils	AFun, ACra	Grapes	[38]
PLA	SBS	Essential oils	AFun, ACra	Grapes	[39]
PLA	SBS	Sucuk spice mix essential oils	ABac, AOx	–	[40]
PLA, PCL	ES	Oregano essential oil	ABac, AFun	Blackberries	[41]

PLA/PEO	ES	Salicylaldehyde	ABac	–	[42]
PLA/PCL	ES	Lactobionic acid	ABac, AOx	–	[43]
PLA/PCL	ES	Plant essential oils and growth regulators, phenolics, natamycin, nisin	ABac AFun	Blackberries	[44]
PLA/PEG	SBS	Peppermint essential oil	ABac, AOx	Strawberries	[45]
PLA/PHB-co-HV[c]	ES	Iron-doped zinc oxide nanoparticles	ABac	–	[46]
PLA/PVP	ES	Thymol	ABac, AOx	–	[47]
Polyacrylonitrile	ES	Titanium dioxide	Removal of ethylene	Bananas, tomatoes	[48]
Polyacrylonitrile	ES	Potassium permanganate	Removal of ethylene	Avocados	[49]
Poly(vinyl alcohol-co-vinyl acetate)/	ES	Silver nanoparticles	ABac, AFun	–	[50]
PS	SBS	Orange oil	ABac	–	[51]
PU/PVP	SBS	Natamycin, ascorbic acid	AFun	Apples	[52]
PU, PVP	SBS	Chlorogenic acid	ABac, AOx	–	[53]
PVA	ES	*Laurus nobilis* and *Rosmarinus officinalis* essential oils	AOx	Meat	[54]
PVA	ES	Bacteriophages	ABac	–	[55]
PVA	ES	Hemp extract	ABac	–	[56]
PVA	SBS	Extracts of carrot, cocoa, Jaboticaba, acai	AOx	Strawberries	[57]
PVA	ES	Cobalt phthalocyanine–titanium dioxide	ABac, removal of ethylene	Strawberries, bananas	[58]
PVA	ES	Rosemary extract	AOx	–	[59]
PVA	ES	Rosemary extract	AOx	–	[60]
PVA	ES	Algae extract	ABac	Meat	[61]
PVA, PLA	ES	Ethyl lauroyl arginate	ABac	–	[62]
PVA	ES	Permutite, cinnamaldehyde	ABac	Strawberries	[63]
PVA	ES	Lignin	ABac, AOx		[64]
PAV	ES	Poly(hexamethylene biguanide)	ABac	Fish	[65]

(*Continued*)

TABLE 8.1 (Continued)
Synthetic Polymer Nanofibre Films with Active Agents

Polymers	Preparation	Agents	Activities[a]	Food	References
PVA	ES	Clove oil	ABac	–	[66]
PVA	ES	Jujube extract	ABac	Strawberries	[67]
PVA	ES	*Coptis chinensis* extract	ABac, AFun	–	[68]
PVA	ES	Caffeic acid	ABac	–	[69]
PVA	ES	Pomegranate peel extract, sodium dehydroacetate	ABac, AOx	–	[70]
PVA	ES	Cinnamon essential oil	ABac	Shrimp	[71]
PVA	ES	*Zanthoxylum bungeanum* essential oil	AFun	Strawberries, cherries	[72]
PVA	ES	Lysozyme, cinnamon essential oil	ABac, AFun	–	[73]
PVA	ES	Graphene oxide, zinc oxide nanoparticles	ABac	–	[74]
PVA	ES	Cinnamon essential oil	ABac	Mushrooms	[75]
PVA/PCL	ES	Glucose oxidase enzyme	ABac	–	[76]
PVA/PLA	ES	Algae extract	ABac, AOx	–	[77]
PVB[d]	ES	Camellia oil, zinc oxide–titanium dioxide	ABac, removal of ethylene	Strawberries, Tomatoes	[78]
PVC/PDMS[e]	ES	Nanodiamond	ABac	–	[79]
PVP	CS	Lavender essential oil	ABac, AOx, AYst	Meat	[80]
PVP, PEO	CS	Copper nanoparticles	ABac	–	[81]
PVP/PVB[d]	ES	Allicin	ABac	Meat	[82]

[a] ABac, antibacterial; ACra, antiocratoxigenic; AFun, antifungal; AOx, antioxidant; AYst, anti-yeast.
[b] Release of ethylene for accelerated ripening and photocatalytic removal of ethylene of increased shelf life.
[c] Poly(3-hydroxybutyrate-co-3-hydroxyvalerate).
[d] Poly(vinyl butyral).
[e] Polyvinyl chloride/polydimethylsiloxane.

TABLE 8.2
Synthetic Polymer Nanofibre Films with Smart Agents

Polymers	Preparation	Agents	Response or Property Indicated[a]	Food	References
PCL	SBS	Anthocyanins	pH	Fish	[83]
PCL	ES	Natural fatty acids	Thermal buffering	–	[84]
PCL	SBS	Thymol microencapsulated in a covalent organic framework	Temperature-responsive release of thymol	–	[85]
PCL	ES	Anthocyanins	Ammonia	Meat	[86]
PCL	ES	Red cabbage extract	Amines	Fish	[87]
PCL, PLA	ES	Commercial phase change material	Thermal buffering	–	[88]
PCL, PS	ES	Commercial phase change material	Thermal buffering	–	[89]
PCL, PLA, PS	ES	Commercial phase change material	Thermal buffering	–	[90]
PCL/PEO	ES	Anthocyanins, silver nanoparticles	pH, trimethylamine, microbes	Shrimp	[91]
PCL/PEO	ES	Phycocyanin	pH	–	[92]
PLA	ES	Anthocyanins, cinnamaldehyde	pH	–	[93]
PLA	ES	Pectin, thymol	Microbe-responsive release of antimicrobial agent	Oranges	[94]
PLA	ES	Blueberry anthocyanins, quaternised chitosan	pH	Blueberries	[95]
PLA	ES	Bromocresol Purple, narigin	pH	Fish	[96]
PLA/PEO	ES	Phycocyanin	pH	–	[97]
PLA/PEO	ES	Spirulina	pH	–	[98]
PLA/PVA/PEO		Thyme essential oil	Humidity-responsive release of antimicrobial agent	Strawberries	[99]
PMMA[b]	CS	Spiropyran	Light	–	[100]
PVA	ES	Black carrot extract, tin oxide nanoparticles	pH, ammonia, hydrogen sulphide, formaldehyde	Fish	[101]
PVA	ES	Curcumin, gardenia blue	pH, ammonia	Shrimp	[102]
PVA	ES	Anthocyanins	pH	Shrimp	[103]
PVA	ES	Red cabbage extract	pH	Meat	[104]
PVA	ES	Red cabbage extract	pH	Fruit	[105]

(Continued)

TABLE 8.2 (Continued)
Synthetic Polymer Nanofibre Films with Smart Agents

Polymers	Preparation	Agents	Response or Property Indicated[a]	Food	References
PVA	ES	Curcumin, indigo	pH, ammonia	–	[106]
PVA	ES	Eggplant skin extract	pH	Fish	[107]
PVA/PEI[c]	ES	Methylene Blue	Bacteria	Various fruit, intact & cut	[108]
PVA/PS	ES	Methylene Blue/ titanium dioxide	Oxygen	Meat balls	[109]
PVF[d]	ES	Anthocyanins, cinnamon essential oil	pH, ammonia	Meat	[110]
PVF[d]	ES	Fluorescein isothiocyanate	Amines	Fish	[111]
PVF[d]	ES	Carbon quantum dots	Amines	Meat	[112]
PVP	ES	Curcumin	Amines	–	[113]

[a] Colour change unless indicated otherwise.
[b] PMMA: poly(methyl methacrylate).
[c] PEI: polyethyleneimine.
[d] PVF: polyvinylidene fluoride.

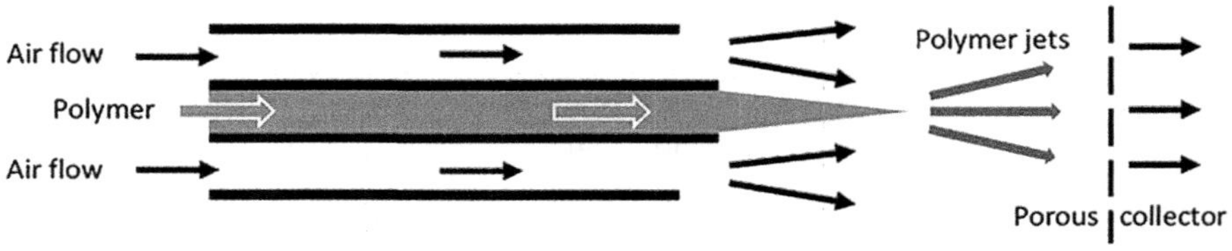

FIGURE 8.4 Solution blow spinning.

Several critical processing parameters influence the efficacy of this type of spinning. These include the ejection rate of the polymer solution from the nozzle, and the shape and internal diameter of the nozzle (typically 0.2–0.7 mm). Maintaining an optimal separation between the nozzle and the collector is essential to ensure the complete stretching and drying of the fibres. Air pressure is a pivotal factor in this process; insufficient pressure results in inadequate polymer stretching, whereas excessive pressure induces turbulence, causing the airborne fibres to tangle and adhere to one another. Spinning is also influenced by environmental conditions, particularly temperature, relative humidity and atmospheric pressure. SBS systems can be configured with multiple nozzles for faster production and control of fibre alignment through mechanisms such as electric fields [125].

In SBS, the most critical characteristics of the polymer solution are its surface tension and the volatility of the solvents, which are determined by the specific materials and concentrations used. For instance, if the viscosity of the polymer is too high, the jets of polymer cannot be drawn into fibres. Nanofibre-based food packaging has been developed using SBS, incorporating both synthetic polymers, biopolymers and various actives, see Section Active and Smart Additives. SBS has also been utilised in at least one study of food packaging with smart functionality, where an antimicrobial agent was released from PCL fibres at a specific temperature [85].

Centrifugal Spinning

CS is also known as centrifugal jet spinning, rotor spinning, rotary force spinning, rotary jet spinning and force spinning. The term 'centrifugal spinning' arises from the fictitious or inertial 'centrifugal force' that appears to act on jets of polymer expelled from a rotating spinneret, see Figure 8.5. CS represents another technique for fabricating nanofibres, which has garnered substantial research interest concerning both its fundamental principles and its diverse applications. These applications encompass edible fibres, wound dressings and drug delivery systems [126]. However, relatively few studies have focused on the preparation of films for packaging applications, positioning CS as the third most extensively investigated method for producing nanofibrous food packaging.

The basic principle of CS is that polymer is pumped to a spinneret, and when the rotational velocity of the spinneret is sufficiently high, the inertial force acting upon the polymer exceeds its surface tension, causing it to be ejected from the spinneret. The polymer solution or melt dries or solidifies and elongates while traversing through the air, ultimately being deposited onto a collector. The trajectory of the fibres can be controlled using airflow, either through blowing or vacuum techniques [1]. Common configurations typically employ spinneret nozzle diameters ranging from 100 to 400 μm, rotational speeds between 2,000 and 6,000 revolutions per minute, and a spinneret-to-collector separation of 5–50 cm.

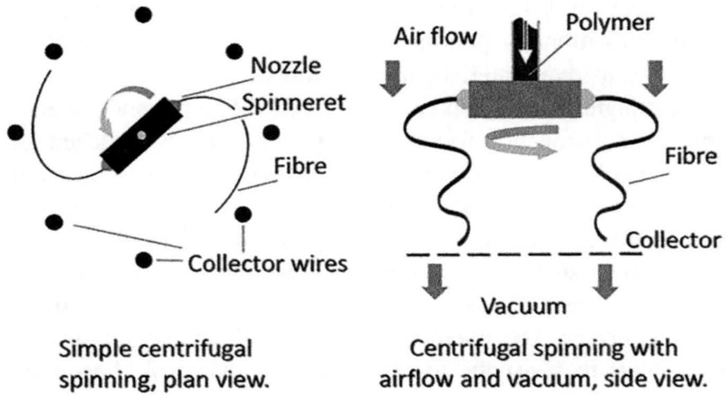

FIGURE 8.5 Centrifugal spinning.

To obtain homogeneous, low-diameter fibres through CS, many parameters need to be in accord, and some of these parameters are common to other spinning methods. These parameters can be categorised into three groups:

1. **Properties of the Polymer Solution/Melt**: This encompasses factors such as viscosity, surface tension and the rate of solidification. These properties are influenced by the nature of the polymer(s), their concentrations and the solvents employed (if any).
2. **Spinning Equipment and Operating Conditions**: This includes the number, shape and diameter of the spinneret nozzles, the rotational speed of the spinneret, the distance between the nozzle and the axis of rotation and the separation between the spinneret and the collector.
3. **Environmental Factors**: Variables such as ambient temperature and relative humidity significantly impact the rate at which the polymer solidifies.

CS is regarded by some experts as a cost-effective and highly productive method for producing nanofibre films with excellent fibre alignment [126]. The range of polymers that can be utilised for CS can be expanded by applying high-pressure gas to the polymer as it enters the spinneret, a process known as pressurised gyration spinning [127]. Studies have reported the use of centrifugally spun nanofibres in food packaging and related fields, employing polymers such as PLA, polyvinyl pyrrolidone (PVP) and polyethylene oxide (PEO) [30, 36, 80, 81], and biopolymers. These polymers incorporated active agents such as lavender essential oil and nanosilver, see Table 8.1. Additionally, spiropyran has been combined with poly (methyl methacrylate) and then subjected to CS to create a photochromic smart film [128].

Microfluidic Spinning

This process has been less extensively studied for the preparation of nanofibre food packaging films than ES, SBS and CS. However, it merits consideration here, as it represents, in some respects, the simplest route to nanofibres, yet it can produce very complex nanofibres [100, 129–131]. A notable advantage of MFS is that it employs only very mild conditions of pressure, shear, and electrical fields, allowing for the incorporation of sensitive materials into the fibres.

In the most straightforward implementation of MFS, a polymer solution is pumped through a needle, subsequently forming a single fibre that is stretched (drawn) on its way to a moving collector, thus reducing the diameter and aligning the molecules of polymer, see Figure 8.6. The final solidification of the film takes place after collection and typically involves solvent evaporation, photo-polymerisation, or cross-linking. To fully exploit the potential of this fibre spinning approach, the polymer solution is pumped through narrow (100–500 μm) channels in a microfluidic chip before being directed to the collector. The microfluidic chips can be designed with progressively narrowing channels to assist the formation of fine-diameter fibres and to combine

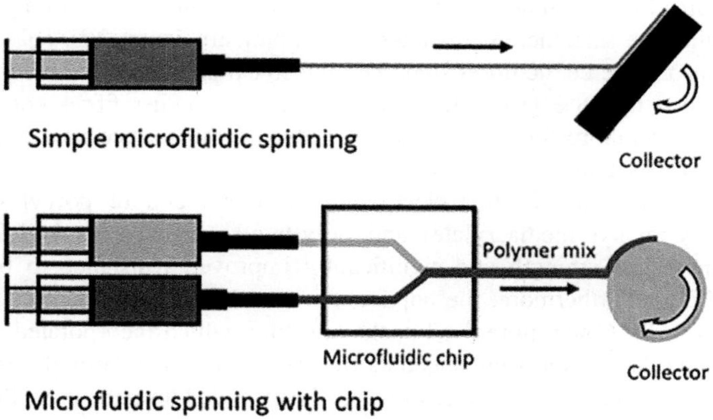

FIGURE 8.6 Microfluidic spinning.

different polymers. All reported studies on MFS of packaging materials have utilised the simple, non-chip approach to produce fibres. In the packaging area, MFS has predominantly been applied to bio-based polymers to date, particularly glucomannan, see Smart Active Biopolymer Nanofibre Films: The Future of Food Packaging chapter in this book [10, 132]. Only one study has been identified involving synthetic polymers in MFS of food packaging, specifically polyvinylpyrrolidone blended with ethyl cellulose loaded with natamycin [133].

Post-spinning Processing

Nanofibre films generally exhibit lower mechanical strength compared to films produced by methods such as extrusion, calendaring and casting, which limits their application in food packaging. However, the robustness of nanofibre films can be enhanced by improving the mechanical properties of individual fibres and by strengthening inter-fibre bonding in post-spinning processes. This can be achieved through various chemical, thermal and mechanical techniques, including cross-linking, exposure to solvents, cold and hot drawing, hot pressing, thermal treatment/annealing and exposure to plasma, UV light and electron beams [2, 134]. Post-spinning processes have been extensively explored in the context of nanofibre films for applications such as water filtration and wound dressings, but their utilisation in food packaging has received significantly less attention. This is likely to change in the future. Therefore, some illustrative studies will be briefly discussed here.

Exposing films to the vapours of solvents chosen to soften the polymer can significantly enhance their mechanical strength. For example, a 48-hour exposure to formic acid vapour increased the tensile strength of an electrospun nylon film, intended for water filtration, by 91% [135]. However, the optimal exposure duration was determined to be 5 hours, as this resulted in a 62% increase in strength without enlarging the pore size, thereby maintaining filtration efficiency. This solvent vapour treatment also rendered the fibres smoother, which extended the time before the filter became fouled by 38% [135]. It is plausible that smoother fibres could also be beneficial in nanofibre food packaging by reducing the surface area available for microbial colonisation.

Photo-cross-linking of films electrospun from a blend of polyvinyl alcohol (PVA), poly(methyl methacrylate) and poly(butyl acrylate) resulted in a 31% increase in tensile strength and significantly improved resistance to both water and heat [136]. Furthermore, the application of a thermal treatment at 170°C for 10 minutes to PVA nanofibre food packaging film, which incorporated antimicrobial essential oils, markedly enhanced its water resistance and thermal stability [54]. The annealing process of nanofibre films can be extended until the nanofibres merge into continuous films [27].

SYNTHETIC POLYMERS FOR NANOFIBROUS FILMS

Polylactide

PLA, also known as poly(lactic acid) and polylactic acid, is a thermoplastic polyester derived from lactic acid or lactide. These precursors are produced through the fermentation of starch. PLA has a glass transition temperature of 60°C–65°C and a melting point ranging from 130°C to 180°C. This polymer is hydrophilic, with good flexibility and resistance to abrasion. However, it is expensive and mechanically weak in comparison to most of the other polymers used for packaging. PLA is compostable under industrial composting conditions, making it one of the most biodegradable of the polymers reviewed here [137]. However, it is only poorly biodegradable in landfills, soils and aquatic environments [138, 139].

PLA is an effective polymer for the microencapsulation of highly water-soluble antimicrobials for various applications [140]. The mechanical properties of a PLA nanofibrous film, incorporating smart and active agents, were enhanced through cross-linking with tannic acid and the inclusion of cellulose nanocrystals [93]. This study demonstrated that the film biodegraded in soil, releasing nutrients that promoted the growth of plant roots.

PLA was utilised to fabricate an enzyme-responsive packaging film. The PLA was spun using a protocol that rendered it highly porous. Subsequently, it underwent a post-spinning treatment with polyethylenimine, followed by the application of pectin infused with thymol [94]. The pectin was employed so that it would be hydrolysed by the pectinase released by any microbes contaminating the food and thus release the thymol to counter the microbes. The film demonstrated significant activity against bacteria and fungi and prolonged the shelf life of mandarin oranges.

A PLA nanofibrous film encapsulating 1-methylcyclopropene has been shown to retard the ripening process of bananas [37]. PLA nanofibre films prepared by SBS, devoid of active agents, have been utilised as porous sachets to contain active substances that extend the shelf life of tomatoes [36]. These active substances were potassium permanganate, which decomposes ethylene, and a proprietary humidity absorbent.

Polycaprolactone

This synthetic polyester is hydrophobic, with a relatively low melting point, ranging from 59°C to 64°C, and a glass transition temperature of approximately −60°C. It biodegrades more rapidly than most synthetic polymers, although at a slower rate than biopolymers [84]. This polymer has been extensively studied for its applications in nanofibre active and smart packaging, owing to its excellent spinnability, good mechanical properties, high biocompatibility and substantial capacity for being loaded with smart or active agents [85]. A representative selection of studies is included in Tables 8.1 and 8.2.

A PCL nanofibre film incorporating β-carotene was prepared on a solid starch film and subsequently annealed to create an antioxidant composite suitable for food packaging applications. The addition of the PCL film resulted in an approximately 85% improvement in oxygen and water barrier properties. However, the annealing process markedly diminished the antioxidant efficacy of the β-carotene [141].

Most food packaging is designed to extend the shelf life of food. However, for climacteric fruits that undergo post-harvest ripening, such as bananas, it is desirable to accelerate ripening initially and subsequently slow it to extend shelf life. To achieve this, a PCL nanofibre film that incorporated ethylene (encapsulated in cyclodextrin–potassium metal–organic frameworks) and photocatalytic titanium dioxide nanoparticles was prepared and used to package bananas [22]. It was observed that the release of ethylene was influenced by humidity and temperature. At 90% relative humidity and 20°C, approximately 90% of the ethylene was released from the film over three days. When exposed to UV light, the titanium dioxide decomposed ethylene into carbon dioxide, thereby prolonging the shelf life of the now ripe bananas.

PCL nanofibre films have been fabricated using SBS, incorporating anthocyanins derived from eggplant and purple cabbage and utilised to monitor the quality of stored fish fillets [83]. Additionally, anthocyanins have been employed in films spun from blends of PCL and PEO [91, 142]. The PCL component imparts moisture resistance to the film, while the inclusion of PEO is essential to render the films wettable, thereby facilitating interaction with packaged food. The film exhibited a colour change in response to variations in pH, trimethylamine levels and microbial counts, effectively indicating the freshness of shrimp. Furthermore, PCL has been combined with thymol encapsulated in covalent organic frameworks and processed into a film via SBS. The release of the thymol antimicrobial agent is temperature-sensitive, occurring at twice the rate at 37°C (the optimal temperature for the proliferation of foodborne pathogens such as *Escherichia coli*) compared to 4°C [85].

The natural fatty acids, linoleic acid and oleic acid, were incorporated into PCL nanofibres at concentrations ranging from 10% to 30% by weight of the polymer. This incorporation aimed to utilise the phase change properties of these fatty acids as a means of thermal buffering for frozen or chilled products, thereby aiding in the maintenance of the cold chain [84]. However, it was observed that the formation of hydrogen bonds between the fatty acids and the PCL matrix disrupted the freezing and melting processes of the acids. This disruption nullified their effectiveness as thermal buffers. Nonetheless, the hydrogen bonding significantly reduced the migration of the fatty acids out of the films, with less than 2% migration observed after 14 days in a food simulant.

Polyvinyl Alcohol

PVA, also known as poly(vinyl alcohol), is a water-soluble polymer characterised by a glass transition temperature of approximately 85°C and decomposition above 200°C. PVA exhibits excellent fibre and film-forming properties, and the films are notable for their high flexibility, biocompatibility, chemical resistance and barrier properties against oxygen and aromas. Due to its hydrophilic nature, many of the properties of PVA films are highly dependent on moisture content and ambient humidity.

PVA's water solubility renders it unsuitable for certain types of food packaging unless its solubility is reduced through methods such as thermal or chemical cross-linking, blending with hydrophobic polymers or copolymerising with hydrophobic monomers [50, 54]. Water-resistant nanofibrous films have been synthesised by combining aqueous dispersions of poly(methyl methacrylate), poly(butyl acrylate) and PVA, followed by spinning and photo-cross-linking [136]. PVA was selected due to its high water-solubility. This preparation method, which utilised aqueous spinning and residual photo-cross-linker from the polymerisation of methyl methacrylate and butyl acrylate, was considered by the researchers to be a relatively sustainable approach.

PVA nanofibrous films have been prepared which incorporated various active agents, including essential oils, photocatalysts and bacteriophages, and are employed in the packaging of various foods, see Table 8.1 [54–58]. Several smart materials have also been integrated into PVA films, see Table 8.2. For example, a PVA film was spun that incorporated two smart components: black carrot anthocyanins and tin oxide. This film was able to monitor the freshness of stored smoked fish by visually indicating the presence of compounds such as ammonia, dimethylamine and trimethylamine. Although the film was not cross-linked, its lack of direct contact with the fish mitigated any potential sensitivity to water [101]. A PVA nanofibrous film has been developed with an ammonia detection limit of less than 20 parts per million and a response time of approximately one minute [102]. PVA was selected, in part, due to its high transparency which is important for packaging intended to show a visual response. The film was rendered moisture-resistant by exposure to glutaraldehyde and trifluoroacetic acid vapours. The film's response, marked by a colour change, was facilitated by the inclusion of curcumin and gardenia blue natural pigment. This visual change was interpreted by software on a smartphone.

Polyethylene Oxide

PEO, also referred to as polyethylene glycol or polyoxyethylene depending on its molecular weight, is a hydrophilic, biocompatible and biodegradable polymer [91, 142, 143]. Its good spinnability has led to it being employed to assist the spinning of polymers that are otherwise challenging to spin [91, 142–144]. PEO was selected for a comparative study of direct current and alternating current ES, because it can be electrospun at moderate potentials [118]. Additionally, PEO has been utilised in conjunction with active and smart additives, see Tables 8.1 and 8.2.

T4 bacteriophage, a virus that specifically infects *E. coli* bacteria, maintained full viability in PEO nanofibres during storage for several weeks at 4°C [31]. A blend of PEO and PCL was developed to produce a smart nanofibrous film with the optimum hydrophilicity for interaction with food and retention of incorporated anthocyanins [91]. A post-spinning treatment with cold plasma enhanced the antimicrobial activity of a PEO–tea tree oil nanofibre film by improving the release of the active compounds [33].

Polyvinyl Pyrrolidone

PVP, also known as polyvidone and povidone, is a water-soluble polymer synthesised from the monomer N-vinylpyrrolidone, with a glass transition temperature around 90°C [81]. PVP is available in a variety of molecular weights and corresponding viscosities, allowing for selection based on specific application requirements. Its water solubility is advantageous for applications such as drug delivery [81]. However, water solubility can be reduced through cross-linking or blending with other polymers. For instance, thermal cross-linking [80] was employed to create a sandwich structure consisting of water-resistant outer layers, PU nanofibres and an internal layer of PVP nanofibres containing the active compound, chlorogenic acid [53]. Additionally, the blending of PVP with poly(vinyl butyral) and incorporating the antimicrobial agent allicin, resulted in a film with strong water absorption properties, which extended the shelf life of chicken breasts stored at room temperature [82].

Polystyrene

This hydrophobic polymer is characterised by its hardness and brittleness. The glass transition temperature of the most common form, atactic polystyrene, is approximately 90°C, while its melting point is around 240°C [51, 109, 145]. It exhibits poor barrier properties against air and water vapour. It has been employed with active and smart additives in a limited number of studies, as detailed in Tables 8.1 and 8.2. For example, a study was conducted to compare the efficacy of different solvents, specifically toluene and orange oil (comprising 97% D-limonene), in the SBS of PS [51]. It was observed that residues of the orange oil acted as plasticisers for the PS, imparting flexibility and antimicrobial properties. Additionally, the hydrophobic nature of PS nanofibre films has been exploited to minimise the leaching of oxygen indicators from PVA nanofibre films [109].

Certain proteins exhibit considerable potential as active agents in food packaging, thereby prompting researchers to explore their incorporation into nanofibrous films and the engineering of their release kinetics for specific requirements. For instance, the protein bovine serum albumin has been utilised as a model protein for integration into PS fibres [145]. By ES, an emulsion of PS and the protein, a core–shell structure was produced. As the protein resided within the core of the fibres, an initial burst release was circumvented, and the release was sustained for a period of 50 days. Furthermore, it was shown that the release rate could be modulated by selecting the molecular weight of the PS.

Other Polymers

The efficacy and stability of water-soluble smart agents within hydrophilic polymers present significant challenges, as these agents tend to leach out of the polymer under the very conditions required for interaction with food. A promising mitigation strategy is the utilisation of hydrophobic polymers. For instance, polyvinylidene fluoride was electrospun with cinnamon essential oil and anthocyanins, resulting in a composite material that not only indicated the freshness of pork but also extended its shelf life [110].

Ethylene vinyl alcohol is a copolymer of ethylene and vinyl alcohol, which exhibits good barrier properties against gases, odours and fragrances. However, under conditions of high humidity, it becomes more permeable to gases. This characteristic can be advantageous for the storage of packaged food, as it facilitates the release of volatile active compounds from the packaging into the food when the humidity is high [20]. Coaxial ES has been used to fabricate nanofibre films with an ethylene vinyl alcohol shell and a core of the same polymer, incorporating the volatile compound cuminaldehyde. These films extended the shelf life of minced chicken meat [20].

PU has been utilised in a limited number of studies that are pertinent to this review, typically in conjunction with other polymers. For example, a blend of PU and PVP produced nanofibre films exhibiting superior tensile properties and increased hydrophobicity compared to the corresponding PVP films [52]. These fibres incorporated natamycin and ascorbic acid, thereby enhancing the preservation of freshly cut apples. Additionally, nanofibre films composed of PU and PVP were integrated into a sandwich composite, featuring outer layers of hydrophobic PU and an inner layer of PVP containing chlorogenic acid as an antimicrobial agent. This composite film demonstrated high mechanical strength and a sustained release of chlorogenic acid over a period of 144 hours [53].

Polyacrylonitrile is another polymer which has only been used in a limited number of studies on nanofibre food packaging films. For instance, it was impregnated with a titanium dioxide photocatalyst, which, under ultraviolet light, decomposed ethylene, thereby delaying the ripening of bananas and tomatoes [48]. Carbon nanofibres containing alumina nanoparticles loaded with potassium permanganate have been prepared which decomposed ethylene and enhanced the storage longevity of avocados. These fibres were synthesised by carbonising electrospun polyacrylonitrile fibres [146].

A blend of polyvinyl chloride and polydimethylsiloxane containing nanodiamonds has been electrospun. The nanodiamonds offer a compelling approach to improving a crucial yet often overlooked aspect of food packaging: the reduction of bacterial adhesion to prevent cross-contamination. The best performing film reduced the adhesion of *Salmonella* and *Listeria* by 99.8% and 99.6%, respectively, compared to simple polyvinyl chloride nanofibrous films [79].

ACTIVE AND SMART ADDITIVES

Incorporation and Functionalisation Methods

The most prevalent method for incorporating active and smart agents into films involves directly adding them to the polymer solution or melt prior to spinning. To enhance compatibility between the agents and the polymer, as well as among different polymers. Surfactants can be employed to assist with emulsification [147]. These emulsions can be tailored to produce core–shell nanofibres, where the active material is concentrated in the core of the nanofibre, allowing for sustained release. Core–shell fibres can be produced using a simple ES configuration with a standard needle [145]. Alternatively, coaxial spinnerets, fed by different solutions or melts, can also be employed [20]. Active agents can be encapsulated before being combined with the solution or melt. For instance, bioactive anthocyanidins and cinnamaldehyde have been encapsulated in cyclodextrin and subsequently added to a solution of PLA before electrospinning [93].

Active Agents

A wide variety of active agents were utilised in the studies listed in Table 8.1. Where an agent is smart as well as active, or a combination of smart and active agents was employed, they are included in Table 8.2. Microorganisms are the primary agents responsible for food spoilage and foodborne illnesses; therefore, controlling them is a critical function of food packaging, and antimicrobial activity is one of the most extensively researched aspects of active packaging. Various types of microorganisms, including bacteria, fungi and yeasts, must be controlled. Antioxidants contribute to food preservation by neutralising free radicals that would otherwise degrade the food. Many antioxidants also exhibit antimicrobial properties.

Since active materials can migrate from packaging into food [49], their toxicity is a critical consideration, significantly influencing the selection of materials. A balance must be struck between efficacy and safety. This is particularly challenging with antimicrobial compounds, as for these compounds to be effective, they must eliminate or inhibit the proliferation of living organisms, specifically microbes. Naturally occurring organic molecules, such as plant extracts, are the most extensively studied antimicrobials and antioxidants for packaging applications, see Table 8.1. These compounds function by disrupting the membranes of bacteria, fungi and yeasts [148]. Antimicrobials with surfactant properties, such as ethyl lauroyl arginate and saponins, may prove particularly useful due to their ability to emulsify incompatible polymers and active substances. These properties are currently being explored in the development of nanofibre packaging [62, 149, 150]. A notable advantage of certain antifungal agents (anticratoxigens) is their ability to reduce the production of the mycotoxin ochratoxin A through multiple mechanisms. These mechanisms extend beyond simply inhibiting the fungi responsible for its production and include reducing the synthesis of ochratoxin A, suppressing its release and decomposing it into less toxic compounds.

Nanoparticles of silver, zinc oxide and diamond antimicrobials have been utilised in nanofibrous films, and several examples are included in Table 8.1. Certain nanoparticles can eliminate or deter insects by disrupting their waterproof cuticle and gut microbiome. The ability to repel insects has obvious benefits for food packaging [151–153]. Photocatalysts are a class of active agents with multifaceted effects. The light-induced hydroxyl radicals and reactive oxygen species they generate exhibit strong antimicrobial properties and can also decompose ethylene. The most well-known photocatalyst, titanium dioxide, is non-toxic.

Bacteriophage viruses are another type of antimicrobial being investigated for packaging, as discussed in a recent review [154]. Certain bacteriophages are approved for direct application to meat to control *E. coli*, and they have been incorporated into PVA and PEO nanofibre food packaging films, where they have demonstrated antimicrobial activity [31, 55]. Enzymes are an emerging class of antimicrobials with significant potential in food packaging. To date, lysozyme and glucose oxidase have been studied in nanofibrous synthetic polymer films. Lysozyme, which is abundant in animals, specifically catalyses the hydrolysis of bacterial cell walls while having little to no effect on animal cells, thereby ensuring its safety [73]. Glucose oxidase is antimicrobial by virtue of producing hydrogen peroxide, a potent antimicrobial agent [76].

Specific activities can be employed in packaging to regulate the ripening of some fruits. For example, removing ethylene, a compound responsible for fruit ripening, can slow the ripening process. Another approach involves the release of encapsulated 1-methylcyclopropene, which blocks ethylene receptors on fruit [37], thereby delaying ripening. Conversely, active packaging can accelerate ripening by releasing encapsulated ethylene into the package [22]. Potassium permanganate has been utilised in packaging to decompose ethylene. Due to its high reactivity, it must be carefully encapsulated and is typically employed within a pouch of nanofibres. Ethylene itself has been incorporated as an active component in nanofibre films, where its release accelerates the ripening of fruit. Zeolites and silica are examples of active materials employed due to their excellent adsorption properties, which enable them to remove unwanted compounds from packaged food, such as water vapour and histamine [27, 36]. Other activities investigated in biopolymer nanofibre films, but not yet extensively studied in their synthetic counterparts, include oxygen absorption, reduction of ultraviolet light transmittance and carbon dioxide emission.

Smart and Smart/Active Agents

Smart agents can be categorised into three distinct types based on their response to changes in packaged food and the surrounding environment, as well as any additional functionalities they may possess:

- **Indication**: For example, a halochromic agent that changes colour in response to pH.
- **Indication with Activity**: For instance, an agent that is both halochromic and antimicrobial, providing a dual function of indicating pH and inhibiting microbial growth.
- **Activity in Response to Changes**: Such as a phase change material that absorbs a large amount of heat when the temperature rises to a particular value.

Examples of smart and smart/active agents that have been used in nanofibrous food packaging are given in Table 8.2. The freshness of packaged food can be indicated by changes in the appearance of smart packaging associated with pH levels, microbial counts, ammonia, amines, ethylene and oxygen. Additionally, monitoring time–temperature history provides valuable information. These indicators can inform food processors, consumers and retailers about the current state of food freshness. Anthocyanins are the most widely used indicators of pH because they occur naturally in a variety of fruits and vegetables, are edible and have some antimicrobial activity. Electronic components with sensing, data storage and communication capabilities can be printed onto packaging to provide traceability, which is particularly important for food traveling along international supply chains [95]. Of course, it is possible to add multiple compounds to a nanofibre film, subject to compatibility with each other, the polymers and the spinning method. For instance, halochromic Bromocresol Purple and the antioxidant naringin have been combined in PLA nanofibres [155].

Packaging that contains phase change materials exemplifies the concept of sensing and responding to conditions, by undergoing a phase transition at a particular temperature, thereby exerting a temperature-buffering effect. Dodecane and linoleic acid have been incorporated into nanofibres due to their melting points, which are suitable for buffering at approximately −10°C. The melting of these materials requires a substantial amount of energy (~200 kJ/kg), slowing undesirable warming in some circumstances. These materials have been added to polymer solutions before spinning, at concentrations of up to 20% of the weight of the polymer [84, 88].

The release of antimicrobial agents from nanofibre films onto food has been engineered to respond to specific factors like temperature, humidity and the presence of microbes. For instance, pectin, which is susceptible to hydrolysis by microbial pectinase, was utilised in nanofibres loaded with the antimicrobial agent, thymol [94]. Encapsulating the thymol within a covalent organic framework and subsequently integrating it into a nanofibre film meant that the thymol was only released in warm conditions. These conditions coincide with the highest risk of microbial proliferation [85].

Carbon materials are starting to attract attention in food packaging and may prove to be important in the development of smart and active packaging. For instance, a carbon nanofibre biosensor has demonstrated the capability of swiftly detecting adulterants, such as pork, in non-pork meat products [156]. Carbon quantum dots are particularly noteworthy due to their multifaceted functionalities, including antioxidant, antimicrobial and UV light-blocking properties, in addition to their biocompatibility. Their application in food safety sensing has recently been reviewed [157]. Carbon quantum dots have been incorporated into nanofibres fabricated from polyvinylidene fluoride, as well as from a blend of polyvinyl alcohol and soy protein, to detect amines and assess the freshness of packaged meat [112, 158].

ECOLOGICAL CONSIDERATIONS

The negative perceptions of plastic packaging are largely due to its high visibility in terrestrial and marine environments [159]. Plastic's high tensile strength allows for the use of very lightweight, thin films in food packaging. These films are disproportionately visible relative to their mass compared to denser and thicker materials like glass and metals. However, these negative perceptions are often unsubstantiated when a comprehensive analysis is conducted, taking into account all relevant ecological factors [160].

The entire lifecycle of food, including cultivation, transportation, storage and processing, exerts a substantial ecological impact. Packaging plays an indispensable role in minimising food wastage, thereby conferring ecological benefits that must be weighed against its negative impacts. Innovations in packaging, like integrating smart and active agents, help reduce food wastage and consequently reduce the ecological footprint of food production.

Packaging is the largest application of plastic, with a total usage of 143 million tonnes in 2019 [161]. Consequently, the production and disposal of packaging materials have a significant and escalating ecological impact. Environmental concerns are fostering interest in food packaging derived from biopolymers, such as polysaccharides and proteins, in both conventional and nanofibrous forms. This includes the incorporation of natural, smart and active agents [10]. However, biopolymer-based packaging faces substantial challenges in matching the economy and performance of synthetic polymer packaging, particularly in terms of sensitivity to humidity and moisture, and availability in large quantities.

A comprehensive review has been published on life cycle assessment studies of various packaging materials. It concluded that synthetic polymer packaging did not generally exhibit a greater environmental impact than alternative materials [160]. When comparing biopolymers to synthetic polymers, biopolymers were rated superior in terms of climate change mitigation and fossil resource depletion, although no significant differences were observed in other impact categories. The substantial weight of glass packaging negatively influenced its environmental performance compared to the much lighter plastics. However, the potential for reusing glass significantly reduced its impact, though not always sufficiently to outperform single-use plastics. Plastics and metals were rated similarly overall, with plastics being the most favourable for food packaging and aluminium excelling in beverage packaging. Paper was generally preferred over plastic. The relatively favourable rating of plastics was partly due to their light weight, and their environmental profile could be significantly improved through recycling.

Long-standing concerns about microplastics' effects on both ecological systems and human health are now well established in the public consciousness [162]. Microplastics originate from a multitude of sources, including packaging, vehicle tyres, the production and laundering of synthetic polymer textiles and the degradation of plastics in the environment [163, 164]. Concerningly, microplastics can migrate from packaging into foods [165].

Most food packaging manufactured from synthetic polymers exhibits significant resistance to biodegradation in both terrestrial and aquatic environments, in contrast to their biopolymer counterparts [166–168]. The success of PLA can be attributed, in part, to its superior biodegradability compared to other common synthetic polymers, as it is compostable under industrial composting conditions [137]. However, PLA degrades at a markedly slower rate in domestic composting systems, landfills, soils and marine environments [138, 139]. When synthetic polymers undergo biodegradation in industrial composting, anaerobic digestion and uncontrolled degradation in the environment, they can generate undesirable microplastics before being fully mineralised into compounds such as carbon dioxide, water, methane, nitrates and sulphates [137]. Recent work with PCL films containing sage extract prepared by ES and annealing to fuse the nanofibres together showed good biodegradability in the soil [169]. Complete biodegradation occurred within three months, and this could be reduced to four weeks by inoculating the soil with *Pseudomonas aeruginosa*.

Concerns have been raised regarding the leaching of chemical additives and residues from food packaging and other plastic products into ecosystems and foodstuffs. These chemicals include plasticisers and residual monomers [170–173]. Such concerns have led to the implementation of legislation. For instance, the European Union mandates that the chemical concentration in food must not exceed 60 mg per kg of food or 10 mg per square decimetre of packaging area [174]. Consequently, researchers developing new packaging materials must consider the migration of these substances into food and the toxicity of any smart and active agents used. In response, researchers are investigating more ecologically sustainable additives. For example, l-lactide oligomer, PEG and epoxidised soybean oil have been evaluated as plasticisers for PLA, with the latter significantly reducing the glass transition temperature and increasing extensibility [175].

There are a range of approaches that can be taken to reduce the ecological impact of synthetic polymer packaging:

1. **Utilisation of Renewable Sources**: For instance, PLA is synthesised from monomers derived through the fermentation of starch. Additionally, blending synthetic polymers with biopolymers can enhance sustainability.
2. **Enhancement of Packaging Performance**: Advanced packaging technologies can extend the shelf life of food products, thereby minimising waste.
3. **Implementation of Down-Cycling and Recycling**: Effective down-cycling and recycling processes help in repurposing materials, reducing the demand for virgin polymer and reducing the amount of packaging that must be disposed of.
4. **Promotion of Biodegradability**:
   - **Nanofibrous Films**: The high surface area of nanofibrous films is anticipated to accelerate biodegradation compared to their non-nanofibrous counterparts.
   - **Microbial Susceptibility**: Engineering packaging materials to be more susceptible to microbial degradation in composting systems, soils and marine environments can enhance their biodegradability.
   - **Microplastic Prevention**: Designing polymers to avoid the formation of microplastics during biodegradation will reduce environmental contamination.

FUTURE PROSPECTS AND CONCLUSIONS

To attempt to make predictions about future research directions, it is prudent to first assess the current packaging materials research landscape, as outlined in the preceding text, and to then consider the factors that may precipitate change. The demand for enhanced performance to further improve packaging's role in extending the shelf life of food and reducing wastage will persist indefinitely. Additional drivers of change are summarised in Figure 8.7; however, these are too numerous and diverse to be reviewed here. Therefore, only a select few, which are particularly insightful, will be discussed.

Technology Innovation

- Smart phones
- AI-assisted research

Materials innovations

- Polymers
- Active agents
- Smart agents

Advances in other applications

- Medical
- Energy storage
- Filtration

Reducing environmental impact

- Enhanced biodegradability
- Reduction of microplastics
- Downcycling and recycling

Changing consumer needs

- Single-serve and convenience
- Edible packaging
- Microwaveable

FIGURE 8.7 Drivers of innovation in synthetic polymer nanofibre-based food packaging.

In the future, a diverse array of polymers, which have hitherto received minimal or no attention, are likely to be investigated for their potential applications in nanofibrous packaging. One such polymer is poly(ethylene adipate), notable for its low cost and excellent biodegradability [176]. To date, poly(ethylene adipate) has been utilised in a single nanofibre study. In this instance, it was blended with PLA and subjected to electrospinning, with the objective of enhancing both the biodegradability and crystallinity compared to PLA alone [177].

The excellent physical properties of synthetic polymers will make them the first choice in the development of mycelium composites [178]. Currently, limited research has been conducted on the utilisation of enzymes and bacteriophages in nanofibrous packaging, with even less focus on peptides such as LfcinB (21-25) Pal [85, 179]. Nevertheless, their highly specific activities and low toxicities are poised to make them the subjects of intense study in the future. Progress in the incorporation of active and smart agents into films via encapsulation in nanoliposomes, nano-niosomes, solid lipid nanoparticles and nano-micelles is likely to expand the range of actives that can be utilised, by overcoming limitations of stability and solubility [180].

Future research in smart packaging is expected to place greater emphasis on the human-packaging interface, as well as materials development. For example, indicating the freshness of food by a colour change is not useful for individuals with colour blindness. Researchers are addressing this issue by investigating systems that display changes in contrast rather than colour [181]. Smartphones can assist with the interpretation of the appearance of smart packaging. For instance, a combination of PVA nanofibre smart packaging film and smartphone software has been developed that can identify amines [70]. Ionic liquids are only starting to receive attention in advanced packaging and have potential as plasticisers. An amine-responsive fluorescent ionic liquid has been utilised in smart packaging to indicate the freshness of seafood [182]. Carbon nanomaterials, such as graphene, nanotubes and dots, have significant potential to enhance the mechanical and barrier properties of polymers and are likely to attract interest in advanced packaging [183]. Aerogels are being investigated for many applications, including additives to fibres, and their extremely high porosities make them an interesting addition to nanofibrous films [184]. The high cost of many of the smart and active agents, especially those sourced from natural sources, is a significant barrier to the broader adoption of smart active packaging [185]. This will lead to a focus on cost-effective, smart and active materials.

In the domain of spinning technology, the low production rate of nanofibrous films presents a significant constraint. This limitation is anticipated to increase interest in needleless electrospinning techniques. Furthermore, the application of alternating current as opposed to direct current is expected to garner increased attention. The direct coating of individual food items, especially fruits and vegetables, constitutes a valuable goal, which can be most effectively realised through solution blow spinning. As the spinning of nanofibrous films becomes more established and well understood, it is foreseen that more attention will be given to preparing fibres with smaller diameters, to increase their specific areas—at present, most fibres are generally between 100 and 400 nm in diameter.

In conclusion, the future of food packaging technology is set for significant advancements, driven by the need for enhanced performance and sustainability. Nanofibrous packaging, utilising innovative smart and active materials, will be at the forefront of future packaging solutions.

Abbreviations

Spinning

CS Centrifugal spinning
ES Electrospinning
MFS Microfluidic spinning
SBS Blow spinning

Polymers

PCL Polycaprolactone
PEG Polyethylene glycol
PEO Polyethylene oxide

PLA Polylactide
PS Polystyrene
PU Polyurethane
PVA Polyvinyl alcohol
PVP Polyvinyl pyrrolidone

REFERENCES

1 Chang, H., Xu, J., Macqueen, L. A., et al. (2022). High-throughput coating with biodegradable antimicrobial pullulan fibres extends shelf life and reduces weight loss in an avocado model. *Nature Food*, 3(6), 428–436. https://doi.org/10.1038/s43016-022-00523-w

2 McNeil, S. J. & Standard, O. C. (2017). Increased bending rigidity of wool fabric imparted by hybrid organic-inorganic sol-gel coatings. *Textile Research Journal*, 87(5), 607–616. https://doi.org/10.1177/0040517516635990

3 Hami, S. S. B. M., Affandi, N. D. N., Indrie, L., et al. (2023). Enhancing mechanical properties and flux of nanofibre membranes for water filtration. *Polymers*, 15, 3281. https://doi.org/10.3390/polym15153281

4 Nayak, R., Padhye, R., Kyratzis, I. L., et al. (2012). Recent advances in nanofibre fabrication techniques. *Textile Research Journal*, 82(2), 129–147. https://doi.org/10.1177/0040517511424524

5 Krifa, M., & Prichard, C. (2020). Nanotechnology in textile and apparel research: An overview of technologies and processes. *Journal of the Textile Institute*, 111, 1778–1793. https://doi.org/10.1080/00405000.2020.1721696

6 Nurfaizey, A. H., Tucker, N., Stanger, J., & Staiger, M. P. (2012). Functional nanofibers in clothing for protection against chemical and biological hazards. In Q. Wei (Edt.), *Functional Nanofibers and their Applications*, (pp. 236–261). Woodhead. https://doi.org/10.1533/9780857095640.2.236

7 Hassan, M. M., Heins, K., & Zheng, H. (2024). Wound dressing based on silver nanoparticle embedded wool keratin electrospun nanofibers deposited on cotton fabric. *ACS Applied Bio Materials*, 7, 2164–2174. https://doi.org/10.1021/acsabm.3c01111

8 Hammami, M. A., Krifa, M., & Harzallah, O. (2014). Centrifugal force spinning of PA6 nanofibers–processability and morphology of solution-spun fibers. *Journal of The Textile Institute*, 105, 637–647. https://doi.org/10.1080/00405000.2013.842680

9 Schabikowski, M., Tomaszewska, J., Kata, D., & Graule, T. (2015). Rotary jet-spinning of hematite fibers. *Textile Research Journal*, 85, 316–324. https://doi.org/10.1177/0040517514542969

10 McNeil, S. J. (in press). Smart and active biopolymer nanofibre films: The future of food packaging. In M. Montazer (Edt.), *Eco-friendly Smart Functional Wearables: From Nanofibers to Textile*, (pp. 125–148). CRC Press. *doi to be assigned*

11 Hu, X., Zhang, X., Shen, X., et al. (2014). Plasma-induced synthesis of CuO nanofibers and ZnO nanoflowers in water. *Plasma Chemistry and Plasma Processing*, 34, 1129–1139. https://doi.org/10.1007/s11090-014-9546-0

12 Nayak, R., Padhye, R., Arnold, L., et al (2012). Mechanism of nanofibre fabrication by meltblowing. *Applied Mechanics and Materials*, 217, 207–212. https://doi.org/10.4028/www.scientific.net/AMM.217-219.207

13 Li, H., Ke, Y., & Hu, Y. (2006). Polymer nanofibers prepared by template melt extrusion. *Journal of Applied Polymer Science*, 99, 1018–1023. https://doi.org/10.1002/app.22597

14 Lopez-Polo, J., Muñoz-Shugulí, C., Vidal, M. P., & Vidal, C. P. (2023). Electrospun edible films and coatings. *Trends in Food Science & Technology*, 143, 104253. https://doi.org/10.1016/j.tifs.2023.104253

15 Tucker, N., Stanger, J. J., Staiger, M. P., et al. (2012). The history of the science and technology of electrospinning from 1600 to 1995. *Journal of Engineered Fibers and Fabrics*, 7(2-suppl), 63–73. https://doi.org/10.1177/155892501200702s10

16 Stanger, J., Tucker, N., Wallace, A., et al. (2009). The effect of electrode configuration and substrate material on the mass deposition rate of electrospinning. *Journal of Applied Polymer Science*, 112, 1729–1737. https://doi.org/10.1002/app.29663

17 Mazari, A. (2014). Relation between needle and needle-less electrospinning using poly (ethylene oxide). *Vlakna a Textil*, 21(1), 21–30.

18 Nurfaizey, A. H., Stanger, J., Tucker, N., et al. (2014). Control of spatial deposition of electrospun fiber using electric field manipulation. *Journal of Engineered Fibers and Fabrics*, 9(1), 155–164. https://doi.org/10.1177/155892501400900118

19 Zhu, Z., Yu, M., Ren, R., et al. (2024). Thymol incorporation within chitosan/polyethylene oxide nanofibers by concurrent coaxial electrospinning and in-situ crosslinking from core-out for active antibacterial packaging. *Carbohydrate Polymers*, 323, 121381. https://doi.org/10.1016/j.carbpol.2023.121381

20 Li, H., Wong, S. Y., Zhang, Y., et al. (2024). Humidity-responsive antimicrobial properties of EVOH nanofibers loaded with cuminaldehyde/HβCD inclusion complexes and its application in chicken preservation. *Food Hydrocolloids*, 150, 109749. https://doi.org/10.1016/j.foodhyd.2024.109749

21 Liao, M., Pan, Y., Fu, X., et al. (2023). Electrospun polylactic acid nanofiber film modified by silver oxide deposited on hemp fibers for antibacterial fruit packaging. *International Journal of Biological Macromolecules*, 253, 126569. https://doi.org/10.1016/j.ijbiomac.2023.126569

22 Qin, Z., Jiang, Q., Chen, M., et al. (2024). Food packaging for ripening and preserving banana based on ethylene-loaded nanofiber films deposited with nanosized cyclodextrin metal-organic frameworks and TiO_2 nanoparticles. *Food Packaging and Shelf Life*, 45, 101332. https://doi.org/10.1016/j.fpsl.2024.101332

23 Huang, Y., Huang, H., Wu, J., et al. (2024). Preparation and properties of PCL coaxial electrospinning films with shell loaded with CEO and core coated LEO nanoemulsions. *Food Research International*, 194, 114817. https://doi.org/10.1016/j.foodres.2024.114817

24 Piri, H., Moradi, S., & Amiri, R. (2021). The fabrication of a novel film based on polycaprolactone incorporated with chitosan and rutin: Potential as an antibacterial carrier for rainbow trout packaging. *Food Science and Biotechnology*, *30*, 683–690. https://doi.org/10.1007/s10068-021-00898-9

25 Salević, A., Prieto, C., Cabedo, L., et al. (2019). Physicochemical, antioxidant and antimicrobial properties of electrospun poly (ϵ-caprolactone) films containing a solid dispersion of sage (*Salvia officinalis* L.) extract. *Nanomaterials*, 9, 270. https://doi.org/10.3390/nano9020270

26 Hu, S., Peng, F., Wang, C., et al. (2025). Preparation and application of polycaprolactone/β-cyclodextrin/gamma-Decalactone nanofiber composites in citrus postharvest diseases. *Food Chemistry*, 463, 141476. https://doi.org/10.1016/j.foodchem.2024.141476

27 Alp-Erbay, E., Figueroa-Lopez, K. J., Lagaron, J. M., et al. (2019). The impact of electrospun films of poly(ϵ-caprolactone) filled with nanostructured zeolite and silica microparticles on *in vitro* histamine formation by *Staphylococcus aureus* and *Salmonella Paratyphi* A. *Food Packaging and Shelf Life*, 22, 100414. https://doi.org/10.1016/j.fpsl.2019.100414

28 Nabati, S., Aminzare, M., Roohinejad, S., et al. (2023). Electrospun polycaprolactone nanofiber containing *Ganoderma lucidum* extract to improve chemical and microbial stability of rainbow trout fillets during storage at 4°C. *Food Control*, 150, 109777. https://doi.org/10.1016/j.foodcont.2023.109777

29 Cao, Y., Shen, C., Yang, Z., et al. (2022). Polycaprolactone/polyvinyl pyrrolidone nanofibers developed by solution blow spinning for encapsulation of chlorogenic acid. *Food Quality and Safety*, 6, 1–10. https://doi.org/10.1093/fqsafe/fyac014

30 Hasan, M. T., Gonzalez, R., Chipara, M., et al. (2021). Antibacterial activities of centrifugally spun polyethylene oxide/silver composite nanofibers. *Polymers for Advanced Technologies*, 32, 2327–2338. https://doi.org/10.1002/pat.5261

31 Korehei, R., & Kadla, J. (2013). Incorporation of T4 bacteriophage in electrospun fibres. *Journal of Applied Microbiology*, 114, 1425–1434. https://doi.org/10.1111/jam.12158

32 Lin, L., Dai, Y., & Cui, H. (2017). Antibacterial poly (ethylene oxide) electrospun nanofibers containing cinnamon essential oil/beta-cyclodextrin proteoliposomes. *Carbohydrate Polymers*, 178, 131–140. https://doi.org/10.1016/j.carbpol.2017.09.043

33 Cui, H., Bai, M., & Lin, L. (2018). Plasma-treated poly(ethylene oxide) nanofibers containing tea tree oil/beta-cyclodextrin inclusion complex for antibacterial packaging. *Carbohydrate Polymers*, 179, 360–369. https://doi.org/10.1016/j.carbpol.2017.10.011

34 Cui, H., Wu, J., Li, C., & Lin, L. (2017). Improving anti-listeria activity of cheese packaging via nanofiber containing nisin-loaded nanoparticles. *LWT*, 81, 233–242. https://doi.org/10.1016/j.lwt.2017.04.003

35 Solaberrieta, I., Jiménez, A., Cacciotti, I., & Garrigós, M. C. (2020). Encapsulation of bioactive compounds from aloe vera agrowastes in electrospun poly (ethylene oxide) nanofibers. *Polymers*, 12, 1323. https://doi.org/10.3390/polym12061323

36 Miranda, K. W. E., Bresolin, J. D., Natarelli, C. V. L., et al. (2021). Potential use of poly (lactic acid) nanofibers mats as nano-sachets in postharvest of climacteric fruits and vegetables. *Journal of Applied Polymer Science*, 138, 50735. https://doi.org/10.1002/app.50735

37 de Barros, H. E. A., Natarelli, C. V. L., Santos, I. A., et al. (2024). A new nano-sachet manufactured by solution blow spinning to retard ripening of bananas using biodegradable nanostructured polymeric mats emitting 1-methylcyclopropene. *Postharvest Biology and Technology*, 214, 112993. https://doi.org/10.1016/j.postharvbio.2024.112993

38 Brandão, R. M., Batista, L. R., de Oliveira, J. E., et al. (2023). In vitro and in vivo efficacy of poly(lactic acid) nanofiber packaging containing essential oils from *Ocimum basilicum* L. and *Ocimum gratissimum* L. against *Aspergillus carbonarius* and *Aspergillus niger* in table grapes. *Food Chemistry*, 400, 134087. https://doi.org/10.1016/j.foodchem.2022.134087

39 Brandão, R. M., Batista, L. R., de Oliveira, J. E., et al. (2022). Active packaging of poly (lactic acid) nanofibers and essential oils with antifungal action on table grapes. *FEMS Microbiology Letters*, 369(1), 116. https://doi.org/10.1093/femsle/fnac116

40 Ozcan, I., Saricaoglu, F. T., Parlak, M. E., et al. (2023). Characterization of solution blow spun poly(lactic) acid based nanofibers containing sucuk spice mix essential oils. *Journal of Polymers and the Environment*, 31, 2334–2346. https://doi.org/10.1007/s10924-023-02761-w

41 Shi, C., Zhou, A., Fang, D., et al. (2022). Oregano essential oil/β-cyclodextrin inclusion compound polylactic acid/polycaprolactone electrospun nanofibers for active food packaging. *Chemical Engineering Journal*, 445, 136746. https://doi.org/10.1016/j.cej.2022.136746

42 Zaitoon, A., Anguraj, V., Singh, S. S., et. al. (2023). Salicylaldehyde-functionalized polyethyleneimine precursor: Synthesis, characterization, and encapsulation in electrospun nonwoven for moisture-triggered release applications. *Chemical Engineering Journal*, 476, 146462. https://doi.org/10.1016/j.cej.2023.146462

43 Vencato, A. A., Clerici, N. J., Juchem, A. L. M., et al. (2024). Electrospun nanofibers incorporating lactobionic acid as novel active packaging materials: Biological activities and toxicological evaluation. *Discover Nano*, 19, 135. https://doi.org/10.1186/s11671-024-04084-8

44 Shi, C., Fang, D., Xia, S., et al. (2024). Poly (lactic acid)/polycaprolactone nanofibrous packaging containing different functional agents for blackberry postharvest preservation. *International Journal of Biological Macromolecules*, 279, 134544. https://doi.org/10.1016/j.ijbiomac.2024.134544

45 Mendes, J. F., Norcino, L. B., Corrêa, T. Q., et al. (2023). Obtaining poly(lactic acid) nanofibers encapsulated with peppermint essential oil as potential packaging via solution-blow-spinning. *International Journal of Biological Macromolecules*, 230, 123424. https://doi.org/10.1016/j.ijbiomac.2023.123424

46 Râpă, M., Stefan, M., Popa, P. A., et al. (2021). Electrospun nanosystems based on PHBV and ZnO for ecological food packaging. *Polymers*, 13, 2123. https://doi.org/10.3390/polym13132123

47 Mohammad, A. A., Oladzadabbasabadi, N., Kareem, H. R., et al. (2024). Investigating the impact of thymol/ɤ-cyclodextrin concentrations on the properties of electrospun PLA/PVP nanofibers for packaging applications. *Colloids and Surfaces A: Physicochemical and Engineering Aspects*, 701, 134923. https://doi.org/10.1016/j.colsurfa.2024.134923

48 Zhu, Z., Zhang, Y., Shang, Y., & Wen, Y. (2019). Electrospun nanofibers containing TiO_2 for the photocatalytic degradation of ethylene and delaying postharvest ripening of bananas. *Food and Bioprocess Technology*, 12, 281–287. https://doi.org/10.1007/s11947-018-2207-1

49 Thakur, V., & Satapathy, B. K. (2024). Migration concerns of biopolymer-based food packaging. In S. Kumar, A. Mukherjee & V. Katiyar (Eds.), *Agro-Waste Derived Biopolymers and Biocomposites: Innovations and Sustainability in Food Packaging*, (pp. 421–450). Scrivener Publishing https://doi.org/10.1002/9781394175161.ch15

50 Rzayev, Z. M., Erdönmez, D., Erkan, K., et al. (2015). Functional copolymer/organo-MMT nanoarchitectures. XXII. Fabrication and characterization of antifungal and antibacterial poly (vinyl alcohol-co-vinyl acetate)/ODA-MMT/AgNPs nanofibers and nanocoatings by e-spinning and c-spinning methods. *International Journal of Polymeric Materials and Polymeric Biomaterials*, 64(5), 267–278. https://doi.org/10.1080/00914037.2014.936599

51 Miranda, K. W. E., Mattoso, L. H. C., Bresolin, J. D., et al. (2019). Polystyrene bioactive nanofibers using orange oil as an ecofriendly solvent. *Journal of Applied Polymer Science*, 136, 47337. https://doi.org/10.1002/app.47337

52 Shen, C., Rao, J., Wu, D., & Chen, K. (2024). Multifunctional fibrous films fabricated by microfluidic blow spinning for the preservation of fresh-cut apples. *Postharvest Biology and Technology*, 211, 112807. https://doi.org/10.1016/j.postharvbio.2024.112807

53 Shen, C., Yang, Z., Rao, J., et al. (2022). Chlorogenic acid-loaded sandwich-structured nanofibrous film developed by solution blow spinning. *Food Packaging and Shelf Life*, 32, 100854. https://doi.org/10.1016/j.fpsl.2022.100854

54 Göksen, G., Fabra, M. J., Pérez-Cataluña, A., et al. (2021). Biodegradable active food packaging structures based on hybrid cross-linked electrospun polyvinyl alcohol fibers containing essential oils and their application in the preservation of chicken breast fillets. *Food Packaging and Shelf Life*, 27, 100613. https://doi.org/10.1016/j.fpsl.2020.100613

55 Costa, M. J., Pastrana, L. M., Teixeira, J. A., et al. (2021). Characterization of PHBV films loaded with FO1 bacteriophage using polyvinyl alcohol-based nanofibers and coatings. *Innovative Food Science & Emerging Technologies*, 69, 102646. https://doi.org/10.1016/j.ifset.2021.102646
56 Dudley, A., Kassama, L., Jackson-Davis, A., et al. (2023). Electrospinning of northern Alabama grown hemp extract in PVA to develop an active packaging film. In *2023 ASABE Annual International Meeting* (p. 1). American Society of Agricultural and Biological Engineers. https://doi.org/10.13031/aim.202300754
57 de Barros, H. E. A., Natarelli, C. V. L., Santos, I. A., et al. (2024). Development of poly (vinyl alcohol) nanofibers incorporated with aqueous plant extracts by solution blow spinning and their application as strawberry coatings. *Journal of Food Engineering*, 363, 111761. https://doi.org/10.1016/j.jfoodeng.2023.111761
58 Bian, X., Sun, X., Min, T., et al. (2023). Functionalized polyvinyl alcohol nanofibers with visible light-triggered antibacterial and ethylene scavenging capabilities for food packaging. *Food Packaging and Shelf Life*, 36, 101056. https://doi.org/10.1016/j.fpsl.2023.101056
59 Estevez-Areco, S., Guz, L., Candal, R., & Goyanes, S. (2018). Release kinetics of rosemary (*Rosmarinus officinalis*) polyphenols from polyvinyl alcohol (PVA) electrospun nanofibers in several food simulants. *Food Packaging and Shelf Life*, 18, 42–50. https://doi.org/10.1016/j.fpsl.2018.08.006
60 Estevez-Areco, S., Guz, L., Candal, R., & Goyanes, S. (2020). Active bilayer films based on cassava starch incorporating ZnO nanorods and PVA electrospun mats containing rosemary extract. *Food Hydrocolloids*, 108, 106054. https://doi.org/10.1016/j.foodhyd.2020.106054
61 Reboleira, J., Adão, P., Guerreiro, S. F., et al. (2020). Poultry shelf-life enhancing potential of nanofibers and nanoparticles containing *Porphyra dioica* extracts. *Coatings*, 10, 315. https://doi.org/10.3390/coatings10040315
62 Vidal, C. P., Velásquez, E., Galotto, M. J., & de Dicastillo, C. L. (2021). Antimicrobial food packaging system based on ethyl lauroyl arginate-loaded core/shell electrospun structures by using hydrophilic and hydrophobic polymers. *Polymer Testing*, 93, 106937. https://doi.org/10.1016/j.polymertesting.2020.106937
63 Shao, P., Yan, Z., Chen, H., & Xiao, J. (2018). Electrospun poly(vinyl alcohol)/permutite fibrous film loaded with cinnamaldehyde for active food packaging. *Journal of Applied Polymer Science*, 135, 46117. https://doi.org/10.1002/app.46117
64 Yan, J., Bai, T., Lu, J., et al. (2024). Pea-like inspired design: Lignin nanospheres-mediated PVA film with high strength, low water evaporation permeability, strong UV resistance and anti-oxidation properties for extending food freshness. *Chemical Engineering Journal*, 494, 152993. https://doi.org/10.1016/j.cej.2024.152993
65 Zhang, Y., Yang, L., Dong, Q., & Li, L. (2021). Fabrication of antibacterial fibrous films by electrospinning and their application for Japanese sea bass (*Lateolabrax japonicus*) preservation. *LWT*, 149, 111870. https://doi.org/10.1016/j.lwt.2021.111870
66 Yu, D., Feng, Y. Y., Xu, J. X., et al. (2021). Fabrication, characterization, and antibacterial properties of citric acid crosslinked PVA electrospun microfibre mats for active food packaging. *Packaging Technology and Science*, 34, 361–370. https://doi.org/10.1002/pts.2566
67 Zeinali, T., Alemzadeh, E., Zarban, A., et al. (2021). Fabrication and characterization of jujube extract-loaded electrospun polyvinyl alcohol nanofiber for strawberry preservation. *Food Science & Nutrition*, 9, 6353–6361. https://doi.org/10.1002/fsn3.2601
68 Yang, S. B., Kim, E. H., Kim, S. H., et al. (2018). Electrospinning fabrication of poly (vinyl alcohol)/*coptis chinensis* extract nanofibers for antimicrobial exploits. *Nanomaterials*, 8, 734. https://doi.org/10.3390/nano8090734

69 Narayanan, V., Alam, M., Ahmad, N., et al. (2021). Electrospun poly (vinyl alcohol) nanofibers incorporating caffeic acid/cyclodextrins through the supramolecular assembly for antibacterial activity. *Spectrochimica Acta Part A: Molecular and Biomolecular Spectroscopy*, 249, 119308. https://doi.org/10.1016/j.saa.2020.119308

70 He, L., Lan, W., Ahmed, S., et al. (2019). Electrospun polyvinyl alcohol film containing pomegranate peel extract and sodium dehydroacetate for use as food packaging. *Food Packaging and Shelf Life*, 22, 100390. https://doi.org/10.1016/j.fpsl. 2019.100390

71 Nazari, M., Majdi, H., Milani, M., et al. (2019). Cinnamon nanophytosomes embedded electrospun nanofiber: Its effects on microbial quality and shelf-life of shrimp as a novel packaging. *Food Packaging and Shelf Life*, 21, 100349. https://doi.org/10.1016/j.fpsl. 2019.100349

72 Zhang, H., Zhang, C., Wang, X., et al. (2022). Antifungal electrospinning nanofiber film incorporated with *Zanthoxylum bungeanum* essential oil for strawberry and sweet cherry preservation. *LWT*, 169, 113992. https://doi.org/10.1016/j.lwt.2022.113992

73 Feng, K., Wen, P., Yang, H., et al. (2017). Enhancement of the antimicrobial activity of cinnamon essential oil-loaded electrospun nanofilm by the incorporation of lysozyme. *RSC Advances*, 7, 1572–1580. https://doi.org/10.1039/C6RA25977D

74 Zhang, R., Wang, Y., Ma, D., et al. (2019). Effects of ultrasonication duration and graphene oxide and nano-zinc oxide contents on the properties of polyvinyl alcohol nanocomposites. *Ultrasonics Sonochemistry*, 59, 104731. https://doi.org/10.1016/j. ultsonch.2019.104731

75 Pan, J., Ai, F., Shao, P., et al. (2019). Development of polyvinyl alcohol/β-cyclodextrin antimicrobial nanofibers for fresh mushroom packaging. *Food Chemistry*, 300, 125249. https://doi.org/10.1016/j.foodchem.2019.125249

76 Alonso-González, M., Corral-González, A., Felix, M., et al. (2020). Developing active poly (vinyl alcohol)-based membranes with encapsulated antimicrobial enzymes via electrospinning for food packaging. *International Journal of Biological Macromolecules*, 162, 913–921. https://doi.org/10.1016/j.ijbiomac.2020.06.217

77 Arrieta, M. P., de Dicastillo, C. L., Garrido, L., et al. (2018). Electrospun PVA fibers loaded with antioxidant fillers extracted from *Durvillaea antarctica* algae and their effect on plasticized PLA bionanocomposites. *European Polymer Journal*, 103, 145–157. https://doi.org/10.1016/j.eurpolymj.2018.04.012

78 Cheng, J., Li, D., Mai, Z., et al. (2024). In-situ electrospinning PVB/Camellia oil/ZnO-TiO2 nanofibrous membranes with synergistic antibacterial and degradation of ethylene applied in fruit preservation. *Food Chemistry*, 460, 140629. https://doi.org/10.1016/ j.foodchem.2024.140629

79 Liu, S., Iepure, M., Zhou, W., et al. (2024). Electrospun superhydrophobic polyvinyl chloride/polydimethylsiloxane-nanodiamond nanocomposite with enhanced antifouling and mechanical properties for fresh produce packaging. *Applied Food Research*, 4(1), 100417. https://doi.org/10.1016/j.afres.2024.100417

80 Doğan, C., Doğan, N., Gungor, M., et al. (2022). Novel active food packaging based on centrifugally spun nanofibers containing lavender essential oil. *Food Packaging and Shelf Life*, 34, 100942. https://doi.org/10.1016/j.fpsl.2022.100942

81 Hasan, M. T., Gonzalez, R., Munoz, A. A., et al. (2022). Forcespun polyvinylpyrrolidone/copper and polyethylene oxide/copper composite fibers and their use as antibacterial agents. *Journal of Applied Polymer Science*, 139, 51773. https://doi.org/10.1002/ app.51773

82 Song, Y., Zhang, H., Huang, H., et al. (2022). Allicin-loaded electrospun PVP/PVB nanofibrous films with superior water absorption and water stability for antimicrobial food packaging. *ACS Food Science & Technology*, 2(5), 941–950. https://doi.org/ 10.1021/acsfoodscitech.2c00080

83 de Oliveira Filho, J. G., de Oliveira Silva, C., do Canto, R. A., et al. (2024). Fast and sustainable production of smart nanofibers mats by solution blow spinning for food quality monitoring. *Food Chemistry*, 140057. https://doi.org/10.1016/j.foodchem.2024. 140057
84 Drago, E., Campardelli, R., Barbucci, A., & Perego, P. (2023). Polycaprolactone submicrometric fibers optimization for primary packaging loaded with fatty acids as natural phase change materials. *Journal of Food Engineering*, 358, 111680. https://doi.org/10.1016/j.jfoodeng.2023.111680
85 Shen, C., Ma, Y., Wu, D., et al. (2023). Preparation of covalent organic framework-based nanofibrous films with temperature-responsive release of thymol for active food packaging. *Food Chemistry*, 410, 135460. https://doi.org/10.1016/j.foodchem.2023.135460
86 Alavijeh, D. K., Heli, B., & Ajji, A. (2024). Development of a sensitive colorimetric indicator for detecting beef spoilage in smart packaging. *Sensors*, 24, 3939. https://doi.org/10.3390/s24123939
87 Kim, S. W., Kim, E. S., Park, B. J., et al. (2023). Polycaprolactone/anthocyanin-based electrospun volatile amines gas indicator with improved visibility by varying bi-solvent ratio: A case of intelligent packaging of mackerel. *Foods*, 12, 3850. https://doi.org/10.3390/foods12203850
88 Perez-Masia, R., López-Rubio, A., Fabra, M. J., & Lagaron, J. M. (2013). Biodegradable polyester-based heat management materials of interest in refrigeration and smart packaging coatings. *Journal of Applied Polymer Science*, *130*, 3251–3262. https://doi.org/10.1002/app.39555
89 Chalco-Sandoval, W., Fabra, M. J., López-Rubio, A., & Lagaron, J. M. (2015). Development of polystyrene-based films with temperature buffering capacity for smart food packaging. *Journal of Food Engineering*, 164, 55–62. https://doi.org/10.1016/j.jfoodeng.2015.04.032
90 Pérez-Masiá, R., López-Rubio, A., Fabra, M. J., & Lagaron, J. M. (2014). Use of electrohydrodynamic processing to develop nanostructured materials for the preservation of the cold chain. *Innovative Food Science & Emerging Technologies*, 26, 415–423. https://doi.org/10.1016/j.ifset.2014.10.010
91 Jovanska, L., Chiu, C. H., Yeh, Y. C., et al. (2022). Development of a PCL-PEO double network colorimetric pH sensor using electrospun fibers containing *Hibiscus rosa sinensis* extract and silver nanoparticles for food monitoring. *Food Chemistry*, 368, 130813. https://doi.org/10.1016/j.foodchem.2021.130813
92 Terra, A. L. M., Moreira, J. B., Costa, J. A. V., et al. (2021). Development of pH indicators from nanofibers containing microalgal pigment for monitoring of food quality. *Food Bioscience*, 44, 101387. https://doi.org/10.1016/j.fbio.2021.101387
93 Guan, Y., Li, F., Wang, Y., et al. (2024). "Reservoir-law" synergistic reinforcement of electrostatic spun polylactic acid composites with cellulose nanocrystals and 2-hydroxypropyl-β-cyclodextrin for intelligent bioactive food packaging. *International Journal of Biological Macromolecules*, 274, 133405. https://doi.org/10.1016/j.ijbiomac.2024.133405
94 Min, T., Zhou, L., Sun, X., et al. (2022). Enzyme-responsive food packaging system based on pectin-coated poly (lactic acid) nanofiber films for controlled release of thymol. *Food Research International*, 157, 111256. https://doi.org/10.1016/j.foodres.2022.111256
95 Ranford, S. L. (2009). Traceability of packaging and packaged products. In *Proceedings of the 63rd Appita Annual Conference*, Melbourne, Australia (pp 221–228). https://search.informit.org/doi/10.3316/informit.734039623274159
96 Li, T., Wang, D., Ren, L., et al. (2021). Preparation of pH-sensitive polylactic acid-naringin coaxial electrospun fiber membranes for maintaining and monitoring salmon freshness. *International Journal of Biological Macromolecules*, 188, 708–718. https://doi.org/10.1016/j.ijbiomac.2021.08.087

97 Moreira, J. B., Terra, A. L. M., Costa, J. A. V., & de Morais, M. G. (2018). Development of pH indicator from PLA/PEO ultrafine fibers containing pigment of microalgae origin. *International Journal of Biological Macromolecules*, 118, 1855–1862. https://doi.org/10.1016/j.ijbiomac.2018.07.028
98 Kuntzler, S. G., Costa, J. A. V., Brizio, A. P. D. R., & de Morais, M. G. (2020). Development of a colorimetric pH indicator using nanofibers containing Spirulina sp. LEB 18. *Food Chemistry*, 328, 126768. https://doi.org/10.1016/j.foodchem.2020.126768
99 Min, T., Sun, X., Yuan, Z., et al. (2021). Novel antimicrobial packaging film based on porous poly (lactic acid) nanofiber and polymeric coating for humidity-controlled release of thyme essential oil. *LWT*, 135, 110034. https://doi.org/10.1016/j.lwt.2020.110034
100 Zhang, M., Peng, X., Fan, P. et al. (2022). Recent progress in preparation and application of fibers using microfluidic spinning technology. *Macromolecular Chemistry and Physics*, 223, 2100451. https://doi.org/10.1002/macp.202100451
101 Cetinkaya, T. (2024). Evaluation of intelligent packaging functions of black carrot extract-infused polyvinyl alcohol nanofibers. *Applied Surface Science Advances*, 19, 100571. https://doi.org/10.1016/j.apsadv.2023.100571
102 Luo, M., Liu, J., Zhang, Y., et al. (2024). Amine response smartphone-based portable and intelligent polyvinyl alcohol films for real-time detection of shrimp freshness. *Food Chemistry*, 450, 139347. https://doi.org/10.1016/j.foodchem.2024.139347
103 Forghani, S., Zeynali, F., Almasi, H., & Hamishehkar, H. (2023). Accurate monitoring of shrimp freshness via cornflower anthocyanins-loaded bio-based nanofiber mats and casted films. *Journal of Polymers and the Environment*, 31, 4258–4273. https://doi.org/10.1007/s10924-023-02867-1
104 Thakur, V., & Satapathy, B. K. (2023). Electrospun nanofibrous mat-on-film bilayered halochromic freshness marker-enabled intelligent food packaging film based on red cabbage anthocyanins. *ACS Food Science & Technology*, 3, 2094–2107. https://doi.org/10.1021/acsfoodscitech.3c00352
105 Maftoonazad, N., & Ramaswamy, H. (2019). Design and testing of an electrospun nanofiber mat as a pH biosensor and monitor the pH associated quality in fresh date fruit (*Rutab*). *Polymer Testing*, 75, 76–84. https://doi.org/10.1016/j.polymertesting.2019.01.011
106 Li, L., Li, W., Sun, W., et al. (2024). A visual ratio PVA/curcumin/indigo nanofibre membrane based on natural pigments for detecting mutton freshness. *International Journal of Food Science & Technology*, 59, 5742–5753. https://doi.org/10.1111/ijfs.17313
107 Jebel, F. S., Roufegarinejad, L., Alizadeh, A., & Amjadi, S. (2025). Development and characterization of a double-layer smart packaging system consisting of polyvinyl alcohol electrospun nanofibers and gelatin film for fish fillet. *Food Chemistry*, 462, 140985. https://doi.org/10.1016/j.foodchem.2024.140985
108 Zhang, W., An, H., Sun, X., et al. (2023). Dual-functional smart indicator for direct monitoring fruit freshness. *Food Packaging and Shelf Life*, 40, 101192. https://doi.org/10.1016/j.fpsl.2023.101192
109 Yılmaz, M., & Altan, A. (2021). Optimization of functionalized electrospun fibers for the development of colorimetric oxygen indicator as an intelligent food packaging system. *Food Packaging and Shelf Life*, 28, 100651. https://doi.org/10.1016/j.fpsl.2021.100651
110 Zhang, J., Huang, X., Zhang, J., et al. (2022). Development of nanofiber indicator with high sensitivity for pork preservation and freshness monitoring. *Food Chemistry*, 381, 132224. https://doi.org/10.1016/j.foodchem.2022.132224
111 Zhai, X., Xue, Y., Song, W., et al. (2024). A ratiometric fluorescent electrospun film with high amine sensitivity and stability for visual monitoring of livestock meat freshness. *Food Chemistry: X*, 101801. https://doi.org/10.1016/j.fochx.2024.101801

112 Song, W., Zhai, X., Shi, J., et al. (2024). A ratiometric fluorescence amine sensor based on carbon quantum dot-loaded electrospun polyvinylidene fluoride film for visual monitoring of food freshness. *Food Chemistry*, 434, 137423. https://doi.org/10.1016/j.foodchem.2023.137423
113 Luo, X., & Lim, L. T. (2020). Curcumin-loaded electrospun nonwoven as a colorimetric indicator for volatile amines. *LWT*, *128*, 109493. https://doi.org/10.1016/j.lwt.2020.109493
114 Kim, K. W., Lee, K. H., Khil, M. S., et al. (2004). The effect of molecular weight and the linear velocity of drum surface on the properties of electrospun poly (ethylene terephthalate) nonwovens. *Fibers and Polymers*, 5, 122–127. https://doi.org/10.1007/BF02902925
115 Jha, B. S., Colello, R. J., Bowman, J. R., et al. (2011). Two pole air gap electrospinning. *Acta Biomaterialia*, 7(1), 203–215. https://doi.org/10.1016/j.actbio.2010.08.004
116 Kılıç, A., Yıldırım, B., İçoğlu, H. İ., et al. (2024). Production of continuous nanofiber bundles by multi parallel electrodes in needleless electrospinning. *Materials Today Communications*, 39, 109025. https://doi.org/10.1016/j.mtcomm.2024.109025
117 LeCorre-Bordes, D., Tucker, N., Huber, T., et al. (2016). Shear-electrospinning. *Journal of Materials Science*, 51, 6686–6696. https://doi.org/10.1007/s10853-016-9955-y
118 Kessick, R., Fenn, J., & Tepper, G. (2004). The use of AC potentials in electrospraying and electrospinning processes. *Polymer*, 45, 2981–2984. https://doi.org/10.1016/j.polymer.2004.02.056
119 Pokorny, P., Kostakova, E., Sanetrnik, F., et al. (2014). Effective AC needleless and collectorless electrospinning for yarn production. *Physical Chemistry Chemical Physics*, 16, 26816–26822. https://doi.org/10.1039/C4CP04346D
120 Medeiros, E. S., Glenn, G. M., Klamczynski, A. P., et al. (2009). Solution blow spinning: A new method to produce micro- and nanofibers from polymer solutions. *Journal of Applied Polymer Science*, 113, 2322–2330. https://doi.org/10.1002/app.30275
121 Dias, F. T. G., Rempel, S. P., Agnol, L. D., & Bianchi, O. (2020). The main blow spun polymer systems. *Journal of Polymer Research*, 27, 205. https://doi.org/10.1007/s10965-020-02173-7
122 Song, J., Li, Z., & Wu, H. (2020). Blowspinning: A new choice for nanofibers. *ACS Applied Materials & Interfaces*, 12, 33447–33464. https://doi.org/10.1021/acsami.0c05740
123 Li, L., Li, W., Sun, W., et al. (2024). Preparation of solution blow spinning nanofibers and its application in the food field. *International Journal of Food Engineering*, 20(5), 291–302. https://doi.org/10.1515/ijfe-2024-0053
124 Dadol, G. C., Kilic, A., Tijing, L. D., et al. (2020). Solution blow spinning (SBS) and SBS-spun nanofibers. *Materials Today Communications*, 25, 101656. https://doi.org/10.1016/j.mtcomm.2020.101656
125 Dayisoylu, K. S., Akboğa, Z., Doğan, C., et al. (2023). Rapid fabrication of micro-nanofibers from grapevine leaf extract and gelatine via electroblowing. *International Journal of Biological Macromolecules*, 253, 127309.
126 Priyanto, A., Hapidin, D. A., Suciati, T., & Khairurrijal, K. (2022). Current developments on rotary forcespun nanofibers and prospects for edible applications. *Food Engineering Reviews*, 14(3), 435–461. https://doi.org/10.1007/s12393-021-09304-w
127 Mahalingam, S., & Edirisinghe, M. (2013). Forming of polymer nanofibers by a pressurised gyration process. *Macromolecular Rapid Communications*, 34, 1134–1139. https://doi.org/10.1002/marc.201300339
128 Doan, H. N., Tsuchida, H., Iwata, T., et al. (2017). Fabrication and photochromic properties of Forcespinning® fibers based on spiropyran-doped poly (methyl methacrylate). *RSC Advances*, 7, 33061–33067. https://doi.org/10.1039/C7RA03794E
129 Du, X. Y., Li, Q., Wu, G., & Chen, S. (2019). Multifunctional micro/nanoscale fibers based on microfluidic spinning technology. *Advanced Materials*, 31, 1903733. https://doi.org/10.1002/adma.201903733

130 Li, R., Feng, Y., Zhang, H., et al. (2023). Recent advances in fabricating, characterizing, and applying food-derived fibers using microfluidic spinning technology. *Food Hydrocolloids*, 108947. https://doi.org/10.1016/j.foodhyd.2023.108947
131 Li, S., Hang, Y., Ding, Z., et al. (2020). Microfluidic silk fibers with aligned hierarchical microstructures. *ACS Biomaterials Science & Engineering*, 6, 2847–2854. https://doi.org/10.1021/acsbiomaterials.0c00060
132 Lin, W., Ni, Y., Liu, D., et al. (2019). Robust microfluidic construction of konjac glucomannan-based micro-films for active food packaging. *International Journal of Biological Macromolecules*, 137, 982–991. https://doi.org/10.1016/j.ijbiomac.2019.07.045
133 Yang, X., Rao, J., Shen, C., et al. (2023). Natamycin-loaded ethyl cellulose/PVP films developed by microfluidic spinning for active packaging. *Foods*, 13(1), 132. https://doi.org/10.3390/foods13010132
134 Han, W., Wang, Q., Kang, Y., et al. (2023). Cross-linking electrospinning. *Nanoscale*, 15, 15513–15551. https://doi.org/10.1039/D3NR03956K
135 Abd Halim, N. S., Wirzal, M. D. H., Bilad, M. R., et al. (2019). Improving performance of electrospun nylon 6, 6 nanofiber membrane for produced water filtration via solvent vapor treatment. *Polymers*, 11, 2117. https://doi.org/10.3390/polym11122117
136 Giebel, E., & Greiner, A. (2012). Water-stable nonwovens composed of electrospun fibers from aqueous dispersions by photo-cross-linking. *Macromolecular Materials and Engineering*, 297(6), 532–539. https://doi.org/10.1002/mame.201100401
137 Lara-Topete, G. O., Castanier-Rivas, J. D., Bahena-Osorio, M. F., et al. (2024). Compounding one problem with another? A look at biodegradable microplastics. *Science of the Total Environment*, 944, 173735. https://doi.org/10.1016/j.scitotenv.2024.173735
138 Karamanlioglu, M., Preziosi, R., & Robson, G. D. (2017) Abiotic and biotic environmental degradation of the bioplastic polymer poly (lactic acid): A review. *Polymer Degradation and Stability*, 137, 122–130. https://doi.org/10.1016/j.polymdegradstab.2017.01.009
139 Shekhar, N., & Mondal, A. (2024). Synthesis, properties, environmental degradation, processing, and applications of Polylactic Acid (PLA): An overview. *Polymer Bulletin*, 81, 11421–11457. https://doi.org/10.1007/s00289-024-05252-7
140 Hassan, M. M., & Sunderland, M. (2015). Antimicrobial and insect-resist wool fabrics by coating with microencapsulated antimicrobial and insect-resist agents. *Progress in Organic Coatings*, 85, 221–229. https://doi.org/10.1016/j.porgcoat.2015.04.016
141 Fabra, M. J., López-Rubio, A., Sentandreu, E., & Lagaron, J. M. (2016). Development of multilayer corn starch-based food packaging structures containing β-carotene by means of the electro-hydrodynamic processing. *Starch-Stärke*, 68, 603–610. https://doi.org/10.1002/star.201500154
142 Guclu, N., Duzyer Gebizli, S., & Orhan, M. (2023). Development of polycaprolactone-based electrospun pH-sensitive sensors as instant colorimetric indicators for food packaging. *Coloration Technology*, 139, 527–543. https://doi.org/10.1111/cote.12701
143 Lin, L., Mahdi, A. A., Li, C., et al. (2022). Enhancing the properties of *Litsea cubeba* essential oil/peach gum/polyethylene oxide nanofibers packaging by ultrasonication. *Food Packaging and Shelf Life*, 34, 100951. https://doi.org/10.1016/j.fpsl.2022.100951
144 Aydogdu, A., Sumnu, G., & Sahin, S. (2019). Fabrication of gallic acid loaded hydroxypropyl methylcellulose nanofibers by electrospinning technique as active packaging material. *Carbohydrate Polymers*, 208, 241–250. https://doi.org/10.1016/j.carbpol.2018.12.065
145 Wang, X., Yuan, Y., Huang, X., & Yue, T. (2015). Controlled release of protein from core–shell nanofibers prepared by emulsion electrospinning based on green chemical. *Journal of Applied Polymer Science*, 132, 41811. https://doi.org/10.1002/app.41811
146 Tirgar, A., Han, D., & Steckl, A. J. (2018). Absorption of ethylene on membranes containing potassium permanganate loaded into alumina-nanoparticle-incorporated alumina/carbon nanofibers. *Journal of Agricultural and Food Chemistry*, 66, 5635–5643. https://doi.org/10.1021/acs.jafc.7b05037

147 Güler, H. K., & Çallioğlu, F. C. (2023). Suspension electrospinning of azithromycin loaded nanofibers. *Journal of Drug Delivery Science and Technology*, 88, 104947. https://doi.org/10.1016/j.jddst.2023.104947
148 Prajapati, U., Ksh, V., & Joshi, A. (2022). Extraction and use of bioactive components from underutilized horticultural crops. In T. Belwal & M. Thakur (Edts.), *Bioactive Components: A Sustainable System for Good Health and Well-Being* (pp. 535–570). Springer Nature. https://doi.org/10.1007/978-981-19-2366-1
149 Leighs, S. J., McNeil, S. J., & Ranford, S. L. (2019). The application of biosurfactants for scouring wool. *Coloration Technology*, 135(1), 48–52. https://doi.org/10.1111/cote.12370
150 Jiang, N., Hu, Y., & Cheng, Y. (2024). TEMPO-oxidized nanocellulose films modified by tea saponin derived from *Camellia oleifera*: Physicochemical, mechanical, and antibacterial properties. *Polymers*, 16(7), 1016. https://doi.org/10.3390/polym16071016
151 Sunderland, M. R., Cruickshank, R. H., & Leighs, S. J. (2014). The efficacy of antifungal azole and antiprotozoal compounds in protection of wool from keratin-digesting insect larvae. *Textile Research Journal*, 84, 924–931. https://doi.org/10.1177/0040517513515312
152 McNeil, S. J., & Sunderland, M. R. (2016). The nanocidal and antifeedant activities of titanium dioxide desiccant toward wool-digesting *Tineola bisselliella* moth larvae. *Clean Technologies and Environmental Policy*, 18, 843–852. https://doi.org/10.1007/s10098-015-1060-4
153 Sunderland, M. R., & Cruickshank, R. H. (2016). Investigations into the toxic and repellent effects of propiconazole on the wool-digesting carpet beetle larvae *Anthrenocerus australis* (Coleoptera: Dermestidae). *Journal of Insect Behavior*, 29, 57–68. https://doi.org/10.1007/s10905-015-9542-0
154 Narayanan, K. B., Bhaskar, R., & Han, S. S. (2024). Bacteriophages: Natural antimicrobial bioadditives for food preservation in active packaging. *International Journal of Biological Macromolecules*, 276, 133945. https://doi.org/10.1016/j.ijbiomac.2024.133945
155 Xu, M., Fang, D., Shi, C., et al. (2024). Anthocyanin-loaded polylactic acid/quaternized chitosan electrospun nanofiber as an intelligent and active packaging film in blueberry preservation. *Food Hydrocolloids*, 158, 110586. https://doi.org/10.1016/j.foodhyd.2024.110586
156 Jiang, G., Ramachandraiah, K., & Zhao, C. (2023). Advances in the development and applications of nanofibers in meat products. *Food Hydrocolloids*, 146, 109210. https://doi.org/10.1016/j.foodhyd.2023.109210
157 Khan, A., Ezati, P., Kim, J. T., & Rhim, J. W. (2023). Biocompatible carbon quantum dots for intelligent sensing in food safety applications: Opportunities and sustainability. *Materials Today Sustainability*, 21, 100306. https://doi.org/10.1016/j.mtsust.2022. 100306
158 Qin, W., Zou, L., Hou, Y., et al. (2024). Characterization of novel anthocyanins film @ carbon quantum dot nanofiber intelligent active double-layer film, physicochemical properties and fresh-keeping monitoring in *Ictalurus punctatus* fish. *Chemical Engineering Journal*, 496, 154041. https://doi.org/10.1016/j.cej.2024.154041
159 Grosso, M. (2022). It's all about plastics. *Waste Management & Research*, 40, 607–608. https://doi.org/10.1177/0734242X221093821
160 Dolci, G., Puricelli, S., Cecere, G., et al. (2024). How does plastic compare with alternative materials in the packaging sector? A systematic review of LCA studies. *Waste Management & Research: The Journal for a Sustainable Circular Economy*, in press. https://doi.org/10.1177/0734242X241241606
161 OECD (2022). Global Plastics Outlook. Accessed on 1 October 2024. https://stats.oecd.org/viewhtml.aspx?datasetcode=PLASTIC_USE_10&lang=en
162 Indrie, L., Gherghel, S., & McNeil, S. Nanomaterials in textiles: Performance, health, and environmental aspects. In M. Yusuf & A. Haji (Edts.), *Handbook of Nanofibres and Nanocomposites. Characteristics, Synthesis, and Applications in Textiles*, (pp. 169–192). Jenny Stanford Publishing, 2023. https://doi.org/10.1201/9781003432746

163 Kaur, G., & Dandautiya, R. (2024). Microplastics in synthetic textile industries: Origin, occurrence, routes and their fates on environment and human. In *E3S Web of Conferences, 509, 01010*. https://doi.org/10.1051/e3sconf/202450901010
164 McNeil, S. (2018). Light induced strength loss in jute and polypropylene carpet backing fabrics. *Annals of the University of Oradea Fascicle of Textile, Leatherwork*, 19(1) 65–70.
165 Gupta, R. K., Guha, P., & Srivastav, P. P. (2024). Investigating the toxicological effects of nanomaterials in food packaging associated with human health and the environment. *Journal of Hazardous Materials Letters*, 5, 100125. https://doi.org/10.1016/j.hazl.2024.100125
166 McNeil, S. J., Sunderland, M. R. & Zaitseva, L. I. (2007). Closed-loop wool carpet recycling. *Resources, Conservation and Recycling*, 51, 220–224. https://doi.org/10.1016/j.resconrec.2006.09.006
167 Collie, S. R., Ranford, S. L., Fowler, I. J., & Brorens, P. H. (2019). Microfibre Pollution–What's The Story For Wool? In *Proceedings of the 19th World Textile Conference-Autex 2019*, Ghent.
168 Collie, S., Brorens, P., Hassan, M. M., & Fowler, I. (2024). Marine biodegradation behavior of wool and other textile fibers. *Water, Air, & Soil Pollution*, 235(5), 283. https://doi.org/10.1007/s11270-024-07093-6
169 Salević-Jelić, A., Lević, S., Prieto, C., et al. (2024). Polycaprolactone-based electrospun films incorporating sage extract: From active food packaging application to accelerated biodegradation by Pseudomonas. *Future Foods*, 10, 100465. https://doi.org/10.1016/j.fufo.2024.100465
170 Li, J., Wang, G., Gou, X., et al. (2022). Revealing trace nanoplastics in food packages. *Analytical Chemistry*, 94, 12657–12663. https://doi.org/10.1021/acs.analchem.2c01703
171 Tang, C., Ramos, M. J. G., Heffernan, A., et al. (2023). Evaluation and identification of chemical migrants leached from baby food pouch packaging. *Chemosphere*, 340, 139758. https://doi.org/10.1016/j.chemosphere.2023.139758
172 Indrie, L., McNeil, S., Mutlu, M. M., et al. (2023). The initial development and evaluation of cross-linked casein films for sustainable footwear. *Coatings*, 13(2), 217. https://doi.org/10.3390/coatings13020217
173 Issart, A., Godin, S., Preud'Homme, H., et al. (2019). Direct screening of food packaging materials for post-polymerization residues, degradation products and additives by liquid extraction surface analysis nanoelectrospray mass spectrometry (LESA-nESI-MS). *Analytica Chimica Acta*, 1058, 117–126. https://doi.org/10.1016/j.aca.2019.01.028
174 Commission Regulation (EU) No. 10/2011 of 14 January 2011 on Plastic Materials and Articles Intended to Come into Contact with Food. Accessed on 1 October 2024. https://eur-lex.europa.eu/legal-content/EN/TXT/PDF/?uri=CELEX:32011R0010&from=HR
175 Darie-Niţă, R. N., Vasile, C., Irimia, A., et al. (2016). Evaluation of some eco-friendly plasticizers for PLA films processing. *Journal of Applied Polymer Science*, 133, 43223. https://doi.org/10.1002/app.43223
176 Atkins, T. W. (1998). Biodegradation of poly(ethylene adipate) microcapsules in physiological media. *Biomaterials*. 19, 61–67. https://doi.org/10.1016/s0142-9612(97)00156-7
177 Lee, K. H., & Lo, C. T. (2023). Additives for enhancing the crystallizability and degradability of electrospun poly (l-lactide) fibers. *ACS Applied Polymer Materials*, 5(11), 9525–9535. https://doi.org/10.1021/acsapm.3c01967
178 Iordache, O., Tănăsescu, C., Vărzaru, E., et al. (2023). Mycelium bio-composites: The future of packaging materials. *Annals of the University of Oradea Fascicle of Textiles, Leatherwork*, 24(1), 61–66.
179 Rodríguez-Sánchez, I. J., Rivera-Monroy, Z. J., García-Castañeda, J. E., et al. (2023). Multilayer polycaprolactone-pullulan nanofiber mats incorporated with the antimicrobial palindromic peptide LfcinB (21-25) Pal as a potential application in active packaging. *Food Packaging and Shelf Life*, 38, 101110. https://doi.org/10.1016/j.fpsl.2023.101110

180 Tan, L., Ni, Y., Xie, Y., et al. (2023). Next-generation meat preservation: Integrating nano-natural substances to tackle hurdles and opportunities. *Critical Reviews in Food Science and Nutrition*, 1–24. https://doi.org/10.1080/10408398.2023.2256013
181 Romano, L., Portone, A., Coltelli, M. B., et al. (2020). Intelligent non-colorimetric indicators for the perishable supply chain by non-wovens with photo-programmed thermal response. *Nature Communications*, 11, 5991. https://doi.org/10.1038/s41467-020-19676-y
182 Che, S., Pan, S., Shao, X., et al. (2024). Portable and reversible smart labels for non-destructive detection of seafood freshness via amine-response fluorescent ionic liquids. *Food Chemistry*, 430, 136946. https://doi.org/10.1016/j.foodchem.2023.136946
183 Hui, S., & Das, N. C. (2022). Surface modified carbon nanotubes in food packaging. *ACS Symposium Series*, 1425, 199–233. https://doi.org/10.1021/bk-2022-1425.ch009
184 McNeil, S.J. & Gupta, H. (2022). Emerging applications of aerogels in textiles. *Polymer Testing*, 106, 107426. https://doi.org/10.1016/j.polymertesting.2021.107426
185 Drago, E., Campardelli, R., & Perego, P. (2024). Sustainable development of biodegradable antimicrobial electrospun membranes for active food packaging and economic analysis. *Advanced Sustainable Systems*, 2400360. https://doi.org/10.1002/adsu. 202400360

9 Smart Active Biopolymer Nanofibre Films

The Future of Food Packaging

Steven J. McNeil

INTRODUCTION

PACKAGING

Food packaging plays a crucial role in maintaining the quality of food and reducing spoilage by limiting contamination and oxidation. Packaging is typically composed of polymers, metals, glass, and paper products. Even when preservation techniques such as freezing, refrigeration, salting, and pickling are employed, packaging is still needed. The utility of food packaging can be enhanced by making it active and smart, see Figure 9.1. Active packaging releases substances that interact with microbes, oxygen, and radicals, helping to preserve the freshness of food. Smart packaging monitors food for chemical, physical and biological changes and responds to the changes. The responses range from a colour change that informs retailers and consumers about the freshness of the food, to the release of an antimicrobial agent onto the food when the humidity gets above a particular level.

NANOFIBRE FILMS

Extrusion-produced polymer films, typically continuous and nonporous, are prevalent in packaging. However, nanofibres offer an alternative, yielding what are known as nanofibre films, mats, or membranes. Nanofibres can be combined with non-fibrous polymers to create composite films, see Figure 9.2. Nanofibre films possess some exceptional properties and are under investigation for various applications other than packaging, such as tissue engineering, drug delivery, wound dressings, filters, and textiles [1–6]. Nanofibre films, due to their unique properties, are superior to continuous films for smart active packaging. As it is for many other applications of nanofibres, it is their exceptional fineness and high aspect ratios that contribute to their unique properties in food packaging. Their large surface area per unit mass

DOI: 10.1201/9781003507208-9

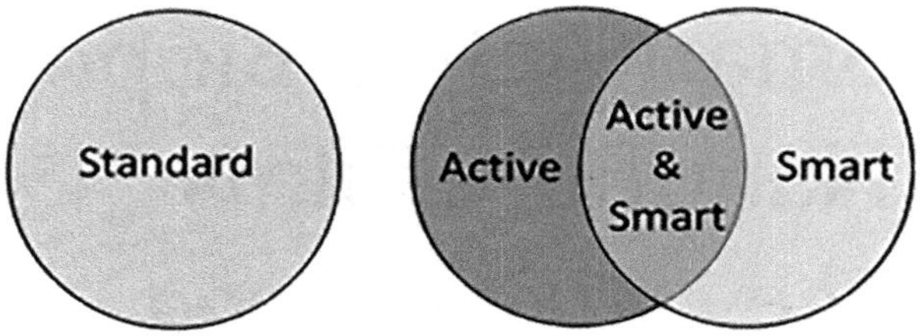

FIGURE 9.1 Types of packaging.

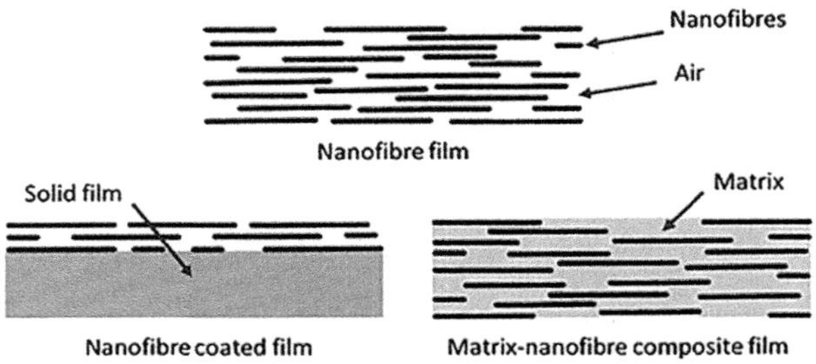

FIGURE 9.2 Types of nanofibre-containing films.

provides a large area of interaction with packaged food, enhancing the effectiveness of the active and smart agents. Nanofibres can be densely packed in films due to their small diameter, giving films with low porosity. To pass through a nanofibre film, molecules must traverse a tortuous path around the fibres. The barrier properties can be engineered to permit the desired amount of movement of water vapour and certain gases away from foods. These properties are illustrated by the studies discussed in this chapter.

Biopolymers

There are widespread concerns regarding the impact of synthetic plastics and polymers on human health and the environment, particularly microplastics in the air, soil, and water. Specifically, in the context of food packaging, it has been established that food can be contaminated by nanoparticles, monomers, plasticisers, and other substances from plastic food packaging, which are known to have adverse effects on health, such as endocrine disruption [7–9]. For instance, a study of baby food pouches found that in 40% of samples, the consumption of one pouch per day may lead to exceeding the threshold of toxicological concern with respect to adipic acid and diethylene glycol [9]. These concerns, along with issues of sustainability, are driving interest in replacing petroleum-derived polymers with biopolymer alternatives.

Scope of This Chapter

Biopolymers are considered here to be polymers that occur in nature, such as proteins, or are formed by modifying naturally occurring polymers, for example, ethylcellulose. Nanofibrous packaging produced from polymers which are prepared by industrial polymerisation of molecules (naturally occurring or otherwise) are not considered here but are the subject of a companion chapter [10]. This chapter examines nanofibre food packaging films fabricated from biopolymers with active and smart additives. Both polysaccharide and protein biopolymers are utilised in their modified and unmodified forms, often in blends with other polymers, including synthetic ones. Nanofibre active and smart food packaging that employs solely synthetic polymers is addressed in a separate chapter in this book [10]. The advancements in nanofibre food packaging films will be contextualised alongside developments in other applications of nanofibrous films where this is informative. The chapter will also discuss the reasons behind the burgeoning interest in this field and speculate on future research directions. Food safety standards and regulations are very important with respect to packaging and will clearly inform the choice of active and smart agents, as they can migrate from the packaging onto the food [11] but fall outside the scope of this review.

PREPARATION OF NANOFIBRES AND NANOFIBRE FILMS BY SPINNING

Nanofibres can be prepared from biopolymers by spinning and non-spinning methods. During spinning, the fibres are generally produced (spun) and immediately assembled into a film. Active and smart materials can be combined with the polymers prior to spinning. The properties of the resulting nanofibre films can subsequently be modified by post-spinning processes. Some techniques for preparing nanofibre films are excluded from this review because they have little apparent potential in nanofibre-based food packaging, for example, melt blowing, melt extrusion and wet spinning.

Certain types of biopolymer nanofibres can be prepared by liberating them from a parent material, such as obtaining nanocellulose from cellulosic sources. Some proteins, notably water-soluble whey proteins, will self-assemble into nanofibres under specific conditions. These protein and polysaccharide nanofibres can be formed into films through solvent casting or incorporated into composite films, see Figures 9.2 and 9.3. These liberated and self-assembled materials are significant in the context of this review, particularly concerning emerging research topics and sustainability. Given that their preparation and conversion into fibrous mats are specific to a limited number of materials, their preparation and application in packaging are discussed in the Biopolymer Materials for Nanofibres section, rather than in the current section.

There are four significant approaches to spinning biopolymers into nanofibres: electrospinning (ES), solution blow spinning (SBS), centrifugal spinning (CS), and microfluidic spinning (MFS). Among these, ES has received by far the most attention for the preparation of nanofibre-based food packaging, followed in turn by SBS, CS, and MFS. Many ingenious variations of these methods have been

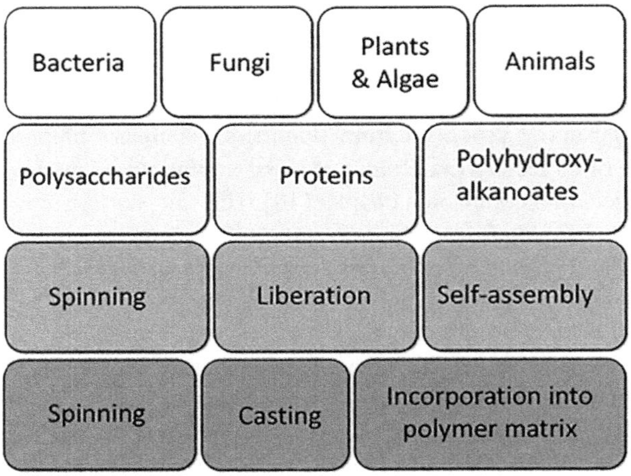

FIGURE 9.3 Routes to biopolymer films. From top, sources of biopolymers, types of biopolymers, fibre preparation methods, and film-forming methods.

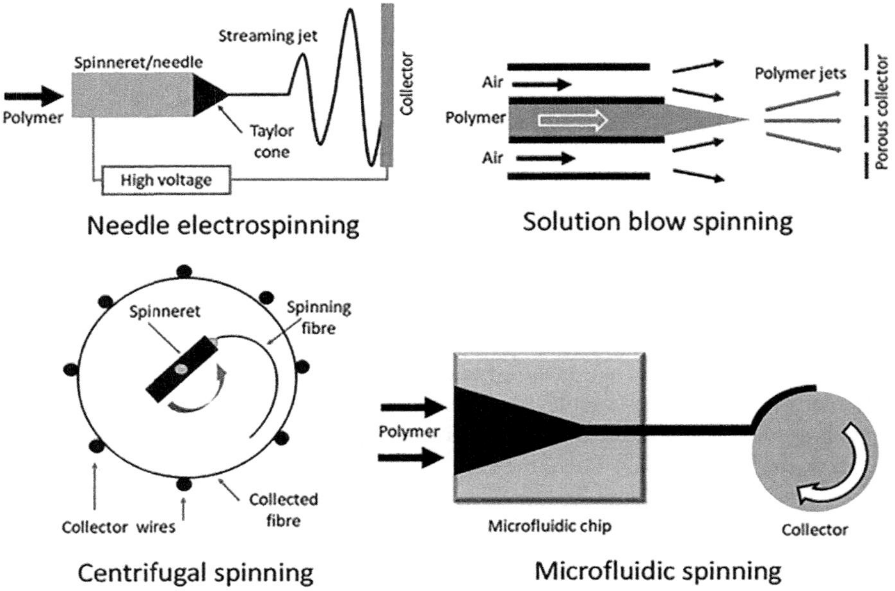

FIGURE 9.4 Methods for spinning nanofibres and nanofibre films.

explored, including some that combine elements of multiple techniques. These methods are illustrated in Figure 9.4 and discussed in greater detail in other chapters in this book and various informative reviews [10–17].

In ES, a strong electric field induces jets of polymer to emerge from a droplet on a spinneret and travel towards a collector maintained at a different voltage than the spinneret. During this flight, the polymer either dries (in the case of solutions) or solidifies (in the case of melts), while electrostatic forces stretch the jets into nanofibres. Typically, 0.5–2 mL/h of polymer passes through a needle-type spinneret, with a voltage difference of 5–30 kV maintained between the spinneret and the collector, which are separated by 10–20 cm. Many proteins and polysaccharides have been spun using this method, either alone or in combination with other polymers and a wide array of active and smart agents, see Tables 9.1–9.5.

TABLE 9.1
Biopolymer Nanofibre Films with Inherent Activity

Polymers	Preparation	Activities[a]	Food	References
Carboxymethyl chitosan/ polycaprolactone	SBS	AFun	Tomato	[18]
Gelatine	SBS	AOx, ABac	–	[19]
Nanocellulose in matrix of silk protein	Liberation, casting	AOx	–	[20]
Soy protein fibrils/pullulan	Self-assembly, ES	ABac	–	[21]

[a] ABac, antibacterial; AFun, antifungal; AOx, antioxidant.

TABLE 9.2
Biopolymer Nanofibre Films with Active Agents

Polymers	Preparation of Fibre and Film	Agents	Activities[a]	Food	References
Carboxymethyl cellulose/ gelatine	ES	*Mentha longifolia* essential oil	ABac	Prawns	[22]
Casein/gelatine	ES	*Mentha spicata* essential oil, MgO nanoparticles	ABac	Fish	[23]
Collagen/zein	ES	Gallic acid	ABac, AOx	Fish	[24]
Chitosan/gelatine	ES	*Zataria multiflora* extract	ABac	Mushrooms	[25]
Cylcodextrin[b]	ES	Octyl gallate	ABac	Meat	[26]
Gelatine	SBS	Grape leaf extract	ABac, AFun, AOx	Cheese	[27]
Gelatine	CS	Lemon peel essential oil	ABac, AFun	Cheese	[28]
Gelatine	CS	Caffeic acid	AOx	Olive oil	[29]
Gelatine	ES	Polylysine, thyme essential oil	ABac	Meat	[30]
Gelatine	ES	Lactoferrin	ABac	–	[31]
Gelatine	SBS	Carvacrol	ABac	–	[32]

(*Continued*)

TABLE 9.2 (Continued)
Biopolymer Nanofibre Films with Active Agents

Polymers	Preparation of Fibre and Film	Agents	Activities[a]	Food	References
Gelatine/zein	Air-assisted ES	Dihydromyricetin, eugenol	ABac, AOx	–	[33]
Gliadin	ES	Ferulic acid	AOx	–	[34]
Gliadin	ES	Phycocyanin, Spirulina extract	ABac, AOx	Walnut kernels	[35]
Glutelin, hordein	ES	*Oliveria decumbens* essential oil	ABac, AOx	–	[36]
Myofibrils	Liberated, cast	Phenolics	UV	–	[37]
Nanocelluose	Liberated, cast	Sakacin	ABac	Fish	[38]
Nanocelluose	Liberated, cast	Pine seed extract	ABac, AOx	Shrimp	[39]
Nanocellulose	Cast	Tea saponin, zinc oxide	ABac, AOx	–	[40]
Pea protein fibrils	Self[c], cast	Epigallocatechin gallate	AOx	–	[41]
Pea protein isolate, pullulan	ES	Allicin	ABac	–	[42]
Pullulan	CS	Thyme oil, nisin	ABac, AFun	Avocados	[43]
Pullulan, Pullulan/whey	ES	Clove extract	ABac, AOx	–	[44]
Pullulan/gum arabic	ES	*Lactobacillus bulgaricus*	ABac	Meat	[45]
Starch	ES	Carvacrol	AFun	Bread	[46]
Starch	ES	Ginger essential oil	ABac	–	[47]
Starch	ES	Carvacrol	ABac	Soy products	[48]
Starch/pullulan	ES	Cinnamon essential oil	ABac	–	[49]
Starch/pullulan	ES	Catechin	ABac, AOx	Strawberries	[50]
Whey fibrils	Self[c], cast	Carvacrol	ABac, AOx	Cheese	[51]
Whey fibrils	Self[c], cast	TiO_2 nanotubes	ABac, AOx	Meat	[52]
Whey fibrils	Self[c], cast	TiO_2 nanoparticles	ABac, AFun, AOx	Meat, cheese	[53]
Whey/pullulan	ES	Clove extract	ABac, AOx	–	[44]
Zein	ES	Saffron extract	AOx	Fish	[54]
Zein	ES	Allyl isothiocyanate	AOx	Strawberries	[55]
Zein	ES	Nisin	ABac, AOx	Peaches	[56]
Zein	ES	Rose hip extract	–	Cumquats, bananas	[57]
Zein	ES	Sakacin	ABac	Meat	[58]

[a] ABac, antibacterial; AFun, antifungal; AOx, antioxidant; UV, reducing transmission of UV light.
[b] Included here for convenience only, the fibre did not contain any polymer.
[c] Self, self-assembled.

TABLE 9.3
Biopolymer–Synthetic Polymer Nanofibre Films with Active Agents

Polymers[a]	Preparation of Fibre and Film	Agents	Activities[b]	Food	References
Alginate/PEO	ES	Bacteriophage	ABac	Meat, cucumbers, tomatoes	[59]
Casein/PEO	ES	Thymol	AOx	Meat	[60]
Casein/PVA	ES	Rosemary extract	ABac	Apple pulp	[61]
Chitosan/PCL	SBS	Thymol	AFun	Tomatoes	[62]
Chitosan/PCL/PEO	ES	Thymol	ABac	Meat	[63]
Chitosan/PEO	ES	*Melissa officinalis* extract	ABac, AFun	Mushrooms	[64]
Chitosan/PEO	ES	Thymol	ABac, AOx	–	[65]
Chitosan/PVA	ES	Baicalin	ABac	Mushrooms	[66]
Ethyl cellulose/PVP	MFS	Natamycin	AFun	–	[67]
Glucomannan/PCL	MFS	Nanosilver	ABac	–	[68]
Glucomannan/PMA	MFS	Chlorogenic acid	ABac	–	[69]
Glucomannan/PLA	MFS	Chlorogenic acid	ABac	–	[70]
Hydroxypropyl methylcellulose/PEO	ES	Gallic acid	AOx, OSc	Walnut kernels	[71]
Kefiran/PVA/PCL	ES	Flexirubin	AOx	Apples	[72]
Lentil flour/PEO	ES	Gallic acid	AOx, OSc	Walnut kernels	[73]
Peach gum polysaccharide/PEO	ES	*Litsea cubeba* essential oil	ABac, AOx	Meat	[74]
Pullulan/PVA	ES	Thymol, photocatalyst[c]	ABac	Grapes, strawberries	[75]
PVP/glucomannan	MFS	Anthocyanins	ABac, AOx	–	[76]
Silk fibroin/PEO	ES	Thyme essential oil	ABac	Meat	[77]
Soy/PVA	ES	Ginger essential oil	ABac	Cheese	[78]
Whey/PVA	ES	Probiotic bacteria	ABac	–	[79]
`Zein/PCL	ES	Caryophyllene	ABac, AOx	Strawberries	[80]

[a] PCL, polycaprolactone; PEO, polyethylene oxide; PLA, polylactic acid; PMMA, poly(methyl methacrylate); PVA, polyvinyl alcohol; PVP, polyvinyl pyrrolidone.

[b] ABac, antibacterial; AFun, antifungal; AOx, antioxidant; OSc, oxygen scavenging; UV, reducing transmission of UV light.

[c] Porphyrin–zirconium-organic framework nanoparticles.

TABLE 9.4
Biopolymer Nanofibre Composite Films with Active Agents

Nanofibrous (Non-fibrous) Polymers	Preparation of Fibre and Film	Agents	Activities[a]	Food	References
Chitosan (methylcellulose)	Liberated, cast	Lactoferrin, nanosilver	ABac, UV	Apples	[81]
Gelatine, PCL	SBS	Curcumin	ABac, OSc	–	[82]
Myofibrils (chitosan)	Liberated, cast	Rosemary extract	AOx	Fish	[83]
Nanocellulose (carboxymethyl cellulose)	Liberated, cast	Tea polyphenols	ABac, AOx	Strawberries, blueberries	[84]
Nanocellulose (gelatine)	Liberated, cast	Black pepper essential oil	ABac, AOx	Meat	[85]
Nanocellulose (polyurethane)	Liberated, cast	–	–	Banana peel	[86]
Nanocellulose (soy protein)	Liberated, cast	Tannic acid, TiO_2	ABac	Fish	[87]
Nanocellulose (whey)	Liberated, cast	Rosemary essential oil, TiO_2	ABac	Meat	[88]
Nanocellulose (whey)	Liberated, cast	Limonene, 6 essential oils	ABac	–	[89]
Soybean fibrils (chitosan)	Self[b], cast	–	–	Meat, cherries, strawberries	[90]
Whey fibrils (starch)	Self[b], cast	–	AOx	Apples	[91]
Whey (gluten)	ES, cast	Tocopherol	AOx	–	[92]
Zein (polyhydroxy-butyrate)	ES, cast	Cinnamaldehyde	AVir	–	[93]

[a] ABac, antibacterial; AOx, antioxidant; AVir, anti-viral; OSc, oxygen scavenger; UV, reducing transmission of UV light.
[b] Self, self-assembled.

SBS, also known as energy spinning and blow spinning, involves injecting polymer solutions into fast-moving streams of pressurised gas, typically air, and forcing them through a small nozzle, thereby stretching and drying the polymer into nanofibres. Nanofibre-based food packaging utilising biopolymers such as gelatine, pullulan, and zein has been prepared by SBS, with various active agents including carvacrol, thymol, and natamycin, see Tables 9.1–9.3 [116, 117]. SBS does not appear to have been used to prepare smart nanofibres. An example illustrating the versatility of this method for food packaging is the preparation of a cellulose acetate film with hydrophobic and hydrophilic sides [117]. This was achieved by sequentially applying the polymer dissolved in different mixtures of acetone, acetic acid, and N,N-dimethylformamide.

TABLE 9.5
Biopolymer Nanofibre Films and Matrix–Nanofibre Composite Films with Smart Agents

Nanofibrous (Non-fibrous) Polymers[a]	Preparation of Fibre and Films	Agents	Property Responded to[b]	Food	References
Carrageenan/ PVA	ES	*Prunus domestica* anthocyanins, epigallocatechin gallate	pH, ammonia	Meat	[94]
Chickpea protein (salep)	ES, cast	Black chickpea anthocyanins	pH, amines	Shrimp, fish	[95]
Chitin nanofibre (whey)	Liberated, cast	Poppy anthocyanins	pH, ammonia	Fish	[96]
Chitin nanofibre/ gelatine/ xanthan	Liberated, ES	Black barberry anthocyanins, citral	pH, ammonia	Shrimp	[97]
Chitosan nanofibre (pectin)	Liberated, cast	Sumac anthocyanins	pH, ammonia	Shrimp	[98]
Chitosan/PVA	ES	Cinnamaldehyde	Humidity, fungal volatiles	Strawberries	[99]
Chitosan/PVA	ES	Blueberry anthocyanins, petunia flower extract	pH, ammonia	Meat	[100]
Gelatine	ES	Black elderberry extract, nanoparticles of gold and stannic oxide	pH, ammonia, amines, hydrogen sulphide	Fish	[101]
Gelatine/zein	ES	Curcumin	pH, ammonia, volatiles from bacteria	Meat	[102]
Guar gum/pectin	ES	Poppy anthocyanins	pH, ammonia	Meat	[103]
Myofibrils	Liberated, cast	Sappan wood extract	pH, duration of water immersion	–	[104]
Myofibrils/ nanocrystalline cellulose	Liberated, cast	Myricetin, luteolin, quercetin anthocyanins	pH, ammonia	–	[105]
Nanocellulose (agar, carrageenan)	Liberated, cast	Aronia extract	pH, ammonia	Kimchi	[106]
Nanocellulose (gelatine)	Liberated, cast	Phycoerythrin	pH	Fish	[107]
Nanocellulose (pea protein)	Liberated, cast	Bilberry extract	pH	Meat	[108]

(*Continued*)

TABLE 9.5 (Continued)
Biopolymer Nanofibre Films and Matrix–Nanofibre Composite Films with Smart Agents

Nanofibrous (Non-fibrous) Polymers[a]	Preparation of Fibre and Films	Agents	Property Responded to[b]	Food	References
Nanocellulose (starch)	Liberated, cast	Betalain	pH	Meat	[109]
Silk and PCL layers	ES	Bayberry anthocyanins	pH	Meat, shrimp	[110]
Soy/PVA	ES	Barley anthocyanins, carbon quantum dots	pH, ammonia	Fish	[111]
Starch/PVA	ES	Roselle anthocyanins	pH, ammonia	Shrimp, meat	[112]
Zein/gelatine	SBS	Extracts of purple cabbage and purple potatoes	pH, ammonia	Milk, fish	[113]
Zein, PEO	ES	Sodium bicarbonate, thyme essential oil	Humidity[c]	Strawberries	[114]
Zein	ES	Cinnamic aldehyde, tetradecane	Temperature	Sausages	[115]

[a] PCL, polycaprolactone; PEO, polyethylene oxide; PVA, polyvinyl alcohol.
[b] Colour change, unless indicated otherwise.
[c] High humidity initiates the production of CO_2 which causes the release of antibacterial thyme essential oil.

CS, also known as centrifugal jet spinning, rotor spinning, rotary force spinning, rotary jet spinning, and force spinning, has been extensively researched for its fundamentals and applications, including edible fibres, wound dressings, and drug delivery [16], yet few studies have reported on its use for preparing films applied to food. CS is insensitive to the dielectric constant of polymer solutions and thus applies to a broader range of polymers than ES, so it may become more utilised in the future. For example, starches with an amylopectin content greater than 65% cannot be effectively spun by ES but can be spun by CS [118]. The flight of the fibres in CS can be directed by airflow, a method that has been used to directly coat individual avocados [43]. Centrifugally spun nanofibre food packaging films have utilised gelatine, pullulan, and actives such as thyme oil, lemon peel oil, citric acid, and nisin. Evaluations have been performed on cheese, olive oil and avocados, see Table 9.2. CS does not appear to have been used to make smart packing films.

The MFS process is less studied for preparing nanofibre food packaging films compared to ES, SBS, and CS. However, it is worth considering here because of its simplicity, see Figure 9.4, and ability to produce very complex nanofibres [119, 120]. For instance, MFS was used to prepare silk nanofibres with an aligned and hierarchical structure with mechanical properties superior to those produced by other spinning methods [121]. A review of MFS of food materials highlighted the high spinnability, biocompatibility, and degradability of proteins, glucomannan, chitosan, and alginate. Microfluidic technology is most commonly encountered in inkjet printheads, and the majority of research on MFS is directed towards fibrous films intended for medical applications (such as tissue scaffolds and sutures, drug release, wound healing, and electronics) [120]. Although the production rate of an individual unit is very low, at approximately 0.5 millilitres per hour, scaling up the process with multiple units is feasible. MFS has been used to create nanofibre films for food packaging, sometimes incorporating active materials, often utilising glucomannan or ethyl cellulose combined with synthetic polymers. The active agents included chlorogenic acid, cinnamic acid, natamycin, and silver nanoparticles, see Table 9.3. In one study, MFS was used to prepare polyvinylpyrrolidone/hexadecyl glucomannan fibres containing anthocyanins [76]. The researchers examined properties like strength and antibacterial activity but did not assess the fibres' potential to indicate pH, as would be expected from the anthocyanins. While MFS has not been widely used to produce nanofibre films with smart additives (other than anthocyanins), microfluidic chip devices have been explored in smart food packaging, such as a time–temperature indicator utilising temperature-dependent flow in a microchannel [122], but these devices are beyond the scope of this review.

POST-SPINNING PROCESSING

Nanofibre films can be modified after spinning. For instance, active compounds such as the antimicrobial cinnamaldehyde can be applied [99], but this is not a common approach. The mechanical properties of as-prepared nanofibre films may be insufficient for certain applications and require enhancement by post-spinning processes. These can increase the strength of individual fibres and/or improve inter-fibre bonding, thereby enhancing the overall strength of the film, see Figure 9.5. A variety of methods have been employed, including cold and hot drawing, hot pressing, thermal treatment, exposure to solvent vapour, and cross-linking [1, 123]. Cross-links can be introduced through several techniques, such as cross-linking chemicals and enzymes, exposure to solvents, plasma, UV light, and electron beams [1, 64, 123]. While these techniques have been extensively studied in the context of nanofibre films for wound dressings and filtration, their application in food packaging remains relatively unexplored.

Significant property enhancements can be achieved through straightforward post-spinning processes. For example, soaking an electrospun film of cellulose acetate in a mixture of ethanol and acetone increased its tensile strength by 97%, by fusing fibres at junction points and increasing the polymer's molecular orientation [124]. Wongkrongsak et al. described electron beam-induced cross-linking of an electrospun silk fibroin–polyethylene oxide film, which resulted in tensile strength and Young's modulus increases of up to 270% and 130%, respectively [125].

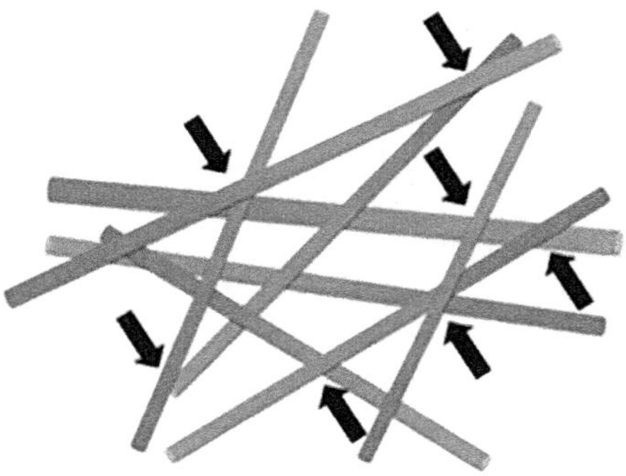

FIGURE 9.5 Inter-fibre bonding points in nanofibre films.

The hydrophilicity of certain biopolymer films is an obstacle to their use in food packaging. One way of overcoming this is the application of glow discharge plasma, which has been demonstrated to reduce both the water vapour permeability and solubility of a myofibrillar protein film [126]. These improvements were ascribed to the removal of weakly bound, low-molecular-weight protein fragments from the fibre surfaces and the formation of new cross-links.

The post-spinning modification of polysaccharide nanofibre films for food packaging has attracted particular attention due to the abundance of raw materials, relative ease of spinning, and the poor mechanical properties of the resulting films [50, 127]. The hydrophobicity and mechanical properties of nanofibre starch films intended for food packaging were significantly improved by cross-linking with glutaraldehyde vapour [127]. The tensile strength increased from approximately 0.7 MPa to nearly 9.7 MPa. Glutaraldehyde reacted with hydroxyl groups in the starch, thereby increasing hydrophobicity, as evidenced by the water contact angle rising from 0° to 94° [127].

NANOFIBRE COMPOSITE FILMS

The use of biopolymer nanofibres in packaging has extended beyond simple nanofibrous films to include packaging where nanofibres are included in a matrix of continuous polymers or applied to one or both faces of a solid film, see Figure 9.2. The nanofibres may be a convenient way of introducing active or smart materials to the film, see Tables 9.4 and 9.5 for examples. For example, whey fibrils have been used inside starch films for reinforcement [91] and nanocellulose has been used to reinforce non-fibrous gelatine and whey films [85, 89]. Applying the nanofibres to the outside of films better preserves the films' flexibility than incorporating them into the film [128]. For example, zein fibres containing phenolic-rich fruit extracts were spun onto one side of a polyhydroxyalkanoate film, so that the activity of the extracts was concentrated on the side of the film intended to be in contact with food [129].

BIOPOLYMER MATERIALS FOR NANOFIBRES

The various biopolymers employed in the preparation of active and smart nanofibre food packaging films are illustrated in Figure 9.6. These biopolymers are utilised in the form of macromolecules, nanofibres, and fibrils. Polysaccharides of both structural and storage types have been utilised in nanofibre film-based food packaging. Proteins derived from plants, such as wheat gluten, zein, and soy proteins, are primarily sourced from crops and are secondary products resulting from the processing of starch, edible oil, or other food items. Nanofibre films spun from proteins generally exhibit inferior mechanical properties and thermal stability but possess superior oxygen barrier properties compared to their polysaccharide counterparts [123, 130].

POLYSACCHARIDES

Cellulose

Cellulose is the most abundant biopolymer and is found in the cell walls of plants, certain algae, microbes, and biofilms. It is hydrophilic yet water-insoluble. In nanofibre-based packaging, it is most often used in the form of nanocellulose fibrils or reconstituted cellulose. The properties of cellulose can be significantly altered by dissolving and reconstituting it, modifying some of its numerous hydroxyl groups,

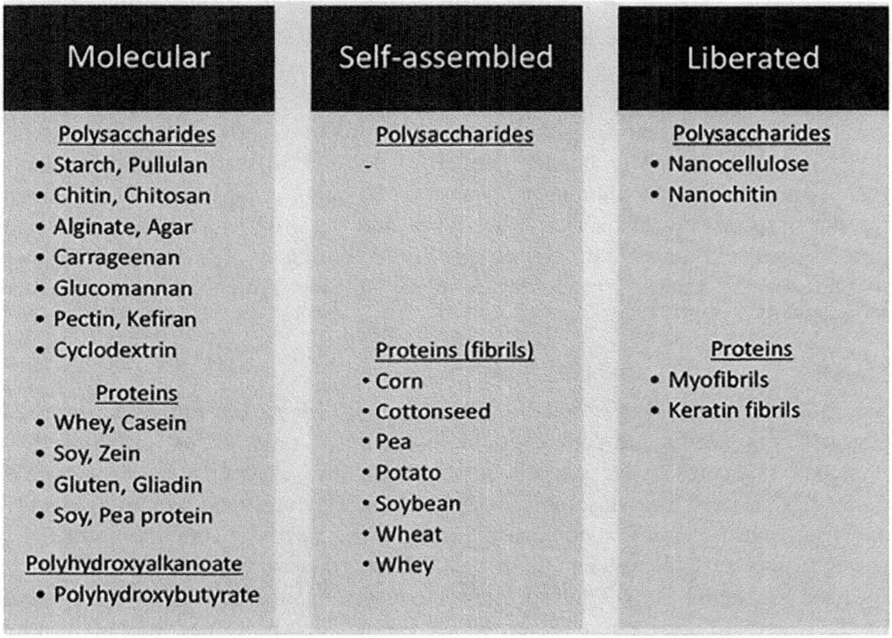

FIGURE 9.6 Biopolymers utilised in nanofibre film-based packaging.

and examples include cellulose acetate and carboxymethyl cellulose [81]. Regenerated cellulose nanofibres have been prepared by ES, SBS, and MFS [22, 67, 117]. For example, a mixture of carboxymethyl cellulose and fish skin gelatine was combined with up to 2% *Mentha longifolia* L. essential oil, electrospun, and evaluated as packaging for prawns [22]. The essential oil reduced the viscosity of the spinning solution and increased its surface tension and electrical conductivity, thereby increasing the diameter of the spun fibres from 394 nm to 455 nm. The films exhibited antibacterial and antioxidant properties, which increased the storage life of the prawns. MFS has been used to prepare films from an ethanolic solution of ethyl cellulose and polyvinylpyrrolidone containing the antifungal agent natamycin. High encapsulation efficiency and resistance to mould were observed with the hydrophobic films [67].

In many types of naturally occurring cellulose, the polysaccharide chains are organised into fibrils. The fibrils have diameters ranging from 5 to 20 nanometres and lengths of several microns. The isolation of nanofibres from natural sources eliminates the need for spinning. Numerous methods for preparing nanocellulose exist, including ball-milling, microfluidisation, enzymatic hydrolysis, and oxidation [85, 131]. The high length-to-width ratio is the primary reason for its use in films to enhance mechanical properties. An illustrative example is provided by a study where nanocellulose fibres and carboxymethyl cellulose (non-fibrous) were cast with tea polyphenols to create a film with antibacterial and antimicrobial properties, which improved the storage life of strawberries and blueberries [84]. The hydrophilicity of nanocellulose presents a disadvantage in packaging applications, but this can be mitigated through some of the post-spinning treatments discussed earlier. Alternatively, incorporating citric acid into a nanocellulose/polyurethane film reduces its hydrophilicity [86]. The cross-linking reaction between citric acid and cellulosic hydroxyl groups increases the water contact angle from 31° to 94°, thereby enhancing the film's water vapour and oxygen barrier properties and improving its ability to prolong the storage of banana peel. By selectively dissolving the amorphous regions of cellulose nanofibres, cellulose nanocrystals can be liberated. These lack the high aspect ratio of nanocellulose fibres and are used to alter the rheological properties of solutions and the properties of films. For instance, they have been used to enhance the physical properties of a myofibrillar protein/anthocyanin smart nanofibre film [105].

Starch

Starch is an abundant polysaccharide found in many plants and fungi, serving as an energy storage medium. It comprises amylose, a linear molecule, and amylopectin, a branched molecule. Starch naturally occurs in granules, where the dense packing of molecules renders them insoluble in cold water. Heating these granules in water and applying mechanical action cause them to burst, releasing starch molecules that strongly interact with water to form gels. These gels can be spun into nanofibres for food packaging films. The significant interest in nanofibrous starch is evidenced by the recent publication of a review on the ES of starch fibres and their application in active and smart food packaging [132]. Starch nanofibres have been loaded with various substances and organisms, for instance, ginger essential oil and live bacteria [47, 133]. Starch has been chemically modified to enhance its spinnability; for example, octenyl succinylated starch has been combined with cinnamon essential oil and spun [49].

Blending starch and modified starches with other polymers can enhance their spinning properties. For example, incorporating pullulan has improved the structural and functional characteristics of an octenyl succinic anhydride starch film containing catechin, thereby reducing nutrient loss in strawberries [50]. Post-spinning modifications of starch nanofibres intended for food packaging have been investigated. For instance, cross-linking with glutaraldehyde vapour has been found to improve hydrophobicity and mechanical properties [127]. The foodborne pathogen *Bacillus cereus* was targeted by a starch-based film containing carvacrol antimicrobial. The carvacrol was immobilised in casein nanoparticles, which were then mixed with cationic starch and spun. The amylases and proteases secreted by any bacteria in the packaged food could hydrolyse the starch and casein, respectively, releasing the carvacrol [48]. The film showed changes in the texture and colour of four soy-based foods, including tofu, during refrigeration. Starch nanofibre films have been utilised for smart as well as active packaging. Modified starch, polyvinyl alcohol, and anthocyanins were spun into pH-sensitive films [112]. The films exhibited a rapid response to pH changes, which was sustained for at least 14 days and provided a visual indication of the freshness of pork and shrimp. Interestingly, the anthocyanins increased the thermal stability of the films, indicating that there was good hydrogen bonding between the anthocyanins and the starch.

Pectin

Pectin is a complex structural polysaccharide found in plants and is particularly abundant in pome and citrus fruits. Due to its strong gelling properties, pectin is utilised in the food industry as a gelling agent, stabiliser, and thickener. It is also employed in solid films and as a matrix for nanofibres in composite food packaging, see Table 9.5. Pectin can be spun into nanofibre films, which have been investigated for applications in medicine and food science. Consequently, extensive information exists on the spinning of pectin and the incorporation and release of active materials for packaging researchers to draw upon. For instance, a probiotic strain of *Lactobacillus rhamnosus* demonstrated a survival rate of 89% when electrospun in a blend of pectin and polyvinyl alcohol [134].

Pectin nanofibres have not yet been utilised for active food packaging, but this is likely to change soon as research progresses towards this application. For instance, mixtures of pectin and chitosan have been spun with curcumin encapsulated in cyclodextrin, and the release of the active compound and the decomposition of the film have been studied [135]. The combination of pectin and chitosan is particularly promising due to the potential bonding between the amine groups of chitosan and the carboxylic acid groups of pectin, which could result in excellent mechanical properties. A smart nanofibre film containing pectin has been evaluated on food. An extract of *Papaver rhoeas* petals was included in an electrospun guar gum–pectin–cellulose nanocrystal film, which exhibited strong colour changes to indicate the spoilage of packaged lamb meat [103].

Pullulan

Pullulan is an extracellular polysaccharide produced by the *Aureobasidium pullulans* fungus, which has gained attention in nanofibrous film packaging because of its good spinnability, biodegradability, and oxygen barrier properties [75]. It has been spun into packaging films by ES, SBS and CS [43, 75, 136], in combination with starch, gelatine, whey, gum arabic, and polyvinyl alcohol, incorporating various active agents, see Tables 9.1–9.4. For instance, pullulan improved the structural and functional properties of a spun starch film containing catechin [50]. In another study, a mixture of pullulan and polyvinyl alcohol, with porphyrin–zirconium-organic framework nanoparticles loaded with thymol, was spun into a film that extended the shelf life of packaged grapes and strawberries [75]. The film's effectiveness was attributed to its dual antimicrobial activity, derived from the thymol and from singlet oxygen that was photochemically generated by the nanoparticles.

Glucomannan

Glucomannan, a water-soluble polysaccharide typically derived from the konjac plant, has attracted particular attention for the MFS of active food packaging. It has been combined with poly(methyl methacrylate) and chlorogenic acid, where the former enhances moisture barrier, thermal, and mechanical properties, while the latter provides antibacterial activity [69]. Additionally, glucomannan has been utilised with polylactic acid/trans-cinnamic acid and poly(caprolactone)/silver nanoparticles, resulting in films with excellent mechanical and gas barrier properties, hydrophobicity, and antibacterial activity [68, 70]. The hexadecyl derivative of glucomannan was shown to be an effective emulsifier for corn oil and corn anthocyanins in polyvinyl pyrrolidone, facilitating the spinning of a film with controlled release of the antibacterial and antioxidant anthocyanins [76].

Chitin and Chitosan

Chitin is the second most abundant natural polysaccharide and occurs in the shells of crustaceans, fungi, insects, molluscs, and diatoms. Chitin nanofibres can be isolated. They have diameters of 10–20 nm and have been used in nanofibrous smart food packaging film, by blending with gelatine, xanthan and anthocyanin [97, 137]. The spun film was able to indicate the freshness of shrimp. Chitin nanofibres have also been incorporated into a matrix of whey and anthocyanin to monitor the freshness of fish [96].

Chitosan is prepared from chitin and is distinguished by its numerous amino groups, which render it water-soluble under acidic conditions and inherently antimicrobial. Chitosan has been more widely used in nanofibrous food packaging than chitin. Due to the high repulsive forces between the ionic groups in chitosan, solutions are challenging to spin. Consequently, chitosan is often blended with other polymers for spinning. Chitosan mixtures have been spun into nanofibre films using all four techniques described earlier. These chitosan films incorporated various active and smart additives such as thymol, boron nitride, 4-terpineol, lactoferrin, rosemary extract, and anthocyanin, have been applied to fruit, mushrooms, shrimp, and meat, see Tables 9.1–9.5. For example, a combination of chitosan, gelatine and antimicrobial *Zataria multiflora* extract was spun into nanofibres. The addition of citric acid cross-linker enhanced the film's mechanical properties, water repellency, and water vapour permeability. The study compared this film with a conventional non-active, non-nanofibre polyethylene film and found that the former was more effective in extending the shelf life of mushrooms [25]. A nanofibre film, composed of chitosan, polyvinyl alcohol, and baicalin, also improved mushroom preservation [66]. A Janus film with directional liquid transport and antibacterial properties has been developed to extract liquid from packaged meat, thereby inhibiting microbial proliferation [63]. The hydrophobic side of the film, designed for contact with the meat, was spun from polycaprolactone, while the hydrophilic side was spun from a blend of chitosan and polyethylene oxide. The film included an antimicrobial agent and extended the shelf life of pork.

Carrageenan, Alginate, and Agar

Polysaccharides isolated from seaweed can exhibit strong gelling properties, making them suitable for nanofibrous food packaging films. For example, a blend of sodium alginate and polyethylene oxide, spun with an active bacteriophage, has been used to package meat, cucumbers, and tomatoes [59]. A blend of carrageenan and polyvinyl alcohol has been spun with smart and active additives, namely anthocyanins and epigallocatechin gallate, respectively, to monitor freshness and extend the shelf life of beef [94]. The packaging indicated changes in freshness and reduced microbial and chemical alterations during storage. Polyvinyl alcohol was needed for the incorporation of the water-soluble anthocyanins and provided good mechanical and thermal properties. A comparison of nanofibre film and cast film, prepared from a mixture of carrageenan and polyvinyl alcohol (1:9) with plant anthocyanins as a smart additive, found that nanofibre films responded faster to pH changes than cast films (10 seconds compared to 15–40 minutes). Agar and carrageenan were used as matrices for nanocellulose and halochromic *Aronia melanocarpa*, and the films successfully indicated both optimal and excessive fermentation of kimchi [106].

Cyclodextrins

Cyclodextrins are renowned for their exceptional encapsulation properties, making them ideal for incorporating active substances into nanofibres. This is particularly beneficial when the active substance is incompatible with the polymer matrix or when controlled release is desired [34, 62]. For instance, a thymol–cyclodextrin inclusion complex was combined with chitosan/polycaprolactam, processed into a nanofibrous film by SBS, and used to package tomatoes [62]. Another application of cyclodextrins in nanofibres is to assist with the spinning of biopolymers. For example, they eliminated the need for carrier polymers and toxic solvents in the spinning of chitosan [135]. The formation of nanofibres by cyclodextrins alone, that is, without polymers, is perhaps less well-known than their use for encapsulation. These are a type of polymer-free nanofibre that can be formed because of the high water solubility of certain cyclodextrins, which allows for the preparation of highly concentrated, viscous solutions suitable for spinning [138, 139]. Polymer-free systems of loaded cyclodextrins enable higher loading of active substances compared to polymeric nanofibres and have been utilised to deliver drugs, pesticides, and flavours [140–142]. Polymer-free cyclodextrin nanofibres have been used in one packaging study where octyl gallate was encapsulated in hydroxypropyl-β-cyclodextrin and then electrospun [26].

The resulting nanofibre film reduced the proliferation of bacteria on stored meat. Cyclodextrins also played a crucial role in an advanced spun polyvinyl alcohol/ chitosan film, which released the active compound 4-terpineol in response to increased humidity, thereby extending the shelf life of chilled shrimp [143].

Kefiran

Kefiran is an exopolysaccharide produced by kefir grains. It is a branched glucogalactan with antimicrobial and film-forming properties. Solid films of kefiran have been investigated for active food packaging, and it has been spun into nanofibres, as highlighted in a recent review [144]. Only one study has reported the preparation and evaluation of kefiran nanofibre films in contact with food, and it involved a double-layered nanofibrous film (kefiran/polyvinyl alcohol and polycaprolactone) prepared by needleless electrospinning [72]. The polymers in both layers contained 15% by weight of flexirubin, an antioxidant and antimicrobial natural pigment. The film enhanced the retention of the colour of sliced apples.

PROTEINS

Gelatine and Collagen

Gelatine is a group of proteins produced by the partial hydrolysis of collagen. It is soluble in warm water and possesses strong gelling properties, making it widely used in food, cosmetics, and other products. Due to its poor mechanical properties and high sensitivity to water, most research on its use in active or smart nanofibre-based packaging films involves blending it with other polymers. Gelatine nanofibres have been prepared by ES, SBS and CS and utilised as the matrix in nanofibre composite films, with various functional additives, see Tables 9.1–9.5. Research on gelatine films has focused on enhancing their properties, often using methods applicable to other biopolymers. For instance, the Maillard cross-linking of gelatine with sugars has been improved through a novel neutralisation-induced approach, resulting in films with superior properties [19]. Pea protein-pullulan blend nanofibres were Maillard crosslinked with glucose [42]. A multilayer film was created by sequentially spinning curcumin-loaded gelatine nanofibres and polycaprolactone nanofibres [82]. This film was stronger than a gelatine-only film. The hydrophobic outer layer of polycaprolactone protected the inner gelatine layer from external moisture, while the gelatine layer could release curcumin in response to moisture, such as from packaged food. A blend of collagen and Eudragit® (a co-polymer of methacrylate and methacrylic) was successfully combined with bitter orange peel essential oil and spun into a nanofibre film which possessed a low rate of water transmission and high antimicrobial and antioxidant activities [145].

Casein

Casein is present in mammalian milk and comprises four phosphoproteins. It is hydrophobic, rendering it poorly soluble in water at neutral pHs, but it readily disperses in alkaline solutions and hence is often encountered in polymer science as sodium and calcium caseinates. Casein products were among the earliest plastics. Due to concerns about petrochemical-based materials, there is renewed interest in using casein for products like textiles and footwear [146, 147]. Casein is not ideally suited for spinning into nanofibres due to the limited entanglement of the protein chains. Consequently, it is typically combined with other polymers. The first instance of spinning casein with an edible co-polymer, pullulan, was reported in 2016, where the randomly ordered chains of pullulan were able to entangle casein proteins, facilitating the spinning of fibres [148]. Additionally, casein has been spun into nanofibres with gelatine, polyvinyl alcohol, and polyethylene oxide, see Tables 9.2 and 9.3.

Casein has been spun with active substances, such as antioxidant rosemary extract, using the minimum amount of polyvinyl alcohol necessary for effective spinning [61]. Prior to spinning, ultrasound was employed to disrupt any protein aggregates. As the concentration of aqueous rosemary extract increased, a greater proportion of PVA was required to spin uniform fibres. The resulting film exhibited radical scavenging activity and slowed the oxidation of apple pulp. In another investigation, a nanofibre film was prepared from sodium caseinate/gelatine incorporating two antimicrobials, *Mentha spicata* L. essential oil and magnesium oxide nanoparticles, which extended the shelf life of trout [23]. A combination of casein and polyethylene oxide with thymol was spun into a nanofibre film and evaluated as packaging for beef [60]. The antimicrobial thymol had been incorporated into cyclodextrin to counter its high volatility and low water solubility. It was found that protease from bacteria on the meat hydrolysed the film, leading to the release of the thymol. This is a good example of an active compound being released where and when it can provide the most benefit to the packaged food, via a pathway that is unique to protein-rich active packaging.

Whey

Whey consists of three water-soluble milk proteins, β-lactoglobulin, α-lactalbumin, and bovine serum albumin. This section specifically focuses on whey in its molecular form, while self-assembled whey fibrils are explored below in the Protein Fibrils section. Whey films exhibit good flexibility and optical qualities, which surpass those of many other proteins and polysaccharides. However, its usage in packaging films faces challenges due to high permeabilities to water vapour and oxygen, as well as low tensile strength [149]. Whey has been utilised as the matrix in composite films incorporating nanofibres. For instance, chitin nanofibres and anthocyanins have been combined with whey to monitor the freshness of fish [96]. Whey has also served as the matrix for bacterial cellulose nanofibres, as well as a combination of chitosan nanofibres and cinnamon oil [89, 150]. Bilayer active packaging films have been prepared by spinning whey, soy and zein nanofibres containing the antioxidant α-tocopherol onto one side of a gluten film [92]. The higher solubilities of the whey and soy proteins, compared to zein, led to a faster release of the active compound. Furthermore, whey has been blended with both polysaccharides and synthetic polymers (pullulan and polyvinyl alcohol) and actives (clove extract and probiotic bacteria) and spun into nanofibre films with good retention of activity [44, 79].

Zein

Zein is a group of proteins derived from maize (*Zea mays*) and has garnered more attention in packaging than many other proteins, and this was recently reviewed [151]. Rich in the amino acids histidine and arginine, zein exhibits insolubility in water, rendering spun fibres resistant to moisture. Significantly, zein can be spun without the need for blending with other polymers. Like many protein nanofibres, those made from zein suffer from poor mechanical properties. In several studies, zein nanofibre films were prepared with active or smart additives and evaluated on food. Zein, in combination with the active compounds saffron extract and allyl isothiocyanate, has been spun and used as packaging to improve the storage of fish and strawberries [54, 55]. Researchers have also developed zein nanofibres containing nisin, which were strengthened by the inclusion of montmorillonite nanoclay and glutaraldehyde and used to package peaches [56].

Coaxial spinning was employed to create fibres with a zein shell and a core of rose hip extract, which served as packaging for cumquats and bananas [57]. The core-shell approach was also employed for a smart, active fibre, where the fibre released its bioactivity selectively at high humidities, addressing the challenge of controlling the release of highly volatile actives [114]. Specifically, the volatile thyme essential oil was encapsulated within a hydrophobic zein–sodium bicarbonate core. The hydrophilic shell of the fibre consisted of polyethylene oxide and citric acid. When exposed to high humidity (such as when in contact with moist food), hydrogen ions from the citric acid diffuse to the bicarbonate core, resulting in the production of carbon dioxide. This gas then exits the fibre through the shell, creating pathways for the release of the active compound. This innovative packaging significantly improved the storage of strawberries.

Soy

The residue left after the extraction of oil from soya beans contains around 50% protein and is mainly used as animal feed. Soy protein can be obtained from this meal. Despite its widespread availability, soy protein has not been extensively studied for food packaging applications. This may be attributed to its high solubility in water and low viscosity, which make it challenging to spin. However, two approaches to address this issue are blending soy protein with other polymers and using solvents in which soy protein is less soluble than in water. Several studies have investigated the development of active/smart nanofibre soy films. However, these films were generally not evaluated on food, for example, studies of a soy protein/polyethylene oxide/raspberry anthocyanin film [152] and a soy/ethyl cellulose/orange peel extract film [153]. In one study involving food applications, a soy protein/polyethylene oxide/ginger essential oil nanofibre film was shown to enhance the storage of cheese [78]. A film incorporating both active and smart additives was spun from a blend of soy protein, polyvinyl alcohol, anthocyanins and carbon quantum dots and applied to fish. The film's colour changed in response to ammonia released from the fish, and when exposed to a strong light, the quantum dots produced reactive oxygen species which inhibited bacteria [111].

Gluten and Gliadin

Gluten occurs in cereal grains and consists of glutenin and gliadin proteins. Its availability, low price, low water solubility, and heat-induced cross-linking make it attractive for food packaging. Wheat gluten films are excellent barriers to carbon dioxide and oxygen, but poor barriers to water vapour [154]. In terms of mechanical properties, these films are comparable to those made from wool keratin and surpass those made from soy and zein proteins [155]. Active nanofibre packaging films have been spun from gliadin and ferulic acid encapsulated in cyclodextrin. This approach gave the antioxidant ferulic acid good resistance to UV light and desirable release rates in dilute ethanol and acetic acid, serving as aqueous and acidic food simulants, respectively [34]. Gliadin can be combined with up to 5% of phycocyanin and Spirulina-extract active compounds, and the antibacterial and antioxidant properties of the film inhibited lipid oxidation in stored walnut kernels [35]. A nanofibre-coated film, see Figure 9.2, was produced by spinning lycopene-infused polylactic acid onto a cast gluten film. The nanofibres resulted in a 14% decrease in water vapour permeability and an enhancement in mechanical properties [156]. An alternative approach for reducing the water vapour permeability of gluten films utilises the Maillard reaction between reducing sugars and proteins. A nanofibrous film was spun from a mixture of gluten, zein, and xylose, then cured. This technique also increased the film's stability in water and increased its rigidity and elasticity due to the augmented intermolecular entanglements and interactions of the polymers [157].

Keratin

The keratin fibre, wool, possesses several properties that make it useful for packaging, including moisture buffering, microbial resistance, and temperature regulation [158]. It holds significant potential in smart, active packaging as it is an excellent substrate for attaching active materials such as nanocides, photocatalysts, and antimicrobials [159–161]. Consequently, it has been utilised in various forms for packaging. For instance, wool fibre packaging is produced by companies like Woolcool [162]. Dissolved wool protein has been both solution blow spun and cast into films which enhanced the shelf life of carrots [163, 164], and wool keratin fibrils have been spun into nanofibre films, see the discussion on protein fibrils below. Waste feathers are another source of keratin, which has been spun into films with inherent antibacterial activity [165].

Silk Fibroin

Fibroin is a protein derived from silk that has extensive applications in biomedicine and food industries due to its excellent biocompatibility [77, 125]. It has been utilised in nanofibrous packaging films. Even without the addition of active compounds, silk's inherent properties meant that a film of nanocellulose in a matrix of silk exhibited antioxidant activity [20]. Silk fibroin and polyethylene oxide can be combined and spun using water as a solvent, and the resultant film, when exposed to an electron beam, becomes water-insoluble due to radical cross-linking [125]. A similar blend of polymers, when spun with thyme essential oil, was treated with cold plasma and significantly reduced the level of *Salmonella Typhimurium* in packaged poultry meat [77]. A Janus film was constructed using silk protein, comprising two distinct nanospun polymer layers designed for directional water transport [110]. The hydrophilic layer was composed of silk fibroin and bayberry anthocyanins, and the hydrophobic layer consisted of polycaprolactone and eucalyptus oil. The anthocyanins provided a pH-sensitive colour response and antioxidant properties, while the eucalyptus oil had antibacterial activity. This multifunctional film effectively prolonged the shelf life of refrigerated pork and room-temperature shrimp.

Protein Fibrils

Protein nanofibrils (fibrils) are variously known as fibrillar protein aggregates, amyloid-like fibrils, and intermediate filaments, and their potential in packaging was first recognised in 2019 [166]. Fibrils have diameters ranging from nanometres to micrometres, lengths of micrometres, and are semi-flexible [51, 167]. Fibrils occur in plants, animals (meat, milk, fibre), bacteria, and fungi, from which they can be liberated. Additionally, they can be synthesised by the self-assembly of certain soluble proteins, such as whey. Films can be produced by casting these fibrils, or they can be mixed with polymers and spun. Fibrils are appealing as food packaging materials due to scalable preparation, lack of requirement for spinning, potential derivation from waste products and their biodegradability.

Fibrils prepared from whey proteins are amongst the most widely studied fibrils for nanofibre films. Their preparation is often based on the 2019 work of Wang et al [51], as this preparation is efficient and only requires common laboratory equipment. The self-assembly of whey proteins into fibrils increases the exposure of amino acids with sulphur and aromatic side chains, resulting in superior hydrophobicity, antioxidant activity, and mechanical properties compared to the precursor proteins [52, 167]. Four studies have investigated the use of whey fibrils in packaging. In one, the fibrils were combined with antimicrobial carvacrol and evaluated as cheese packaging. The results were promising, particularly in terms of hydrophobicity and antioxidant activity [51]. Titanium dioxide nanotubes and nanoparticles have been incorporated into whey fibril coatings to enhance microbial resistance and reduce UV light transmission [52, 53]. These coatings extended the shelf life of meat by reducing microbial growth and moisture loss and by scavenging radicals. Films prepared from whey and titanium dioxide nanoparticles were 29% stronger when the whey was converted into fibrils [53]. Whey fibrils have also been utilised to enhance the mechanical properties of non-fibrous starch films [91].

Fibrils isolated from skeletal muscles are known as myofibrils and have been investigated in several packaging-related studies. They can be isolated from meat off-cuts, with fish meat being a commonly used source. Myofibril films have been prepared with various additives, including antimicrobials, nanoclay, and nanocellulose [168]. Even when actives were present, the thrust of the research was the mechanical and thermal properties, water vapour transport, and water sorption kinetics, reflecting the emerging interest in this area of food packaging and the need to understand the fundamental principles before additives are introduced and testing is conducted on food. For example, cold plasma treatment has been shown to reduce the water sensitivity of films cast from fish myofibrils [126]. A film composed of myofibrils, non-fibrous chitosan and rosemary extract improved the storage of fish [83]. Myofibrils have also been used in combination with (non-fibrous) pectin films, with a myofibril to pectin ratio of 5:3, yielding the optimum physical, mechanical, and thermal properties [169]. One study investigated the addition of a smart additive to a myofibril film, namely an extract of sappan wood, which enabled the film to indicate pH [104].

Fibrils liberated from wool have been both cast and spun into nanofibrous films for wound dressings and pollution absorbents [4, 170, 171]. The incorporation of silver nanoparticles provided significant antibacterial activity while maintaining cytocompatibility. These studies suggest keratin fibrils have potential applications in active food packaging. Fibrils have also been isolated from various plant sources, including peas, potatoes, maize, cottonseed, and wheat [41, 123]. For instance, soybean fibrils enhanced the food packaging properties of chitosan, particularly in terms of strength and barrier selectivity for carbon dioxide and oxygen [90].

A concern with using fibrils in food packaging is their possible role in amyloidosis diseases, in which abnormal amyloid fibrils accumulate in body tissues. For example, fibrils derived from casein have been shown to be toxic to certain cells in vitro [172]. However, other studies have found that most fibrils derived from food proteins are not toxic [166]. As research in this area progresses, the situation will become clearer.

Other Proteins

This review has explored the main proteins used in nanofibrous food packaging. However, due to space constraints, not all can be discussed in full. For instance, pea protein has been used as a matrix in a composite film with modified nanocellulose and bilberry extract to visually track the freshness of pork [173].

Seed Flours and Polyhydroxybutyrate

The flour of certain seeds is rich in proteins and polysaccharides, enabling it to be nanospun and cast as matrices for nanofibre composite films. For instance, pea flour consists of 55% polysaccharides, 22% protein, and 12% dietary fibre, with the remaining 11% comprising fat, moisture, and minerals [108]. It is a cost-effective biopolymer raw material, as it has a more straightforward preparation than most pure polysaccharides and proteins. Flour from lentils, peas, and chickpeas has been employed in active nanofibre films, often blended with polyethylene oxide [73, 108]. An example is the spinning of lentil flour and polyethylene oxide with gallic acid into a film that enhances the oxidative stability of walnut kernels [73]. Many protein-rich seed flours have yet to be investigated for nanofibre film-based packaging such as mustard and canola.

Polyhydroxybutyrates are produced by some microorganisms and are neither a polysaccharide nor a protein. They are interesting to packaging researchers because they are thermoplastic, water-resistant and biodegradable. This polymer has been spun into a nanofibre film with palladium nanoparticles. The film was then annealed at 160°C into a continuous film with oxygen-scavenging activity [174]. In another study, polyhydroxybutyrate and phenolic compounds were extracted from Spirulina and spun into a nanofibrous film which was hydrophobic and antimicrobial [175]. Polyhydroxybutyrate film was used in a rare investigation on anti-viral food packaging. A composite of electrospun zein containing cinnamaldehyde between two layers of polyhydroxybutyrate film packaging material was effective against surrogates of the norovirus [93].

ACTIVE AND SMART SUBSTANCES AND THEIR PERFORMANCE IN FILMS

Inherently Active Nanofibre Films

Organisms benefit significantly from the active properties of their constituents; thus, it is unsurprising that some biopolymers are inherently active. For instance, the antimicrobial properties of chitin and keratin are well-documented [137, 158]. When processed into nanofibre films, these properties may be retained, see Table 9.1. Post-spinning modifications, such as Maillard cross-linking, can further enhance the activity of films [19]. In many of the studies reviewed here, films without additives were used as controls.

Active Agents Incorporated into Nanofibre Films

Many substances have been utilised to impart activity to nanofibrous food packaging films, see Tables 9.2–9.5. These include antimicrobials, antioxidants, absorbers of ethylene, moisture, and UV light, as well as carbon dioxide generators. Some active substances exhibit multiple functionalities. Antimicrobial agents are the active compounds most extensively researched for inclusion in packaging films, reflecting the pivotal role microbes play in the spoilage of food and the risk of food poisoning posed by specific microorganisms, such as *Campylobacter jejuni*. Most of the antimicrobials investigated are organic compounds, often extracted from plants. A typical example is β-caryophyllene, which is found in the essential oils of various plant materials such as cloves. It possesses antioxidant and antibacterial properties and has been used in nanofibrous food packaging, after inclusion in naturally occurring halloysite aluminosilicate nanotubes, to slow its release [80]. Photocatalysts, such as certain types of titanium dioxide, exhibit antimicrobial activity. When exposed to specific wavelengths of light, they generate radicals and reactive oxygen species with potent antimicrobial properties. Incorporated into nanofibrous films, they have been shown to enhance the storage of meat and cheese [52, 53]. Magnesium oxide, another antimicrobial inorganic compound, is an appealing packaging additive due to its ability to improve mechanical properties and thermal stability [23].

Bacteriophages are viruses that infect bacteria, and their activity can be preserved when incorporated into nanofibre films, making the films useful as wound dressings, biosensors, and food packaging. In a study focused on food preservation, T4 bacteriophages, which infect *E. coli*, were incorporated into electrospun polyethylene oxide–cellulose diacetate core/shell fibres and polyethylene oxide fibres, demonstrating promising sustained-release behaviour [176]. Another bacteriophage active against *E. coli* was integrated into electrospun sodium alginate/polyethylene oxide packaging, retaining its activity during storage at both room temperature and refrigeration, and reducing the proliferation of *E. coli* on meat, cucumbers, and tomatoes [59]. Bacteriocins are a group of proteins and peptides that some bacteria produce to inhibit other bacteria, and some of these have been used in nanofibrous films as antimicrobials, such as nisin and sakacin [38, 43, 58].

Several antioxidants have been utilised in nanofibrous packaging to neutralise free radicals. These antioxidants are often derived from plants, for instance, ferulic acid and essential oils, due to their generally favourable safety profiles [34, 74]. Some antioxidants also exhibit antimicrobial activity, for instance, clove extract and titanium dioxide nanotubes [44, 52]. However, antioxidants can diminish the efficacy of photochemical antimicrobial compounds by neutralising the reactive oxygen species they generate [53, 111]. Therefore, these types of active agents should not be deployed together.

Substances that emit carbon dioxide have been utilised to extend the shelf life of food. Carbon dioxide displaces oxygen and exhibits antimicrobial properties, though its application in nanofibres has received little attention. An example is discussed in the section on smart actives below [114]. Ethylene is a phytohormone that promotes the ripening of various fruits. To delay this process, sachets containing potassium permanganate are sometimes placed inside packaging to remove ethylene by oxidation. Currently, there are no reported studies on nanofibre biopolymer packaging films with active agents to remove ethylene. However, a nanofibre sachet has been developed for packaging, incorporating titanium dioxide in zein-spun nanofibres. Under UV light, the titanium dioxide converts ethylene to water and carbon dioxide, thereby extending the storage life of tomatoes [177].

The significance of moisture removal is evidenced by the widespread use of desiccant sachets in packaged goods. The desiccants are materials such as silica gel and calcium oxide. Although limited research has been conducted on the use of desiccants in nanofibre films with synthetic polymers, there has been little exploration with biopolymers. The paucity of research into active moisture removal can be attributed to the simplicity and effectiveness of desiccant pouches and passive methods. Simply enhancing the film's resistance to moisture transport is effective, but a more advanced passive approach involves Janus films with directional water transport. These films draw moisture away from the food and release it into the environment, using combinations of hydrophobic and hydrophilic layers. For instance, nanofibrous films with directional water transport have been prepared from chitosan and silk protein and shown to increase the shelf life of pork and shrimp [63, 110].

Typically, some oxygen is entrapped in the packaging applied to food, and oxygen can diffuse through the packaging. Consequently, compounds that react with oxygen, known as oxygen scavengers, can enhance the performance of packaging. Oxygen scavengers such as curcumin and gallic acid have been incorporated into nanofibrous packaging films, thereby improving food storage [71, 73, 82].

The optical properties of packaging are crucial for consumer appeal and product stability. Packaging should be highly transparent when the food is visible to consumers, if they are making purchasing decisions based on the appearance of the product. However, light, especially UV light, can accelerate the ageing of food. Therefore, packaging should have high transmittance of visible light and low transmittance of UV light. This balance has been achieved by incorporating titanium dioxide into films, such as whey fibril films [52, 53].

Smart Agents Incorporated into Nanofibre Films

The most widely investigated type of smart agents in nanofibre packaging is those which change colour in response to pH (i.e., halochromic) to indicate the condition of food, see Table 9.5. For protein-rich foods, the pH is affected by the release of ammonia due to microbial degradation. A widely used type of halochromic substances is anthocyanins, which are appealing because they are extracted from plants, are edible and possess both antimicrobial and antioxidant activities. The effectiveness of anthocyanins has been enhanced by combining them with nanoparticles of gold and tin oxide in spun gelatine nanofibres, which improved the storage stability of fish [101]. The utility of plant pigments such as anthocyanins in non-nanofibrous smart food packaging has recently been reviewed, and they have significant potential for indicating pH, freshness, time–temperature history, and packaging integrity [178].

Smart compounds that respond to factors other than pH have been investigated. For instance, a combination of citric acid and sodium bicarbonate was used as a smart additive to a zein and polyethylene spun film. At high pH, this combination produced carbon dioxide, which, in turn, released the volatile antimicrobial thyme essential oil from the fibres, thereby enhancing the preservation of strawberries [114]. A smart film has been prepared that responds to two factors—high humidity and the colonisation of fruit by *Botrytis cinerea*—by releasing the antimicrobial agent cinnamaldehyde [99]. The nanofibres were spun from a blend of chitosan and polyvinyl alcohol, and the antimicrobial was applied to the film in a post-spinning vapour process. Phase change materials can be considered smart and active as they respond to a particular temperature by undergoing a phase transition, thereby exerting a temperature buffering effect, for instance, zein nanofibres containing tetradecane [115].

ECOLOGICAL CONSIDERATIONS

The production of food has a substantial ecological impact, yet approximately one-third of the food produced is wasted, according to United Nations estimates [149]. Packaging plays a crucial role in minimising this wastage. Enhancements in packaging can reduce food wastage and, consequently, the ecological impact of food production. Nanofibrous active and smart films are at the forefront of developing advanced packaging solutions. However, food packaging itself has a direct and significant negative environmental impact, as approximately 40% of it is plastic. Packaging, in general, represents the largest use of plastic, amounting to 143 million tonnes in 2019 [179]. Mitigating this impact to some extent is the established recycling of certain types of plastic packaging, such as the recycling and down-cycling of polyethylene terephthalate drink bottles. However, the recycling of food packaging films presents greater challenges compared to drink bottles. These films are often bulky, the specific type of polymer can be difficult to identify, and they may be contaminated with food residues.

The increasing awareness of the necessity for sustainably produced materials, coupled with concerns regarding the adverse effects of plastics on human health and the environment, is propelling research into biopolymers as replacements for oil-based polymers across various applications. These applications include textiles, footwear, filters, and even advanced materials such as aerogels [1, 146, 147, 180]. Packaging is no exception to this trend.

Biopolymers are produced by organisms ranging from bacteria to plants and animals, which make them renewable. There is, of course, an environmental impact from these organisms, but this is mitigated for some biopolymers, as they are derived from by-products, for example, chitin from seafood. A significant advantage of biopolymers over their synthetic counterparts is their biodegradability in both soil and marine environments [181–183]. Additionally, a unique benefit of biopolymers in food packaging is their edibility, which can eliminate the need for disposal.

The environmental impact of packaging is not solely due to the polymers. For instance, the spinning of nanofibres, whether synthetic or bio-based, sometimes requires the use of organic solvents or hazardous chemicals. There is a discernible trend towards the utilisation of water and mild organic acids as solvents for fibre spinning. In certain instances, chemical-free post-spinning processes can be employed to confer the desired properties upon nanofibre films [125]. The environmental impact of the active and smart agents utilised in advanced packaging must also be considered. Most of the agents employed in the studies discussed herein are sustainably produced biodegradable natural products, see Tables 9.2–9.5.

FUTURE DIRECTIONS AND CONCLUSIONS

The intrinsic properties of fibres and films are the foundation upon which active and smart functionalities are built. Continuous advancements will be made in areas such as strength, extensibility, biodegradability, barrier properties, and thermal properties. For example, the thermal conductivity of a chitosan–polycaprolactone nanofibre film was increased by the inclusion of boron nitride sheets, thereby improving its ability to dissipate the heat generated by the respiration of packaged fruit and increasing the storage stability [184].

Numerous biomaterials remain unexplored in nanofibrous packaging, including various proteins and polysaccharides that exhibit desirable properties such as robust film formation and effective barrier capabilities [185]. The most innovative development of biomaterials in food packaging in the future is likely to be the incorporation of lipids into the fibres to modulate water-related properties, such as water repellency and water vapour transfer. Additionally, certain lipids have potential as plasticisers and antimicrobial agents within nanofibrous films [186].

The role of cellulosic nanofibres in packaging applications is well-established and likely to grow. However, the future of protein fibrils remains uncertain as safety concerns have not yet been unequivocally resolved. Future research could usefully expand our understanding of how interactions between different biopolymers can be engineered to enhance the spinnability of their solutions and improve the properties of the resulting films. For instance, chitosan and pectin, which are rich in amines and carboxylic acids, respectively, can exhibit a high level of interaction [135].

Of the four major techniques for spinning nanofibres, electrospinning is likely to remain the most significant in the immediate future. The capability of SBS to form nanofibre films directly on food is expected to make it the most important spinning technique in the long term. Regardless of the spinning technique, constructing films with layers of nanofibres possessing different properties can yield highly beneficial characteristics, such as bidirectional transport of water (both liquid and vapour), and this approach warrants further research.

Food is increasingly being sold to consumers as heat-and-eat products such as frozen meals, which creates specific requirements for packaging, such as stability to heating and transparency to microwaves [187]. Biopolymer films will play an important role in this, as many are edible, and consumers will benefit greatly from smart functionality indicating freshness. Heat-and-eat products will increase the amount of packaging that will need to be produced and disposed of, thereby placing even greater importance on sustainability and ecological considerations, areas where biopolymers rate more highly than their synthetic counterparts.

Artificial intelligence is an increasingly utilised tool across various research fields, including nanofibre films. For instance, machine learning techniques have been employed to analyse published data on spun nanofibres containing curcumin, to identify the effects of polymer density, voltage, and the distance between the spinneret and collector on the rate of release of the active compound and the fibre diameter [130]. Smartphones are poised to play a significant role in the future of smart packaging due to their ubiquity, portability, connectivity, and capability to capture and analyse images of packaged food, thereby providing valuable information to users. A pioneering study in this field utilised a solid gelatine film embedded with red cabbage extract and carbon dots. The release of amines from fish packaged by the film changed the film's colour which was monitored and interpreted in real time by a smartphone running specialised machine learning software [188].

Advancements in biopolymer nanofibre films are anticipated to progress rapidly, particularly within the biomedical, energy and filtration sectors. Consequently, it is reasonable to expect that the swift development of nanofibrous biopolymer active smart food packaging materials that have been described in this chapter will persist, driven by sustainability and ecological concerns. Enhanced knowledge in this field will lead to improved packaging performance, reduced food waste, fewer instances of food poisoning, and lower environmental impacts associated with the production of food and packaging and the disposal of food packaging. These are crucial objectives for the future of our planet and the well-being of all its inhabitants.

REFERENCES

1 Hami, S., Affandi, N., Indrie, L., et al. (2023). Enhancing mechanical properties and flux of nanofibre membranes for water filtration. *Polymers*, 15, 3281. https://doi.org/10.3390/polym15153281

2 Güler, H. K., & Çallioğlu, F. C. (2023). Drug-loaded nanofibers: Production techniques and release behaviors. In M. Yusuf, & A. Haji (Eds.), *Handbook of Nanofibers and Nanocomposites. Characteristics, Synthesis, and Applications in Textiles* (pp. 125–148). Jenny Stanford Publishing. https://doi.org/10.1201/9781003432746

3 Nurfaizey, A. H., Tucker, N., Stanger, J., & Staiger, M. P. (2012). Functional nanofibers in clothing for protection against chemical and biological hazards. In Q. Wei (Ed.), *Functional Nanofibers and their Applications* (pp. 236–261). Woodhead. https://doi.org/10.1533/9780857095640.2.236

4 Hassan, M. M., Heins, K., & Zheng, H. (2024). Wound dressing based on silver nanoparticle embedded wool keratin electrospun nanofibers deposited on cotton fabric. *ACS Applied Bio Materials*, 7, 2164–2174. https://doi.org/10.1021/acsabm.3c01111

5 Hammami, M. A., Krifa, M., & Harzallah, O. (2014). Centrifugal force spinning of PA6 nanofibers–processability and morphology of solution-spun fibers. *Journal of the Textile Institute*, 105, 637–647. https://doi.org/10.1080/00405000.2013.842680

6 Çallioğlu, F. C. & Güler, H. K. (2019). Natural nanofibers and applications. In N. Horzum, M. Demir, R. Muñoz-Espí, & D. Crespy (Eds.), *Green Electrospinning* (pp. 157–188). De Gruyter. https://doi.org/10.1515/9783110581393-007

7 Li, J., Wang, G., Gou, X., et al. (2022). Revealing trace nanoplastics in food packages: An electrochemical approach facilitated by synergistic attraction of electrostatics and hydrophobicity. *Analytical Chemistry*, 94, 12657–12663. https://doi.org/10.1021/acs.analchem.2c01703

8 Withey, J. R. (1976). Quantitative analysis of styrene monomer in polystyrene and foods including some preliminary studies of the uptake and pharmacodynamics of the monomer in rats. *Environmental Health Perspectives*, 17, 125–133. https://doi.org/10.1289/ehp.7617125

9 Tang, C., Ramos, M. J. G., Heffernan, A., et al. (2023). Evaluation and identification of chemical migrants leached from baby food pouch packaging. *Chemosphere*, 340, 139758. https://doi.org/10.1016/j.chemosphere.2023.139758

10 McNeil, S. J. (2026). Synthetic polymer nanofibre films with smart and active agents for food packaging: A review. In M. Montazer (Ed.), *Eco-friendly Smart Functional Wearables: From Nanofibers to Textile*, (pp. 125–148). CRC Press.

11 Thakur, V., & Satapathy, B. K. (2024). Migration concerns of biopolymer-based food packaging. In S. Kumar, A. Mukherjee, V. Katiyar, (Eds.), *Agro-Waste Derived Biopolymers and Biocomposites: Innovations and Sustainability in Food Packaging* (pp. 421–450). Scrivener Publishing https://doi.org/10.1002/9781394175161.ch15

12 Lopez-Polo, J., Muñoz-Shugulí, C., Vidal, M. P., & Vidal, C. P. (2023). Electrospun edible films and coatings. *Trends in Food Science & Technology*, 143, 104253. https://doi.org/10.1016/j.tifs.2023.104253

13 Tucker, N., Stanger, J. J., Staiger, M. P., et al. (2012). The history of the science and technology of electrospinning from 1600 to 1995. *Journal of Engineered Fibers and Fabrics*, 7(2-sup), 63–73. https://doi.org/10.1177/155892501200702s10

14 Song, J., Li, Z., & Wu, H. (2020). Blowspinning: A new choice for nanofibers. *ACS Applied Materials & Interfaces*, 12, 33447–33464. https://doi.org/10.1021/acsami.0c05740

15 Dadol, G. C., Kilic, A., Tijing, L. D., et al. (2020). Solution blow spinning (SBS) and SBS-spun nanofibers. *Materials Today Communications*, 25, 101656. https://doi.org/10.1016/j.mtcomm.2020.101656

16 Priyanto, A., Hapidin, D. A., Suciati, T., & Khairurrijal, K. (2022). Current developments on rotary forcespun nanofibers and prospects for edible applications. *Food Engineering Reviews*, 14, 435–461. https://doi.org/10.1007/s12393-021-09304-w

17 Çallıoğlu, F. C., Güler, H. K., & Çetin, E. S. (2019). Emulsion electrospinning of bicomponent poly(vinyl pyrrolidone)/gelatin nanofibers with thyme essential oil. *Materials Research Express*, 6, 125013. https://doi.org/10.1088/2053-1591/ab5387

18 Shen, C., Yang, X., Wang, D., et al. (2024). Carboxymethyl chitosan and polycaprolactone-based rapid in-situ packaging for fruit preservation by solution blow spinning. *Carbohydrate Polymers*, 326, 121636. https://doi.org/10.1016/j.carbpol.2023.121636

19 Ahmed, S. B., Doğan, N., Doğan, C., & Akgul, Y. (2023). A novel approach to crosslink gelatin nanofibers through neutralization-induced Maillard Reaction. *Food and Bioprocess Technology*, 17, 489–503. https://doi.org/10.1016/j.jfoodeng.2023.111761

20 Kwak, H. W., Lee, H., Lee, M. E., & Jin, H. J. (2018). Facile and green fabrication of silk sericin films reinforced with bamboo-derived cellulose nanofibrils. *Journal of Cleaner Production*, 200, 1034–1042. https://doi.org/10.1016/j.jclepro.2018.07.289

21 Dong, Y., Lan, T., Wang, L., et al. (2023). Development of composite electrospun films utilizing soy protein amyloid fibrils and pullulan for food packaging applications. *Food Chemistry: X*, 20, 100995. https://doi.org/10.1016/j.fochx.2023.100995

22 Shahbazi, Y., Shavisi, N., Karami, N., et al. (2021). Electrospun carboxymethyl cellulose-gelatin nanofibrous films encapsulated with *Mentha longifolia* L. essential oil for active packaging of peeled giant freshwater prawn. *LWT*, 152, 112322. https://doi.org/10.1016/j.lwt.2021.112322

23 Eghbalian, M., Shavisi, N., Shahbazi, Y., & Dabirian, F. (2021). Active packaging based on sodium caseinate-gelatin nanofiber mats encapsulated with *Mentha spicata* L. essential oil and MgO nanoparticles. *Food Packaging and Shelf Life*, 29, 100737. https://doi.org/10.1016/j.fpsl.2021.100737

24 Song, Z., Liu, H., Huang, A., et al. (2022). Collagen/zein electrospun films incorporated with gallic acid for tilapia (*Oreochromis niloticus*) muscle preservation. *Journal of Food Engineering*, 317, 110860. https://doi.org/10.1016/j.jfoodeng.2021.110860

25 Tayebi, L., Bayat, F., Mahboubi, A., et al. (2024). Citric acid cross-linked biopolymeric nanofibers containing *Zataria multiflora* extract, an environmentally friendly active food packaging system. *Journal of Food Measurement and Characterization*, 18, 3458–3473. https://doi.org/10.1007/s11694-024-02417-w

26 Wu, Z. K., Chen, R. K., Shi, Y., et al. (2024). Novel polymer-free antimicrobial system based on octyl gallate/hydroxypropyl-β-cyclodextrin inclusion complex electrospun nanofibers for Chinese Giant Salamander preservation. *Food Biophysics*, 19, 310–320. https://doi.org/10.1007/s11483-023-09824-4

27 Dayisoylu, K. S., Akboğa, Z., Doğan, C., et al. (2023). Rapid fabrication of micro-nanofibers from grapevine leaf extract and gelatine via electroblowing. *International Journal of Biological Macromolecules*, 253, 127309. https://doi.org/10.1016/j.ijbiomac.2023.127309

28 Doğan, N., Doğan, C., Eticha, A. K., et al. (2022). Centrifugally spun micro-nanofibers based on lemon peel oil/gelatin as novel edible active food packaging. *Food Control*, 139, 109081. https://doi.org/10.1016/j.foodcont.2022.109081

29 Guler, G. S., Yildiz, E., Yazicioglu, N., et al. (2024). Reduction of lipid oxidation in olive oil using gelatin-based centrifugally spun fibers loaded with caffeic acid. *Food Bioscience*, 59, 104009. https://doi.org/10.1016/j.fbio.2024.104009

30 Lin, L., Zhu, Y., & Cui, H. (2018). Electrospun thyme essential oil/gelatin nanofibers for active packaging against *Campylobacter jejuni* in chicken. *LWT*, 97, 711–718. https://doi.org/10.1016/j.lwt.2018.08.015

31 Padrão, J., Machado, R., Casal, M., et al. (2015). Antibacterial performance of bovine lactoferrin-fish gelatine electrospun membranes. *International Journal of Biological Macromolecules*, 81, 608–614. https://doi.org/10.1016/j.ijbiomac.2015.08.047

32 Liu, F., Saricaoglu, F. T., Avena-Bustillos, R. J., et al. (2018). Antimicrobial carvacrol in solution blow-spun fish-skin gelatin nanofibers. *Journal of Food Science*, 83, 984–991. https://doi.org/10.1111/1750-3841.14076

33 Liu, S., Chen, Z., Zhang, H., et al. (2023). Comparison of eugenol and dihydromyricetin loaded nanofibers by electro-blowing spinning for active packaging. *Food Bioscience*, 51, 102294. https://doi.org/10.1016/j.fbio.2022.102294

34 Sharif, N., Golmakani, M. T., Niakousari, M., et al. (2018). Active food packaging coatings based on hybrid electrospun gliadin nanofibers containing ferulic acid/hydroxypropyl-beta-cyclodextrin inclusion complexes. *Nanomaterials*, 8, 919. https://doi.org/10.3390/nano8110919
35 Golmakani, M. T., Hajjari, M. M., Kiani, F., et al. (2024). Application of electrospinning to fabricate phycocyanin-and Spirulina extract-loaded gliadin fibers for active food packaging. *Food Chemistry*: X, 22, 101275. https://doi.org/10.1016/j.fochx.2024.101275
36 Zahabi, N., Golmakani, M. T., Fazaeli, M., et al. (2021). Electrospinning of glutelin-hordein incorporated with Oliveria decumbens essential oil: Characterization of nanofibers. *Colloids and Surfaces B: Biointerfaces*, 208, 112058. https://doi.org/10.1016/j.colsurfb.2021.112058
37 Prodpran, T., Benjakul, S., & Phatcharat, S. (2012). Effect of phenolic compounds on protein cross-linking and properties of film from fish myofibrillar protein. *International Journal of Biological Macromolecules*, 51, 774–782. https://doi.org/10.1016/j.ijbiomac.2012.07.010
38 Mapelli, C., Musatti, A., Barbiroli, A., et al. (2019). Cellulose nanofiber (CNF)–Sakacin-A active material: Production, characterization and application in storage trials of smoked salmon. *Journal of the Science of Food and Agriculture*, 99, 4731–4738. https://doi.org/10.1002/jsfa.9715
39 Pirveisi, N., Ariaii, P., Esmaeili, M., & Ahmadi, M. (2023). Investigating active packaging based on cellulose nanofibers oxidized by TEMPO method containing hydrolyzed protein obtained from pine tree fruit on the quality of pacific white shrimp (*Litopenaeus vannamei*) during the storage period. *Journal of Food Measurement and Characterization*, 17(4), 3323–3337. https://doi.org/10.1007/s11694-023-01843-6
40 Jiang, N., Hu, Y., & Cheng, Y. (2024). TEMPO-oxidized nanocellulose films modified by tea saponin derived from camellia oleifera: Physicochemical, mechanical, and antibacterial properties. *Polymers*, 16(7), 1016. https://doi.org/10.3390/polym16071016
41 Chen, X., Yi, J., Wen, Z., & Fan, Y. (2024). Ultrasonic pretreatment and epigallocatechin gallate incorporation enhance the formation, apparent viscosity, and antioxidant activity of pea protein amyloid-like fibrils. *Food Hydrocolloids*, 149, 109630. https://doi.org/10.1016/j.foodhyd.2023.109630
42 Jia, X., Li, X., Zhao, J., et al. (2022). Fabrication and characterization of crosslinked pea protein isolated/pullulan/allicin electrospun nanofiber films as potential active packaging material. *Food Packaging and Shelf Life*, 33, 100873. https://doi.org/10.1016/j.fpsl.2022.100873
43 Chang, H., Xu, J., Macqueen, L. A., et al. (2022). High-throughput coating with biodegradable antimicrobial pullulan fibres extends shelf life and reduces weight loss in an avocado model. *Nature Food*, 3, 428–436. https://doi.org/10.1038/s43016-022-00523-w
44 Zambak, Ö., Özkal, A., & Özkal, S. G. (2022). Production of clove extract loaded pullulan and whey protein nanofibers as antioxidant and antibacterial agent. *Journal of Applied Polymer Science*, 139, e53141. https://doi.org/10.1002/app.53141
45 Huang, X., Du, L., Li, Z., et al. (2024). Lactobacillus bulgaricus-loaded and chia mucilage-rich gum arabic/pullulan nanofiber film. *International Journal of Biological Macromolecules*, 266, 131000. https://doi.org/10.1016/j.ijbiomac.2024.131000
46 Fonseca, L. M., Souza, E. J. D., Radünz, M., et al. (2021). Suitability of starch/carvacrol nanofibers as biopreservatives for minimizing the fungal spoilage of bread. *Carbohydrate Polymers*, 252, 117166. https://doi.org/10.1016/j.carbpol.2020.117166
47 Pires, J. B., Dos Santos, F. N., da Cruz, E. P., et al. (2024). Starch extraction from avocado by-product and its use for encapsulation of ginger essential oil by electrospinning. *International Journal of Biological Macromolecules*, 254, 127617. https://doi.org/10.1016/j.ijbiomac.2023.127617

48 Cui, H., Lu, J., Li, C., et al. (2022). Antibacterial and physical effects of cationic starch nanofibers containing carvacrol@ casein nanoparticles against *Bacillus cereus* in soy products. *International Journal of Food Microbiology*, 364, 109530. https://doi.org/10.1016/j.ijfoodmicro.2022.109530
49 Huang, X., Teng, Z., Xie, F., et al. (2024). Loading of cinnamon essential oil into electrospun octenylsuccinylated starch-pullulan nanofiber mats. *Food Hydrocolloids*, 148, 109426. https://doi.org/10.1016/j.foodhyd.2023.109426
50 Lv, H., Xu, H., Xu, E., et al. (2024). Improving structural and functional properties of starch-catechin-based green nanofiber mats for active food packaging by electrospinning and crosslinking techniques. *International Journal of Biological Macromolecules*, 267, 131460. https://doi.org/10.1016/j.ijbiomac.2024.131460
51 Wang, Q., Yu, H., Tian, B., et al. (2019). Novel edible coating with antioxidant and antimicrobial activities based on whey protein isolate nanofibrils and carvacrol. *Coatings*, 9, 583. https://doi.org/10.3390/coatings9090583
52 Feng, Z., Li, L., Wang, Q., Wu, G., et al. (2019). Effect of antioxidant and antimicrobial coating based on whey protein nanofibrils with TiO_2 nanotubes on the quality and shelf life of chilled meat. *International Journal of Molecular Sciences*, 20, 1184. https://doi.org/10.3390/ijms20051184
53 Hasan, Z., Zeshan, B., Hassan, A., et al. (2023). Preparation and characterization of edible whey protein nanofibrils and efficacy studies on the quality and shelf-life of chilled food products. *Journal of Food Safety*, 43, e13034. https://doi.org/10.1111/jfs.13034
54 Najafi, Z., Cetinkaya, T., Bildik, F., et al. (2022). Nanoencapsulation of saffron (Crocus sativus L.) extract in zein nanofibers and their application for the preservation of sea bass fillets. *LWT*, 163, 113588. https://doi.org/10.1016/J.LWT.2022.113588
55 Colussi, R., da Silva, W. M. F., Biduski, B., et al. (2021). Postharvest quality and antioxidant activity extension of strawberry fruit using allyl isothiocyanate encapsulated by electrospun zein. *LWT*, 143, 111087. https://doi.org/10.1016/j.lwt.2021.111087
56 Khalil, R. K., Elleithy, A. E., Ayoup, M. S., et al. (2024). Zein-based nisin-loaded electrospun nanofibers as active packaging mats for control of *Listeria monocytogenes* on peach. *Food Chemistry*, 140441. https://doi.org/10.1016/j.foodchem.2024.140441
57 Yao, Z. C., Chang, M. W., Ahmad, Z., & Li, J. S. (2016). Encapsulation of rose hip seed oil into fibrous zein films for ambient and on demand food preservation via coaxial electrospinning. *Journal of Food Engineering*, 191, 115–123. https://doi.org/10.1016/j.jfoodeng.2016.07.012
58 Heydari-Majd, M., Shadan, M. R., Rezaeinia, H., et al. (2023). Electrospun plant protein-based nanofibers loaded with sakacin as a promising bacteriocin source for active packaging against *Listeria monocytogenes* in quail breast. *International Journal of Food Microbiology*, 391, 110143. https://doi.org/10.1016/j.ijfoodmicro.2023.110143
59 Cui, H., Yang, X., Li, C., et al. (2022). Enhancing anti-E. coli O157: H7 activity of composite phage nanofiber film by D-phenylalanine for food packaging. *International Journal of Food Microbiology*, 376, 109762. https://doi.org/10.1016/j.ijfoodmicro.2022.109762
60 Dai, J., Hu, W., Yang, H., et al. (2022). Controlled release and antibacterial properties of PEO/casein nanofibers loaded with thymol/β-cyclodextrin inclusion complexes in beef preservation. *Food Chemistry*, 382, 132369. https://doi.org/10.1016/j.foodchem.2022.132369
61 Dumitriu, R. P., Stoleru, E., Rosnes, J. T., et al. (2024). Rheological properties influence on the electrospinning of caseinate for loading with antioxidant rosemary extract. *Food Hydrocolloids*, 151, 109883. https://doi.org/10.1016/j.foodhyd.2024.109883

62 Shen, C., Wu, M., Sun, C., et al. (2022). Chitosan/PCL nanofibrous films developed by SBS to encapsulate thymol/HPβCD inclusion complexes for fruit packaging. *Carbohydrate Polymers*, 286, 119267. https://doi.org/10.1016/j.carbpol.2022.119267

63 Zhu, Z., Meng, L., Gao, Z., et al. (2024). Development of chitosan/polycaprolactone-thymol Janus films with directional transport and antibacterial properties for meat preservation. *International Journal of Biological Macromolecules*, 268, 131669. https://doi.org/10.1016/j.ijbiomac.2024.131669

64 Tayebi, L., Mahboubi, A., Bayat, F., et al. (2024). Photo-cross-linked nanofibers containing melissa officinalis extract as a novel active food packaging. *Journal of Polymers and the Environment*. https://doi.org/10.1007/s10924-024-03209-5

65 Zhu, Z., Yu, M., Ren, R., et al. (2024). Thymol incorporation within chitosan/polyethylene oxide nanofibers by concurrent coaxial electrospinning and in-situ crosslinking from core-out for active antibacterial packaging. *Carbohydrate Polymers*, 323, 121381. https://doi.org/10.1016/j.carbpol.2023.121381

66 Lu, S., Tao, J., Liu, X., & Wen, Z. (2022). Baicalin-liposomes loaded polyvinyl alcohol-chitosan electrospinning nanofibrous films: Characterization, antibacterial properties and preservation effects on mushrooms. *Food chemistry*, 371, 131372. https://doi.org/10.1016/j.foodchem.2021.131372

67 Yang, X., Rao, J., Shen, C., et al. (2023). Natamycin-loaded ethyl cellulose/PVP films developed by microfluidic spinning for active packaging. *Foods*, 13, 132. https://doi.org/10.3390/foods13010132

68 Lin, W., Ni, Y., & Pang, J. (2020). Size effect-inspired fabrication of konjac glucomannan/polycaprolactone fiber films for antibacterial food packaging. *International Journal of Biological Macromolecules*, 149, 853–860. https://doi.org/10.1016/j.ijbiomac.2020.01.242

69 Lin, W., Ni, Y., & Pang, J. (2019). Microfluidic spinning of poly(methyl methacrylate)/konjac glucomannan active food packaging films based on hydrophilic/hydrophobic strategy. *Carbohydrate Polymers*, 222, 114986. https://doi.org/10.1016/j.carbpol.2019.114986

70 Lin, W., Ni, Y., Liu, D., et al. (2019). Robust microfluidic construction of konjac glucomannan-based micro-films for active food packaging. *International Journal of Biological Macromolecules*, 137, 982–991. https://doi.org/10.1016/j.ijbiomac.2019.07.045

71 Aydogdu, A., Sumnu, G., & Sahin, S. (2019). Fabrication of gallic acid loaded hydroxypropyl methylcellulose nanofibers by electrospinning technique as active packaging material. *Carbohydrate Polymers*, 208, 241–250. https://doi.org/10.1016/j.carbpol.2018.12.065

72 Amorim, L. F., Fangueiro, R., & Gouveia, I. C. (2022). Novel functional material incorporating flexirubin-type pigment in polyvinyl alcohol_kefiran/polycaprolactone nanofibers. *Journal of Applied Polymer Science*, 139, e53208. https://doi.org/10.1002/app.53208

73 Aydogdu, A., Yildiz, E., Aydogdu, Y., et al. (2019). Enhancing oxidative stability of walnuts by using gallic acid loaded lentil flour based electrospun. *Food Hydrocolloids*, 95, 245–255. https://doi.org/10.1016/j.foodhyd.2019.04.020

74 Lin, L., Mahdi, A. A., Li, C., et al. (2022). Enhancing the properties of *Litsea cubeba* essential oil/peach gum/polyethylene oxide nanofibers packaging by ultrasonication. *Food Packaging and Shelf Life*, 34, 100951. https://doi.org/10.1016/j.fpsl.2022.100951

75 Min, T., Sun, X., Zhou, L., et al. (2021). Electrospun pullulan/PVA nanofibers integrated with thymol-loaded porphyrin metal–organic framework for antibacterial food packaging. *Carbohydrate Polymers*, 270, 118391. https://doi.org/10.1016/j.carbpol.2021.118391

76 Huang, L., Lin, H., Bu, N., et al. (2023). Robust microfluidic construction of polyvinyl pyrrolidone microfibers incorporated with W/O emulsions stabilized by amphiphilic konjac glucomannan. *International Journal of Biological Macromolecules*, 241, 124563. https://doi.org/10.1016/j.ijbiomac.2023.124563

77 Lin, L., Liao, X., & Cui, H. (2019). Cold plasma treated thyme essential oil/silk fibroin nanofibers against Salmonella Typhimurium in poultry meat. *Food Packaging and Shelf Life*, 21, 100337. https://doi.org/10.1016/j.fpsl.2019.100337
78 da Silva, F. T., da Cunha, K. F., Fonseca, et al. (2018). Action of ginger essential oil (Zingiber officinale) encapsulated in proteins ultrafine fibers on the antimicrobial control in situ. *International Journal of Biological Macromolecules*, 118, 107–115. https://doi.org/10.1016/j.ijbiomac.2018.06.079
79 Panahi, Z., Mohsenzadeh, M., & Hashemi, M. (2023). Fabrication and characterization of PVA/WPI nanofibers containing probiotics using electrospinning technique. *Nanomedicine Journal*, 10, 216–226. https://doi.org/10.22038/nmj.2023.71941.1773
80 Ullah, A., Sun, L., Wang, F. F., et al. (2023). Eco-friendly bioactive β-caryophyllene/halloysite nanotubes loaded nanofibrous sheets for active food packaging. *Food Packaging and Shelf Life*, 35, 101028. https://doi.org/10.1016/j.fpsl.2023.101028
81 Tavassoli, M., Khezerlou, A., Sani, M. A., et al. (2024). Methylcellulose/chitosan nanofiber-based composites doped with lactoferrin-loaded Ag-MOF nanoparticles for the preservation of fresh apple. *International Journal of Biological Macromolecules*, 259, 129182. https://doi.org/10.1016/j.ijbiomac.2023.129182
82 Cai, Z., Shen, C., Deng, Z., et al. (2022). Solution blow spinning of multilayer polycaprolactone/curcumin-loaded gelatin/polycaprolactone nanofilm for slow release and bacterial inhibition. *Food Hydrocolloids for Health*, 2, 100062. https://doi.org/10.1016/j.fhfh.2022.100062
83 Du, H., Liu, C., Unsalan, O., et al. (2021). Development and characterization of fish myofibrillar protein/chitosan/rosemary extract composite edible films. *International Journal of Biological Macromolecules*, 184, 463–475. https://doi.org/10.1016/j.ijbio mac. 2021.06.121
84 Ge, J., Lu, W., Zhang, H., et al. (2024). Exploring sustainable food packaging. *International Journal of Biological Macromolecules*, 129200. https://doi.org/10.1016/j.ijbiomac.2024.129200
85 Acharya, D., Liu, S., Lu, H., et al. (2024). Nanoemulsion-integrated gelatin/bacterial cellulose nanofibril-based multifunctional film. *International Journal of Biological Macromolecules*, 257, 128341. https://doi.org/10.1016/j.ijbiomac.2023.128341
86 Tang, J. S., Kuo, C. T., & Liao, Y. C. (2024). Transparent biodegradable composite plastic packaging film from TEMPO-oxidized cellulose nanofibers. *International Journal of Biological Macromolecules*, 129502. https://doi.org/10.1016/j.ijbiomac.2024.129502
87 Khodanazary, A. (2024). Effect of soybean protein isolated/cellulose nanofibers-based nano-biocomposite films enriched with TiO2 nanoparticle and tannic acid on the quality properties of refrigerated common carp fillets. *Journal of Food Measurement and Characterization*, 18, 4920–4921 https://doi.org/10.1007/s11694-024-02546-2
88 Sani, M. A., Ehsani, A., & Hashemi, M. (2017). Whey protein isolate/cellulose nanofibre/TiO2 nanoparticle/rosemary essential oil nanocomposite film. *International Journal of Food Microbiology*, 251, 8–14. https://doi.org/10.1016/j.ijfoodmicro.2017.03.018
89 Papadaki, A., Lappa, I. K., Manikas, A. C., et al. (2024). Grafting bacterial cellulose nanowhiskers into whey protein/essential oil film composites. *Food Hydrocolloids*, *147*, 109374. https://doi.org/10.1016/j.foodhyd.2023.109374
90 Zhang, Y., Nian, Y., Shi, Q., & Hu, B. (2023). Protein fibrillation and hybridization with polysaccharides enhance strength, toughness, and gas selectivity of bioplastic packaging. *Journal of Materials Chemistry A*, 11, 9884–9901. https://doi.org/10.1039/D3TA00446E
91 Zheng, F., Yang, Q., Yuan, C., et al. (2024). Characterizations of corn starch edible films reinforced with whey protein isolate fibrils. *Food Hydrocolloids*, 147, 109412. https://doi.org/10.1016/j.foodhyd.2023.109412

92 Fabra, M. J., López-Rubio, A., & Lagaron, J. M. (2016). Use of the electrohydrodynamic process to develop active/bioactive bilayer films for food packaging applications. *Food Hydrocolloids*, 55, 11–18. https://doi.org/10.1016/j.foodhyd.2015.10.026

93 Fabra, M. J., Castro-Mayorga, J. L., Randazzo, W., et al (2016). Efficacy of cinnamaldehyde against enteric viruses and its activity after incorporation into biodegradable multilayer systems of interest in food packaging. *Food and Environmental Virology*, 8, 125–132. https://doi.org/10.1007/s12560-016-9235-7

94 Goudarzi, J., Moshtaghi, H., & Shahbazi, Y. (2023). Kappa-carrageenan-poly(vinyl alcohol) electrospun fiber mats encapsulated with *Prunus domestica* anthocyanins and epigallocatechin gallate to monitor the freshness and enhance the shelf-life quality of minced beef meat. *Food Packaging and Shelf Life*, 35, 101017. https://doi.org/10.1016/j.fpsl.2022.101017

95 Amjadi, S., Almasi, H., Gholizadeh, S., & Hamishehkar, H. (2024). Double layer packaging based on active black chickpea protein isolate electrospun nanofibers and intelligent salep film containing black chickpea peel anthocyanins for seafood products. *International Journal of Biological Macromolecules*, 278, 134897. https://doi.org/10.1016/j.ijbiomac.2024.134897

96 Tavassoli, M., Khezerlou, A., Bakhshizadeh, M., et al. (2024). Smart packaging containing red poppy anthocyanins for fish freshness monitoring. *Journal of Food Measurement and Characterization*, 18, 3054–3068. https://doi.org/10.1007/s11694-024-02386-0

97 Heydarian, A., & Shavisi, N. (2023). Multifunctional food packaging materials. *Food Packaging and Shelf Life*, 40, 101219. https://doi.org/10.1016/j.fpsl.2023.101219

98 Tavassoli, M., Khezerlou, A., Moghaddam, T. N., et al. (2023). Sumac (*Rhus coriaria* L.) anthocyanin loaded-pectin and chitosan nanofiber matrices for real-time monitoring of shrimp freshness. *International Journal of Biological Macromolecules*, 242, 125044. https://doi.org/10.1016/j.ijbiomac.2023.125044

99 Jiao, X., Yang, M., Chong, X., et al. (2024). Humidity and pH dual-responsive smart nanofiber antimicrobial packaging. *Journal of Food Engineering*, 380, 112163. https://doi.org/10.1016/j.jfoodeng.2024.112163

100 Zhao, X., Zhang, H., Xiang, H., et al. (2024). A nature pH indicator with high colorimetric response sensitivity for pork freshness monitoring. *Food Bioscience*, 57, 103519. https://doi.org/10.1016/j.fbio.2023.103519

101 Cetinkaya, T., Bildik, F., Altay, F., & Ceylan, Z. (2024). Gelatin nanofibers with black elderberry, Au nanoparticles and SnO2 as intelligent packaging layer used for monitoring freshness of Hake fish. *Food Chemistry*, 437, 137843. https://doi.org/10.1016/j.foodchem.2023.137843

102 Cui, H., Gao, J., Khin, M. N., et al. (2024). Preparation and application of pH-sensitive protein nanofibre membrane loaded with alizarin and curcumin for meat preservation. *Packaging Technology and Science*, 37, 793–807. https://doi.org/10.1002/pts.2823

103 Karambasti, P. R., & Shavisi, N. (2024). Development of guar gum-pectin nanofiber mats containing *Papaver rhoeas* petal anthocyanins and cellulose nanocrystals for real-time visual detection of lamb meat freshness. *LWT*, *194*, 115786. https://doi.org/10.1016/j.lwt.2024.115786

104 Rostini, I., Junianto, & Warsiki, E. (2023). Designing bio-based color sensor from myofibrillar-protein-based edible film incorporated with Sappan wood (*Caesalpinia sappan* L.) extract for smart food packaging. *Applied Sciences*, 13, 8205. https://doi.org/10.3390/app13148205

105 Chen, J., Chai, J., Chen, X., et al. (2023). Development of edible films by incorporating nanocrystalline cellulose and anthocyanins into modified myofibrillar proteins. *Food Chemistry*, 417, 135820. https://doi.org/10.1016/j.foodchem.2023.135820

106 Oun, A. A., Roy, S., Hong, S. J., et al. (2024). Development of smart colorimetric indicators for tracking kimchi freshness by loading aronia extract in agar, κ-carrageenan, and cellulose nanofiber films. *International Journal of Biological Macromolecules*, 270, 132343. https://doi.org/10.1016/j.ijbiomac.2024.132343
107 Ali, M. S., Ho, T. C., Park, J. S., et al. (2024). Fabrication of smart film based on fish gelatin incorporating phycoerythrin and cellulose nanofibrils to monitor fish freshness. *Food Bioscience*, 59, 104111. https://doi.org/10.1016/j.fbio.2024.104111
108 Aydogdu Emir, A., Yildiz, E., Aydogdu, Y., et al. (2021). Gallic acid encapsulated pea flour-based nanofibers produced by electrospinning as a potential active food packaging material. *Legume Science*, 3, e90. https://doi.org/10.1002/leg3.90
109 Surendhiran, D., Roy, V. C., Ho, T. C., et al. (2023). Smart packaging film prepared from subcritical water-modified oat starch and betalain of beetroot extract reinforced with cellulose nanofibrils. *Sustainable Chemistry and Pharmacy*, 36, 101349. https://doi.org/10.1016/j.scp.2023.101349
110 Xia, L., Li, L., Xiao, Y., et al. (2023). Active colorimetric bilayer polycaprolactone-eucalyptus oil@ silk fibroin-bayberry anthocyanins (PCL-EO@ SF-BAs). *Journal of Colloid and Interface Science*, 651, 356–367. https://doi.org/10.1016/j.jcis.2023.07.181
111 Qin, W., Zou, L., Hou, Y., et al. (2024). Characterization of novel anthocyanins film @ carbon quantum dot nanofiber intelligent active double-layer film, physicochemical properties and fresh-keeping monitoring in Ictalurus punctatus fish. *Chemical Engineering Journal*, 496, 154041. https://doi.org/10.1016/j.cej.2024.154041
112 Lv, H., Wang, C., He, D., et al. (2024). Intelligent food tag: A starch-anthocyanin-based pH-sensitive electrospun nanofiber mat for real-time food freshness monitoring. *International Journal of Biological Macromolecules*, 256, 128384. https://doi.org/10.1016/j.ijbiomac.2023.128384
113 de Oliveira Filho, J. G., de Oliveira Silva, C., Egea, M. B., et al. (2024). Efficient and sustainable production of intelligent nonwovens as indicators of food spoilage through solution blow spinning of proteins and natural pigments from agri-food waste. *Food Control*, 166, 110707. https://doi.org/10.1016/j.foodcont.2024.110707
114 Jiao, X., Chong, X., Du, H., et al. (2024). Leveraging gas generation for humidity-responsive release of Thyme essential oil via nanofiber film for fresh fruits. *Chemical Engineering Journal*, 487, 150606. https://doi.org/10.1016/j.cej.2024.150606
115 Karim, M., Fathi, M., Soleimanian-Zad, S., & Spigno, G. (2022). Development of sausage packaging with zein nanofibers containing tetradecane produced via needle-less electrospinning method. *Food Packaging and Shelf Life*, 33, 100911. https://doi.org/10.1016/j.fpsl.2022.100911
116 Shen, C., Cao, Y., Rao, J., et al. (2021). Application of solution blow spinning to rapidly fabricate natamycin-loaded gelatin/zein/polyurethane antimicrobial nanofibers for food packaging. *Food Packaging and Shelf Life*, 29, 100721. https://doi.org/10.1016/j.fpsl.2021.100721
117 Kramar, A., & González-Benito, J. (2023). Preparation of cellulose acetate film with dual hydrophobic-hydrophilic properties using solution blow spinning. *Materials & Design*, 227, 111788. https://doi.org/10.1016/j.matdes.2023.111788
118 Li, X., Chen, H., & Yang, B. (2016). Centrifugally spun starch-based fibers from amylopectin rich starches. *Carbohydrate Polymers*, 137, 459–465. https://doi.org/10.1016/j.carbpol.2015.10.079
119 Du, X. Y., Li, Q., Wu, G., & Chen, S. (2019). Multifunctional micro/nanoscale fibers based on microfluidic spinning technology. *Advanced Materials*, 31, 1903733. https://doi.org/10.1002/adma.201903733
120 Li, R., Feng, Y., Zhang, H., et al. (2023). Recent advances in fabricating, characterizing, and applying food-derived fibers using microfluidic spinning technology. *Food Hydrocolloids*, 108947. https://doi.org/10.1016/j.foodhyd.2023.108947

121 Li, S., Hang, Y., Ding, Z., et al. (2020). Microfluidic silk fibers with aligned hierarchical microstructures. *ACS Biomaterials Science & Engineering*, 6, 2847–2854. https://doi.org/10.1021/acsbiomaterials.0c00060
122 Schmitt, P., Wedrich, K., Müller, L., et al. (2017). Design, fabrication and characterisation of a microfluidic time-temperature indicator. *Journal of Physics: Conference Series*, 922(1), 012004. https://doi.org/10.1088/1742-6596/922/1/012004
123 Chen, D., Jones, O. G., & Campanella, O. H. (2023). Plant protein-based fibers. *Critical Reviews in Food Science and Nutrition*, 63, 4554–4578. https://doi.org/10.1080/10408398.2021.2004991
124 Cai, J., Zhang, Q., Lei, M., et al. (2016). The use of solvent-soaking treatment to enhance the anisotropic mechanical properties of electrospun nanofiber membranes for water filtration. *RSC Advances*, 6, 66807–66813. https://doi.org/10.1039/C6RA11020G
125 Wongkrongsak, S., Pangon, A., Pongsak, N., et al. (2022). Strengthened silk-fibroin/poly (ethylene oxide) nonwoven nanofibers. *ACS Sustainable Chemical Engineering*, 10, 2653–2672. https://doi.org/10.1021/acssuschemeng.1c06965
126 Romani, V. P., Olsen, B., Collares, M. P., et al. (2019). Plasma technology as a tool to decrease the sensitivity to water of fish protein films for food packaging. *Food Hydrocolloids*, 94, 210–216. https://doi.org/10.1016/j.foodhyd.2019.03.021
127 Zhu, W., Zhang, D., Liu, X., et al. (2022). Improving the hydrophobicity and mechanical properties of starch nanofibrous films by electrospinning and cross-linking for food packaging applications. *LWT*, 169, 114005. https://doi.org/10.1016/j.lwt.2022.114005
128 McNeil, S. J., & Standard, O. C. (2017). Increased bending rigidity of wool fabric imparted by hybrid organic-inorganic sol-gel coatings. *Textile Research Journal*, 87, 607–616. https://doi.org/10.1177/0040517516635990
129 Moreno, M. A., Orqueda, M. E., Gómez-Mascaraque, L. G., et al. (2019). Crosslinked electrospun zein-based food packaging coatings containing bioactive chilto fruit extracts. *Food Hydrocolloids*, 95, 496–505. https://doi.org/10.1016/j.foodhyd.2019.05.001
130 Khedri, M., Beheshtizadeh, N., Rostami, M., et al. (2022). Artificial intelligence deep exploration of influential parameters on physicochemical properties of curcumin-loaded electrospun nanofibers. *Advanced NanoBiomed Research*, 2, 2100143. https://doi.org/10.1002/anbr.202100143
131 Jiang, G., Ramachandraiah, K., & Zhao, C. (2023). Advances in the development and applications of nanofibers in meat products. *Food Hydrocolloids*, 146, 109210. https://doi.org/10.1016/j.foodhyd.2023.109210
132 Fonseca, L. M., da Cruz, E. P., Crizel, R. L., et al (2024). New advances of electrospun starch fibers, encapsulation, and food applications. *Trends in Food Science & Technology*, 147, 104467. https://doi.org/10.1016/j.tifs.2024.104467
133 Lancuški, A., Abu Ammar, A., Avrahami, R., et al. (2017). Design of starch-formate compound fibers as encapsulation platform for biotherapeutics. *Carbohydrate Polymers*, 158, 68–76. https://doi.org/10.1016/j.carbpol.2016.12.003
134 Xu, C., Ma, J., Wang, W., et al. (2022). Preparation of pectin-based nanofibers encapsulating *Lactobacillus rhamnosus* 1.0320 by electrospinning. *Food Hydrocolloids*, 124, 107216. https://doi.org/10.1016/j.foodhyd.2021.107216
135 Celebioglu, A., Saporito, A. F., & Uyar, T. (2022). Green electrospinning of chitosan/pectin nanofibrous films by the incorporation of cyclodextrin/curcumin inclusion complexes. *ACS Sustainable Chemistry & Engineering*, 10, 4758–4769. https://doi.org/10.1021/acssuschemeng.2c00650
136 Shen, C., Deng, Z., Rao, J., et al. (2022). Characterization of glycosylated gelatin/pullulan nanofibers fabricated by multi-fluid mixing solution blow spinning. *International Journal of Biological Macromolecules*, 214, 512–521. https://doi.org/10.1016/j.ijbiomac.2022.06.082

137 Fan, Y., Saito, T., & Isogai, A. (2008). Preparation of chitin nanofibers from squid pen β-chitin by simple mechanical treatment under acid conditions. *Biomacromolecules*, *9*(7), 1919–1923. https://doi.org/10.1021/bm800178b
138 Zhang, W., Chen, M., Zha, B., & Diao, G. (2012). Correlation of polymer-like solution behaviors with electrospun fiber formation of hydroxypropyl-β-cyclodextrin and the adsorption study on the fiber. *Physical Chemistry Chemical Physics*, 14, 9729–9737. https://doi.org/10.1039/C2CP41092C
139 Celebioglu, A., & Uyar, T. (2013). Electrospinning of nanofibers from non-polymeric systems: Electrospun nanofibers from native cyclodextrins. *Journal of Colloid and Interface Science*, 404, 1–7. https://doi.org/10.1016/j.jcis.2013.04.034
140 Zhang, W., Gong, X., Cai, Y., et al. (2013). Investigation of water-soluble inclusion complex of hypericin with β-cyclodextrin polymer. *Carbohydrate Polymers*, 95, 366–370.
141 Gao, S., Jiang, J., Li, X., et al. (2021). Electrospun polymer-free nanofibers incorporating hydroxypropyl-β-cyclodextrin/difenoconazole via supramolecular assembly for antifungal activity. *Journal of Agricultural and Food Chemistry*, 69, 5871–5881. https://doi.org/10.1021/acs.jafc.1c01351
142 Saffarionpour, S. (2019). Nanoencapsulation of hydrophobic food flavor ingredients and their cyclodextrin inclusion complexes. *Food and Bioprocess Technology*, 12, 1157–1173. https://doi.org/10.1007/s11947-019-02285-z
143 Cheng, C., Ma, X., Min, T., et al. (2024). Antibacterial nanofibrous film encapsulated with 4-terpineol/β-cyclodextrin inclusion complexes. *Food Hydrocolloids*, 156, 110231. https://doi.org/10.1016/j.foodhyd.2024.110231
144 de Carvalho, A., & Conte-Junior, C. (2021). Food-derived biopolymer kefiran composites, nanocomposites and nanofibers. *Trends in Food Science & Technology*, 116, 370–386. https://doi.org/10.1016/j.tifs.2021.07.038
145 Al-Sudani, B. T., Khoshkalampour, A., Kamil, M. M., et al. (2024). A novel antioxidant and antimicrobial food packaging based on Eudragit®/collagen electrospun nanofiber. *LWT*, 193, 115730. https://doi.org/10.1016/j.lwt.2024.115730
146 Yordanova, D., & Gibbons, S. (2017). Milking it. *Twist*, 18, 36–37.
147 Indrie, L., McNeil, S., Mutlu, M. M., Bota, S., et al. (2023). The initial development and evaluation of cross-linked casein films for sustainable footwear. *Coatings*, 13(2), 217. https://doi.org/10.3390/coatings13020217
148 Tomasula, P. M., Sousa, A. M., Liou, S. C., et al (2016). Electrospinning of casein/pullulan blends for food-grade applications. *Journal of Dairy Science*, 99, 1837–1845. https://doi.org/10.3168/jds.2015-10374
149 Ali, A., Bairagi, S., Ganie, S. A., & Ahmed, S. (2023). Polysaccharides and proteins based bionanocomposites as smart packaging materials. *International Journal of Biological Macromolecules*, 252, 126534. https://doi.org/10.1016/j.ijbiomac.2023.126534
150 Mohammadi, M., Mirabzadeh, S., Shahvalizadeh, R., & Hamishehkar, H. (2020). Development of novel active packaging films based on whey protein isolate incorporated with chitosan nanofiber and nano-formulated cinnamon oil. *International Journal of Biological Macromolecules*, 149, 11–20. https://doi.org/10.1016/j.ijbiomac.2020.01.083
151 Nanda, A., Pandey, P., Rajinikanth, P. S., & Singh, N. (2024). Revolution of nanotechnology in food packaging: Harnessing electrospun zein nanofibers for improved preservation. *International Journal of Biological Macromolecules*, 129416. https://doi.org/10.1016/j.ijbiomac.2024.129416
152 Wang, S., Marcone, M. F., Barbut, S., & Lim, L. T. (2013). Electrospun soy protein isolate-based fiber fortified with anthocyanin-rich red raspberry (*Rubus strigosus*) extracts. *Food Research International*, 52, 467–472. https://doi.org/10.1016/j.foodres.2012.12.036

153 Rashidi, M., Mansour, S. S., Mostashari, P., et al. (2021). Electrospun nanofiber based on ethyl cellulose/soy protein isolated integrated with bitter orange peel extract for antimicrobial and antioxidant active food packaging. *International Journal of Biological Macromolecules*, 193, 1313–1323. https://doi.org/10.1016/j.ijbiomac.2021.10.182

154 García Ibarra, V., Sendón, R., & Rodríguez-Bernaldo de Quirós, A. (2016). Antimicrobial food packaging based on biodegradable materials. In J. Barros-Velázquez (Ed.), *Antimicrobial Food Packaging* (pp. 363–384). Academic Press. https://doi.org/10.1016/B978-0-12-800723-5.00029-2

155 Reddy, N., & Yang, Y. (2007). Novel protein fibers from wheat gluten. *Biomacromolecules*, 8, 638–643. https://doi.org/10.1021/bm0608840

156 Hajikhani, M., Emam-Djomeh, Z., & Askari, G. (2020). Fabrication and characterization of gluten film reinforced by lycopene-loaded electrospun polylactic acid nano-fibers. *Food and Bioprocess Technology*, 13, 2217–2227. https://doi.org/10.1007/s11947-020-02561-3

157 Zhang, Y., Deng, L., Zhong, H., et al. (2022). Impact of glycation on physical properties of composite gluten/zein nanofibrous film. *Food Chemistry*, 366, 130586. https://doi.org/10.1016/j.foodchem.2021.130586

158 Johnson, N. A. G., Wood, E. J., Ingham, P. E., et al. (2003). Wool as a technical fibre. *Journal of The Textile Institute*, 94(3-4), 26–41. https://doi.org/10.1080/00405000308630626

159 McNeil, S. J., & Sunderland, M. R. (2016). The nanocidal and antifeedant activities of titanium dioxide desiccant toward wool-digesting *Tineola bisselliella* moth larvae. *Clean Technologies and Environmental Policy*, 18, 843–852. https://doi.org/10.1007/s10098-015-1060-4

160 Sunderland, M. R., Cruickshank, R. H., & Leighs, S. J. (2014). The efficacy of antifungal azole and antiprotozoal compounds in protection of wool from keratin-digesting insect larvae. *Textile Research Journal*, 84, 924–931. https://doi.org/10.1177/0040517513515312

161 McNeil, S. J., Sunderland, M. R. & Leighs, S. J. (2017). The utilisation of wool as a catalyst and as a support for catalysts. *Applied Catalysis A, General*, 541, 120–140. https://doi.org/10.1016/j.apcata.2017.04.021

162 https://www.woolcool.com/ accessed 28 August 2024.

163 Maurizii, G., Valentini, L., Sotgiu, G., et al. (2024). The dark side of the wool? From wool wastes to keratin microfilaments through the solution blow spinning process. *International Journal of Biological Macromolecules*, 275, 133722. https://doi.org/10.1016/j.ijbiomac.2024.133722

164 Ramirez, D. O. S., Carletto, R. A., Tonetti, C., et al. (2017). Wool keratin film plasticized by citric acid for food packaging. *Food Packaging and Shelf Life*, 12, 100–106. https://doi.org/10.1016/j.fpsl.2017.04.004

165 Ertek, D. A., Sanli, N. O., Menceloglu, Y. Z., & Seven, S. A. (2023). Environmentally friendly, antibacterial materials from recycled keratin incorporated electrospun PLA films. *European Polymer Journal*, 185, 111804. https://doi.org/10.1016/j.eurpolymj.2022.111804

166 Cao, Y., & Mezzenga, R. (2019). Food protein amyloid fibrils. *Advances in Colloid and Interface Science*, 269, 334–356. https://doi.org/10.1016/j.cis.2019.05.002

167 Zhao, Y., Wang, C., Lu, W., et al. (2021). Evolution of physicochemical and antioxidant properties of whey protein isolate during fibrillization process. *Food Chemistry*, 357, 129751. https://doi.org/10.1016/j.foodchem.2021.129751

168 Rostamzad, H., Paighambari, S. Y., Shabanpour, B., et al. (2016). Improvement of fish protein film with nanoclay and transglutaminase for food packaging. *Food Packaging and Shelf Life*, 7, 1–7. https://doi.org/10.1016/j.fpsl.2015.10.001

169 Florentino, G. I. B., Lima, D. A. S., Santos, M. M. F., et al. (2022). Characterization of a new food packaging material based on fish by-product proteins and passion fruit pectin. *Food Packaging and Shelf Life*, 33, 100920. https://doi.org/10.1016/j.fpsl.2022.100920

170 Tissera, N. D., Wijesena, R. N., Ludowyke, N., et al. (2024). Keratin protein nanofibers from merino wool yarns. *RSC Advances*, 14, 6793–6804. https://doi.org/10.1039/D3RA07063H
171 Pervez, M. N., Hassan, M. M., & Naddeo, V. (2024). Separation of cationic methylene blue dye from its aqueous solution by S-sulfonated wool keratin-based sustainable electrospun nanofibrous membrane biosorbent. *Separation and Purification Technology*, 333, 125903. https://doi.org/10.1016/j.seppur.2023.125903
172 Dehle, F. C., Ecroyd, H., Musgrave, I. F., & Carver, J. A. (2010). αB-Crystallin inhibits the cell toxicity associated with amyloid fibril formation by κ-casein and the amyloid-β peptide. *Cell Stress and Chaperones*, 15, 1013–1026. https://doi.org/10.1007/s12192-010-0212-z
173 Zhang, C., Qu, L., Liu, H., et al. (2024). pH-responsive color-indicating film of pea protein isolate cross-linked with dialdehyde carboxylated cellulose nanofibers. *International Journal of Biological Macromolecules*, 257, 128671. https://doi.org/10.1016/j.ijbiomac.2023.128671
174 Cherpinski, A., Gozutok, M., Sasmazel, H. T., et al. (2018). Electrospun oxygen scavenging films of poly (3-hydroxybutyrate) containing palladium nanoparticles for active packaging applications. *Nanomaterials*, 8, 469. https://doi.org/10.3390/nano8070469
175 Kuntzler, S. G., de Almeida, A. C. A., Costa, J. A. V., & de Morais, M. G. (2018). Polyhydroxybutyrate and phenolic compounds microalgae electrospun nanofibers: A novel nanomaterial with antibacterial activity. *International Journal of Biological Macromolecules*, 113, 1008–1014. https://doi.org/10.1016/j.ijbiomac.2018.03.002
176 Korehei, R., & Kadla, J. F. (2014). Encapsulation of T4 bacteriophage in electrospun poly (ethylene oxide)/cellulose diacetate fibers. *Carbohydrate Polymers*, 100, 150–157. https://doi.org/10.1016/j.carbpol.2013.03.079
177 Böhmer-Maas, B. W., Fonseca, L. M., Otero, D. M., et al. (2020). Photocatalytic zein-TiO2 nanofibers as ethylene absorbers for storage of cherry tomatoes. *Food Packaging and Shelf Life*, 24, 100508. https://doi.org/10.1016/j.fpsl.2020.100508
178 Kumar, Y., Bist, Y., Thakur, D., et al. (2024). A review on the role of pH-sensitive natural pigments in biopolymers based intelligent food packaging films. *International Journal of Biological Macromolecules*, 276, 133869. https://doi.org/10.1016/j.ijbiomac.2024.133869
179 OECD (2022). Global Plastics Outlook. https://stats.oecd.org/viewhtml.aspx?datasetcode=PLASTIC_USE_10&lang=en, accessed on 20 August 2024.
180 McNeil, S. J. & Gupta, H. (2022). Emerging applications of aerogels in textiles. *Polymer Testing*, 106, 107426. https://doi.org/10.1016/j.polymertesting.2021.107426
181 Collie, S. R., Ranford, S. L., Fowler, I. J., & Brorens, P. H. (2019). Microfibre Pollution–What's The Story For Wool? In *Proceedings of the 19th World Textile Conference-Autex 2019*, Ghent.
182 McNeil, S. J., Sunderland, M. R. & Zaitseva, L. I. (2007). Closed-loop wool carpet recycling. *Resources, Conservation and Recycling*, 51, 220–224. https://doi.org/10.1016/j.resconrec.2006.09.006
183 Collie, S., Brorens, P., Hassan, M. M., & Fowler, I. (2024). Marine biodegradation behavior of wool and other textile fibers. *Water, Air, & Soil Pollution*, 235(5), 283. https://doi.org/10.1007/s11270-024-07093-6
184 Shen, C., Yang, Z., Rao, J., et al. (2022). Development of a thermally conductive and antimicrobial nanofibrous mat for the cold chain packaging of fruits and vegetables. *Materials & Design*, 221, 110931. https://doi.org/10.1016/j.matdes.2022.110931
185 Razzaq, H. A., Pezzuto, M., Santagata, G., et al, (2016). Barley β-glucan-protein based bioplastic film with enhanced physicochemical properties for packaging. *Food Hydrocolloids*, 58, 276–283. https://doi.org/10.1016/j.foodhyd.2016.03.003

186 Zubair, M., Pradhan, R. A., Arshad, M., & Ullah, A. (2021). Recent advances in lipid derived bio-based materials for food packaging applications. *Macromolecular Materials and Engineering*, 306, 2000799. https://doi.org/10.1002/mame.202000799
187 Thanakkasaranee, S., Sadeghi, K., & Seo, J. (2023). Packaging materials and technologies for microwave applications. *Critical Reviews in Food Science and Nutrition*, 63, 6464–6483. https://doi.org/10.1080/10408398.2022.2033685
188 Doğan, V., Evliya, M., Kahyaoglu, L. N., & Kılıç, V. (2024). On-site colorimetric food spoilage monitoring with smartphone embedded machine learning. *Talanta*, 266, 125021. https://doi.org/10.1016/j.talanta.2023.125021

10 The Future of Diffusion Is Green

Eco-Carriers as Functional Enablers in Sustainable Textile Processing

Ece Kalayci and Ozan Avinc

INTRODUCTION

The textile sector is one of the most resource-intensive sectors worldwide, contributing significantly to the global economy while exerting considerable environmental pressure. Across the value chain—from fiber production and spinning to dyeing, finishing, and garment manufacturing—large amounts of water, energy, and chemicals are consumed [1–3]. Among these processes, dyeing and finishing stand out as particularly critical. They not only determine the aesthetic and functional qualities of textiles but also account for a substantial share of the industry's ecological footprint, including wastewater generation, high chemical oxygen demand (COD), greenhouse gas emissions, and the release of persistent chemical residues [4, 5]. In an era where sustainability is both a regulatory imperative and a consumer-driven demand, innovation in wet processing technologies is indispensable for reducing the environmental burden of textile production.

A central challenge within this context lies in the coloration of hydrophobic synthetic fibers such as polyester and aramids [6, 7]. Among these, polyester dominates the global fiber market, accounting for more than half of all textile fiber consumption. According to recent industry statistics, the annual production of polyester exceeds 60 million metric tons, reflecting its ubiquitous presence in apparel, home textiles, and technical applications [3, 8, 9]. Its widespread use stems from a favorable balance of mechanical strength, durability, cost efficiency, and processability, yet these same attributes—particularly its dense molecular packing and low affinity toward dyes—make it one of the most challenging fibers to color efficiently. As a result, polyester dyeing continues to be a focal point of both industrial innovation and sustainability debates within the textile sector. Their compact, highly oriented molecular structures severely limit the free volume available for dye diffusion [6, 10, 11].

As a result, effective dyeing typically requires elevated temperatures, high-pressure conditions, or the addition of chemical auxiliaries to promote molecular

 DOI: 10.1201/9781003507208-10

mobility [12]. While such methods improve color uptake, they do so at the expense of increased energy use and chemical loading, creating an inherent tension between performance and sustainability [10, 11]. To address these limitations, carrier systems have long been employed as auxiliaries in synthetic fiber dyeing. By temporarily depressing the glass transition temperature (T_g) of the fiber, carriers enhance chain mobility and enable greater penetration of dyestuff molecules into the fiber matrix, thereby facilitating effective coloration under otherwise unfavorable conditions [13–16].

Despite their technical utility, conventional carriers present serious drawbacks that run counter to modern sustainability goals. Typically derived from petrochemical feedstocks, many of these compounds are volatile organic chemicals (VOCs) or halogenated aromatics [17–19]. They pose health risks to dyehouse workers, resist biodegradation, and persist in the environment long after usage. Wastewater effluents containing conventional carriers frequently exhibit elevated COD and biological oxygen demand (BOD), creating challenges for treatment plants and further burdening ecosystems. With tightening global regulations and growing consumer demand for environmentally responsible textiles, continued reliance on such carrier systems is increasingly untenable [6, 20].

These challenges underscore the urgency of rethinking carrier technologies in textile dyeing and finishing. There is an apparent necessity for eco-friendly alternatives that can replicate or surpass the performance of traditional carriers while minimizing environmental harm. Ideally, such eco-carriers should be bio-based, nontoxic, and biodegradable. Beyond facilitating dye diffusion, they also hold potential to impart added functionalities—such as antibacterial activity, ultraviolet (UV) protection, or natural fragrance retention—thus aligning with the market demand for multifunctional and sustainable textiles [6, 17]. In this context, naturally derived aromatic compounds such as vanillin and coumarin represent promising candidates, bridging the gap between high-performance processing and ecological responsibility.

This chapter first provides information about diffusion mechanisms and conventional carrier systems, eco-friendly and bio-based carriers, and then the concept and definition of eco-carriers. Afterward, the structural and functional properties of bio-based aromatic compounds are discussed, followed by detailed information on vanillin and coumarin, and subsequently, information on other potential bio-derived carriers is also given. Afterward, previous studies on chemical structures and functional groups of bio-based carriers, interaction mechanisms with dyes, process parameters, functional outcomes, and value-added functionalities (such as antibacterial activity, UV protection, antioxidant properties, aromatherapeutic and fragrance retention, and application domains) in the literature are reviewed. Finally, the environmental and sustainability perspective and future perspectives issues were discussed.

DIFFUSION MECHANISM AND CONVENTIONAL CARRIER SYSTEMS

Carriers are auxiliary agents employed in the dyeing of synthetic fibers whose compact molecular structures hinder dye adsorption and diffusion, particularly at lower temperatures. By facilitating the mobility of dyestuff molecules into the fiber interior, carriers enable satisfactory coloration of hydrophobic fibers under conditions where conventional aqueous dyeing would otherwise be ineffective [19, 21, 22].

From a chemical perspective, most conventional carriers belong to classes such as phenolics, chlorinated aromatics, aromatic hydrocarbons, and ethers [23, 24]. Based on their solubility and affinity, they are often categorized as either hydrophobic or hydrophilic, with their action mechanisms differing accordingly [25]. The chemical structure of a carrier primarily governs its distribution between the aqueous dye bath and the fiber surface [26]. In both cases, the activity of carriers is believed to rely on their capability to disrupt intermolecular forces within the fiber, thereby promoting dye penetration [26].

Hydrophilic carriers typically contain proton-donating groups (–OH, $-NH_2$), enabling van der Waals interactions with hydrophobic fibers while simultaneously attracting water molecules. When absorbed by the fiber, such carriers may transport a small amount of water into the polymer matrix, enhancing swelling and thus allowing greater fixation of dye molecules [10, 26]. In practice, hydrophilic carriers act as strong dispersing agents, improving the solubility of disperse dyestuff in water but, paradoxically, often reducing their diffusion into polyester fibers [13, 25]. Benzoic acid derivatives exemplify hydrophilic carriers, being more suitable for polar fibers where they enhance water uptake and fiber swelling.

Hydrophobic carriers, by contrast, are generally considered more effective in polyester dyeing, as their poor solubility in aqueous media prevents significant competition with the dye for fiber sites [25]. Their principal mode of action involves decreasing the glass transition temperature (T_g) of the fiber, thereby plasticizing the amorphous regions and increasing the free volume available for dye diffusion [18]. Compounds such as dichlorobenzene and trichlorobenzene are classic hydrophobic carriers, designed specifically to promote dye penetration into polyester by enabling nonaqueous diffusion pathways [13, 14, 27].

The general mechanism of carrier-assisted dyeing can be summarized as follows: carrier molecules, being smaller than dye molecules, first adsorb onto the fiber surface and rapidly diffuse into its amorphous regions. This process induces swelling of the fiber and generates microvoids between macromolecular chains. The resulting increase in free volume reduces the fiber's T_g, effectively plasticizing it and rendering the amorphous domains more accessible to dye molecules [16, 23, 25, 28]. Consequently, carriers significantly accelerate the rate and extent of dyestuff diffusion within hydrophobic fibers such as polyester.

Despite their efficacy, conventional carriers exhibit several undesirable properties. Many are associated with unpleasant odor, poor biodegradability, and high toxicity [29, 30]. During dyeing and subsequent rinsing, a substantial portion is released into wastewater, posing threats to aquatic ecosystems and complicating effluent treatment due to elevated COD and BOD loads. Additionally, a fraction of carrier molecules remains trapped within fibers and may volatilize during heat-based processes such as drying, thermofixation, and ironing, releasing harmful emissions into the atmosphere [11, 27]. These limitations translate into considerable environmental burdens and operational challenges for textile finishing plants, as most conventional carriers are characterized by unpleasant odor, toxicity, poor biodegradability, and substantial wastewater loads—factors that make them increasingly incompatible with modern sustainability targets. In the following part of this chapter, information on eco-friendly and bio-based carriers is given.

ECO-FRIENDLY AND BIO-BASED CARRIERS

In reply to the ecological and health concerns associated with conventional carrier systems, considerable research efforts have shifted toward the development of sustainable alternatives. Biodegradable and bio-based substitutes—often referred to as eco-carriers—have obtained expanding attention for their potential to combine technical efficacy with reduced ecological footprint. Unlike traditional petrochemical carriers, eco-carriers are typically derived from renewable resources, exhibit lower toxicity, and in some cases, provide additional functional properties to the dyed fabric [17, 19, 29, 31].

Recent developments in this field highlight the promise of naturally derived aromatic compounds such as vanillin and coumarin, which possess phenolic and aromatic functionalities that facilitate dye–fiber interactions while offering the advantages of partial biodegradability and derivation from agricultural or lignin-based feedstocks [19, 32, 33]. Literature findings indicate that these compounds not only enhance dye diffusion under milder conditions (100°C–120°C) but also deliver added benefits such as antibacterial activity, UV protection, and fragrance retention, thereby aligning with current trends toward multifunctional textiles [34, 35].

While several commercial eco-carrier formulations are already available, many continue to raise concerns regarding residual wastewater loads and incomplete biodegradability. Nevertheless, the shift from purely petrochemical auxiliaries to bio-based alternatives marks a critical step in reducing the overall environmental impact of textile dyeing and finishing. Moreover, eco-carriers open new pathways for integrating green chemistry principles with advanced functional finishing, bridging sustainability with innovation.

Concept and Definition of Eco-Carriers

The concept of eco-carriers has emerged in answer to the urgent necessity for ecologically benign alternatives to conventional carrier systems traditionally employed in the dyeing of hydrophobic synthetic fibers. Conventional carriers, while effective in promoting dye diffusion by decreasing the glass transition temperature (T_g) of fibers, are often associated with toxicological risks, poor biodegradability, and significant wastewater burdens [11, 23–25]. In this context, eco-carriers are defined as auxiliary agents derived from renewable or environmentally compatible resources that facilitate dye diffusion under milder conditions while minimizing ecological and human health impacts.

Eco-carriers differ fundamentally from petrochemical-based carriers in their origin, environmental behavior, and functional potential. They are frequently obtained from natural or bio-based feedstocks—such as lignin derivatives, essential oil constituents, or plant-derived aromatic compounds—and are often characterized by lower volatility, reduced toxicity, and partial or full biodegradability [18, 33]. Unlike traditional carriers, whose contribution to dyeing is largely limited to fiber plasticization, eco-carriers may also provide added functionalities, including antibacterial activity, antioxidant properties, or fragrance retention, thereby aligning with the current market demand for multifunctional textiles.

From a mechanistic standpoint, eco-carriers act similarly to conventional carriers by reducing intermolecular interactions within the fiber matrix and depressing T_g, which increases the free volume available for dye penetration. However, their ecological footprint is significantly lower, and in many cases, they contribute positively to the circular economy by being derived from renewable by-products of agricultural or forestry processes [19, 21]. For example, vanillin can be synthesized from lignin, a major component of biomass waste streams, while coumarin occurs naturally in a variety of plants and could be extracted from agricultural residues.

In summary, eco-carriers can be described as a new generation of carrier systems that integrate sustainability principles with technical efficiency. They provide a bridge between traditional textile chemistry and green innovation, not only addressing environmental regulations but also contributing to the broader goals of resource efficiency, waste reduction, and product multifunctionality.

Bio-Based Aromatic Compounds: Structural and Functional Features

Bio-based aromatic compounds have recently gained attention as promising candidates for eco-friendly carrier systems in textile dyeing. Their aromatic rings, hydroxyl groups, aldehydes, or lactone structures allow them to interact with both synthetic fibers and dye molecules, thereby influencing diffusion mechanisms in ways similar to conventional carriers. However, unlike petrochemical carriers, these compounds are typically derived from renewable natural sources, such as lignin, essential oils, or plant metabolites, which makes them more aligned with green chemistry principles [10, 36].

The functional effectiveness of bio-based carriers depends largely on two aspects [17, 37, 38]:

- Structural Features—Aromaticity enhances π–π interactions with disperse dyes, while polar substituents (–OH, $–OCH_3$, –CHO, etc.) influence solubility, hydrogen-bonding potential, and fiber affinity.
- Physicochemical Behavior—Their balance between hydrophobic and hydrophilic properties affects how they partition between the aqueous dye bath and the fiber interior. Furthermore, their ability to diminish the glass transition temperature (T_g) of the fiber enhances chain mobility and creates free volume, thereby facilitating dye penetration.

Vanillin and coumarin are the most studied representatives of this class. Vanillin, a naturally occurring phenolic aldehyde, combines moderate hydrophobicity with the capability to form hydrogen bonds, thus functioning both as a mild plasticizer and T_g depressant [32]. Coumarin, a benzopyrone compound, is valued for its conjugated aromatic system and lactone structure, which not only support diffusion processes but also impart secondary functionalities such as UV absorption and antioxidant behavior [37].

In addition to these two well-documented compounds, other naturally derived phenolics and terpenoids—such as thymol, eugenol, and ferulic acid—have been proposed as potential eco-carrier candidates [6, 38, 39]. Their widespread availability from essential oils or agricultural by-products makes them attractive for further exploration. These compounds may also contribute to the development of multifunctional textiles, combining improved dye uptake with properties such as antimicrobial activity or fragrance retention.

Overall, bio-based aromatic carriers represent a convergence of structural versatility, renewable origin, and functional potential, positioning them as a cornerstone of sustainable innovation in textile wet processing. Among these bio-based aromatic compounds, vanillin and coumarin have attracted the most scientific interest, and their structural features, natural occurrence, and potential as eco-carriers are reviewed in detail in the following parts.

Vanillin

Vanillin (4-hydroxy-3-methoxybenzaldehyde) is one of the most broadly recognized flavor and fragrance molecules, extensively employed in the food, beverage, and perfumery industries [40, 41]. Although naturally present in vanilla pods, its extraction is labor-intensive and costly, making natural vanillin a premium product. As a result, the majority of vanillin available on the market is produced through chemical synthesis or biotechnological processes, which provide an economical and scalable alternative [19].

Advanced chromatographic analyses of synthetic vanilla extracts have identified more than 200 compounds contributing to its chemical complexity [18, 42–45]. The nonvolatile fraction consists primarily of polyphenols, tannins, amino acids, and resins, while the volatile fraction includes acids, ethers, alcohols, acetals, heterocycles, hydrocarbons, esters, and carbonyls, all of which contribute to its characteristic aroma and taste [46]. In contrast, synthetic extracts are comparatively simpler and typically composed of six major compounds: vanillin, ethyl vanillin, vanillic acid, o-vanillin, isovanillin, and coumarin [47].

Structurally, vanillin contains aromatic and phenolic groups, making it chemically analogous to traditional carrier agents such as phenols and benzoic acid derivatives. These features enable vanillin to function as a mild plasticizer that reduces the glass transition temperature (Tg) of polyester fibers, facilitating greater dye diffusion and uniform distribution within the amorphous regions of the fiber. In addition, vanillin displays a broad spectrum of biological activities—including antioxidant, antimicrobial, and antimutagenic properties [34, 40, 48]—which provide potential for added-value functionalities in textile applications. Given its structural similarity to conventional carriers, renewable origin, and multifunctional profile, vanillin has emerged as one of the most promising candidates for development as an eco-carrier in synthetic fiber dyeing and finishing.

Coumarin

Coumarin is an aromatic organic compound characterized by a fused benzene and α-pyrone (lactone) ring system, which imparts its distinctive sweet, vanilla-like fragrance. It is a crystalline solid with a molecular weight of 146.14 g/mol, a melting point of 68°C–70°C, and a boiling point of 298°C–302°C [49]. Coumarin exhibits good solubility in organic solvents such as ethanol, chloroform, and diethyl ether, while its aqueous solubility remains limited (1.7 g/L at 20°C) [11, 50]. This compound occurs naturally in a wide variety of plants. Its most prominent sources include tonka beans (*Dipteryx odorata*), which are particularly rich in coumarin; Cassia cinnamon, which contains significantly higher levels than Ceylon cinnamon [51–53]; and floral species such as jasmine and lavender, which owe part of their characteristic aroma to coumarin derivatives [54–56]. Coumarin is also abundant in sweet clover (Melilotus species), where it contributes to the plant's fragrance and has historically been exploited as a flavoring agent [57–59].

Beyond its sensory appeal, coumarin and its derivatives have been widely studied for their pharmacological activities, including antioxidant, antiviral, anti-inflammatory, hepatoprotective, and antithrombotic effects [60–62]. These properties underpin its extensive use in the perfume, cosmetics, and food industries [11]. However, high oral intake has been associated with hepatotoxicity, prompting regulatory restrictions on its use in food and flavoring products [63]. In contrast, its application in textiles is not subject to the same stringent controls, making it more accessible for functional finishing and dyeing processes.

In the textile domain, coumarin derivatives are best known for their fluorescent properties, which are exploited in optical brighteners and photoinitiators [64, 65]. More recently, coumarin itself has been investigated as a bio-based carrier in dyeing processes for hydrophobic fibers. Its conjugated aromatic system and lactone moiety facilitate interactions with disperse dyes and fiber matrices, while its secondary functionalities—such as UV absorption and antioxidant behavior—hold potential for added-value applications in multifunctional textiles.

Other Potential Bio-Derived Carriers

Beyond vanillin and coumarin, a wide range of natural oils and plant-derived phenolics have been investigated as alternative eco-carriers for synthetic fiber dyeing. Kalaycı [6] demonstrated that essential oils such as vanilla, sage, and lavender could function as natural diffusion carriers in the coloration of polyetherimide (PEI) fabrics, with vanilla oil in particular yielding the highest K/S values. Building on this, further research has highlighted additional essential oils—including eucalyptus, oregano, cassia, clove, peppermint, juniper berry, citronella, green tea, and star anise—whose terpenoid and phenolic constituents may promote fiber swelling and T_g reduction, thereby enhancing dye diffusion efficiency. In addition, oils with high aromatic content such as sandalwood oil and argan oil have been applied in disperse printing systems, where they not only improved color yield but also imparted antibacterial activity [17].

Spice- and herb-derived bioactive molecules, including those from cloves, basil, thyme, cinnamon, and tea leaves, have also attracted attention. Their phenolic constituents, such as eugenol and catechins, can form hydrogen bonds and π–π interactions with disperse dyes, potentially enhancing dye–fiber affinity. In supercritical CO_2 dyeing systems, aromatic acids such as benzoic and salicylic acid have also been investigated as eco-carrier candidates, demonstrating the ability to improve color yield under milder dyeing conditions [31].

Although the performance of these natural carriers varies according to their chemical composition, many exhibit dual benefits: they support dye diffusion while simultaneously imparting biofunctional properties such as antimicrobial, antioxidant, or UV-protective effects [38]. Their renewable origin and alignment with consumer demand for eco-conscious, multifunctional textiles position them as promising alternatives to conventional petrochemical carriers, though further optimization and standardization remain necessary for large-scale industrial applications.

Previous research about chemical structures and functional groups of bio-based carriers, dye interaction mechanisms, process parameters, functional outcomes, and value-added functionalities (like antibacterial activity, UV protection, antioxidant properties, aromatherapeutic and fragrance retention, and application domains) in the literature are reviewed in the section that follows in this chapter.

LITERATURE REVIEW

Chemical Structures and Functional Groups

Bio-based carriers such as vanillin, coumarin, and other phenolic derivatives are structurally characterized by the presence of aromatic rings and polar substituents that play a decisive role in their carrier activity. The aromatic core enables π–π stacking interactions with disperse dye molecules, thereby stabilizing dye–carrier complexes within the dye bath. At the same time, functional groups such as hydroxyl (–OH), methoxy (–OCH_3), aldehyde (–CHO), or lactone moieties increase their hydrogen-bonding potential and affinity toward fiber amorphous regions. This dual nature—aromaticity promoting hydrophobic interactions and polar groups enhancing compatibility with aqueous phases—makes these compounds effective at partitioning between the dye liquor and fiber interior. In contrast to petrochemical carriers like chlorobenzenes, bio-based compounds exhibit moderate hydrophobicity, allowing them to swell the fiber structure and lower the glass transition temperature (Tg) without excessive toxicity or volatility. These molecular features underpin their growing adoption as sustainable carrier alternatives in synthetic fiber coloration.

Interaction Mechanisms with Dyes

Numerous studies have searched for the mechanisms through which bio-based carriers such as vanillin and coumarin enhance dye uptake in hydrophobic fibers. Pasquet et al. [19] explored the influence of o-vanillin and p-vanillin on the dyeing performance of polyester at reduced temperatures. Dyeings were carried out at 90°C, 1:50 liquor ratio, and 3% owf dye concentration across a wide pH range (3–11). Both vanillin isomers facilitated fine distribution of disperse dyes within the fiber matrix, comparable to that of conventional carriers such as phenylphenol, dichlorobenzene, and benzoic acid. Notably, o-vanillin provided higher color yields than p-vanillin, and efficient dye uptake was achieved under neutral pH, eliminating the need for chemical pH adjustment.

Carrion-Fite and Radei [66] and subsequent studies [67, 68] developed a microemulsion system incorporating o-vanillin or coumarin with the nontoxic solvent n-butyl acetate. Polyester fabrics dyed at 95°C with this system exhibited K/S values and dye exhaustion comparable to those achieved at 135°C under conventional high-temperature dyeing. The o-vanillin/n-butyl acetate system generally outperformed coumarin, yielding higher dye uptake and better thermodynamic parameters. Importantly, sublimation and wash fastness were not adversely affected, demonstrating that low-temperature coloration with eco-carriers can maintain performance while reducing energy consumption.

Expanding the scope to copolymeric systems, Souissi et al. [10, 20, 69] examined o-vanillin, p-vanillin, and coumarin as carriers for PET/PTT bicomponent fibers dyed with disperse dyes of different molecular weights. Compared to toxic carriers such as dichlorobenzene and o-phenylphenol, the bio-based carriers significantly improved dye kinetics and fastness properties even below 100°C. Among them, p-vanillin showed the most consistent performance and was highlighted as the most practical option because of its lower synthesis cost and simpler production. Further, Souissi and co-workers proposed a single-bath, two-phase dyeing process for PET/PTT–cotton blends using p-vanillin in combination with natural colorants derived from date seeds [10, 20, 69]. This approach yielded not only improved dyeability but also functional properties such as antibacterial, antioxidant, and antifungal activity imparted by the natural extract.

Jalali et al. evaluated vanillin as a natural carrier for disperse dyeing of polyester under boiling conditions, comparing it with commercial carriers CTM (chlorobenzoyl structure) and Levegal PEW (N-alkyl phthalimide structure) [18]. Dye uptake was found to be strongly dependent on bath pH, with maximum values obtained at neutral conditions (at pH 7). Vanillin improved color yield and fastness, while binary mixtures of vanillin with commercial carriers showed synergistic effects, reducing vanillin consumption while enhancing dye uptake (Table 10.1).

Process Parameters and Functional Outcomes

These studies collectively demonstrate that process variables such as temperature, carrier concentration, and bath pH play critical roles in determining the performance of natural carriers. Effective dyeing has been achieved at temperatures as low as 90°C–100°C using vanillin and coumarin, compared to the 130°C–135°C typically required in high-temperature polyester dyeing. Optimal carrier concentrations ranged between 0.04 mol/L [20] and 1–3 g/L [18, 19], while neutral pH conditions proved particularly favorable. Importantly, the adoption of bio-based carriers aligns with sustainability goals by lowering energy demand, chemical load, and effluent toxicity, while in some cases providing added functionalities such as antimicrobial and antioxidant activity [33, 35].

For instance, Gressier et al. [35] demonstrated that polyester fabrics treated with o-vanillin via high-temperature diffusion or microwave-assisted processes exhibited not only coloration but also significant antibacterial activity against *Staphylococcus aureus* and *Klebsiella pneumoniae*. Similarly, Sharma and Ali [32, 33] functionalized polyester with vanillin-derived Schiff bases, reporting over 90% antibacterial activity against *E. coli*, along with enhanced UV protection and antioxidant performance. These findings highlight that eco-carriers may serve a dual role, simultaneously acting as dye diffusion enhancers and multifunctional finishing agents.

Value-Added FunctionalitiesEco-carriers derived from natural aromatics can perform a dual role: they facilitate dye diffusion (via T_g depression and fiber swelling)

TABLE 10.1
Summary of Studies on Bio-Based/Eco-Carriers in Synthetic Fiber Dyeing/Printing

References	Carrier(s)	Fiber Type	Dye Type	Conditions (Temp/Time/Conc./pH)	Key Findings
[19]	o-Vanillin, p-vanillin	PET	Disperse dyeing	90°C, LR: 1:50, 3% owf dye, pH 3–11	o-vanillin > p-vanillin in K/S; neutral pH feasible; fine distribution of dye in fiber
[18]	Vanilla powder	PET	Disperse dyeing	Boiling, LR: 1:50, 3% owf dye, pH 3–11	Maximum uptake at pH 7; improved K/S and fastness; synergistic effect with commercial carriers
[35]	o-Vanillin	PET		HT-HP (130°C, 45 min, LR: 1:20) or microwave (125–450 Watt, 3–7 min, LR: 1:1)	Yellowish color with microwave > HT-HP vanillin uptake; high antibacterial activity against *S. aureus* and *K. pneumoniae*: microwave > HT-HP
[66–68]	o-Vanillin, coumarin + n-butyl acetate (microemulsion)	PET	Disperse dyeing	95°C, microemulsion system	Dye uptake and K/S comparable to 135°C HT dyeing; o-vanillin > coumarin; fastness unaffected
[10, 20]	o-Vanillin, p-vanillin, coumarin	PET/PTT blends	Disperse dyeing	100°C with carrier, 0.04 mol/L carrier, 130°C without carrier	Improved dye kinetics and fastness; p-vanillin is most effective; lower cost synthesis
[32, 33]	Vanillin Schiff base	PET		120 °C, HT-HP, 15–20 min	Excellent antibacterial activity (>90% against *E. coli*), UV protection, antioxidant activity, enhanced coloration
[69]	p-Vanillin	PET/PTT–cotton blends	natural date seed extract	100°C, LR: 1:40, 0.04 mol/L carrier	Good dyeability of blends; added antibacterial, antioxidant, antifungal effects from natural extract
[6]	Vanilla oil, sage oil, lavender oil	PEI	Turmeric extract (microwave-assisted)	Microwave irradiation; aqueous medium	Vanilla oil gave the highest K/S; sage and lavender moderate; eco-carrier concept extended to dyeing

(Continued)

TABLE 10.1 (Continued)
Summary of Studies on Bio-Based/Eco-Carriers in Synthetic Fiber Dyeing/Printing

References	Carrier(s)	Fiber Type	Dye Type	Conditions (Temp/Time/Conc./pH)	Key Findings
[17, 70]	Sandalwood oil, argan oil, selected essential oils	PET	Disperse printing Disperse dyeing	Printing conditions: 40 g/kg disperse dye, 60 g/kg thickening paste, 8 g/kg citric acid (HT steam 180°C, 10 min) Dyeing: 110–130°C, 3% owf dye, %10 owf carrier	Improved color yield for both printing and dyeing with sandalwood oil antibacterial contribution from oils
[31]	Benzoic acid, salicylic acid (aromatic acids)	PET	Disperse dyeing	$ScCO_2$, 120°C	Enhanced dye uptake at milder $ScCO_2$ conditions; potential eco-carrier role
[37]	Coumarin (natural, agro-sourced)	Recycled polyester	Low- and high-molecular-weight disperse dyeing	100°C, 45 min, LR: 1:40, 2–4 g/L, pH 4.5–5.5, HT-HP	At 4 g/L, K/S = 25.8; fastness (washing, rubbing, light, perspiration) = 4–5

K/S: color strength
HT-HP: high temperature, high pressure
$ScCO_2$: supercritical carbon dioxide
UV: ultraviolet
LR: liquor ratio
PET: poly(ethylene terephthalate)
PTT: poly(trimethylene terephthalate)
PEI: polyetherimide

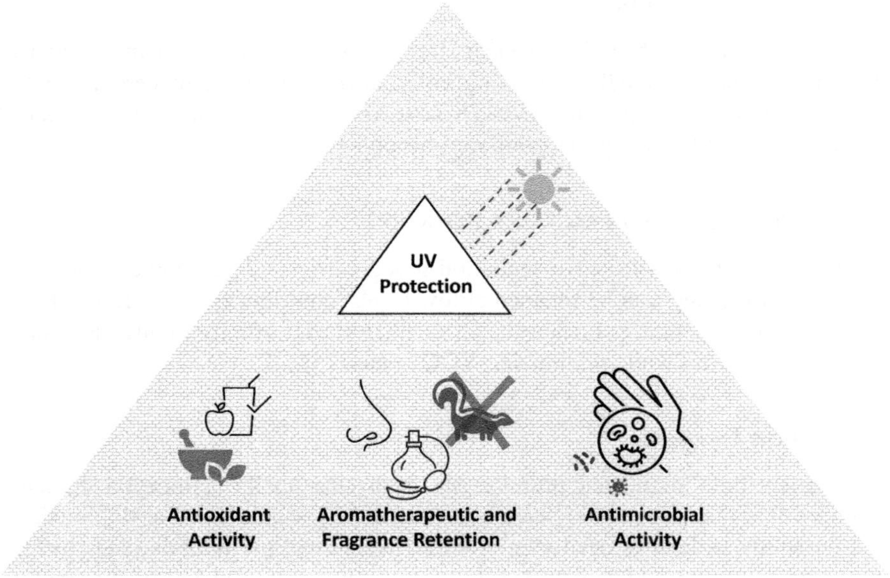

FIGURE 10.1 Functional performance spectrum of eco-carrier-treated textiles.

while imparting biofunctional properties to the substrate (Figure 10.1). This convergence is strategically important for value-added product designs in healthcare, wellness, and performance apparel, where coloration and functionality must coexist without increasing chemical load [6, 17, 37, 38].

Antibacterial Activity

Phenolic and benzopyrone structures (e.g., vanillin derivatives, coumarins, and selected essential oils) are known to disrupt bacterial membranes and/or interfere with cellular redox balance. When applied through dyeing or post-diffusion finishing, they can yield meaningful reductions in common test strains (e.g., *S. aureus*, *K. pneumoniae*, and *E. coli*) while preserving color strength and fastness [17, 38]. Durability depends on uptake level, bonding mode (physical vs. reactive/Schiff base), and post-wash retainment.

UV Protection

Aromatic conjugation and auxochromic substituents provide UV-absorbing capability, raising UPF values without resorting to inorganic particulates. Vanillin-derived finishes and coumarin chromophores can broaden UV attenuation while maintaining optical aesthetics; shade dependence should be reported to decouple color and UPF effects [37, 39].

Antioxidant Properties

Hydrogen-donating phenolics quench free radicals, slowing thermo-oxidative degradation and potentially improving color fastness to light in certain shades. Antioxidant indices (e.g., DPPH/ABTS assays) may be correlated with retained K/S after accelerated aging to quantify co-benefits [39].

Aromatherapeutic and Fragrance Retention

Volatile notes from selected essential-oil-based carriers (e.g., vanilla, clove, and sandalwood profiles) can be retained at low levels, contributing to perceived freshness and user experience. Encapsulation or fiber–carrier affinity tuning can extend fragrance longevity while minimizing VOC release [6].

Application Domains

Eco-carriers hold significant promise across diverse textile application domains, provided that their functional claims are validated by standardized testing methods and durability is demonstrated over multiple laundering, perspiration, and abrasion cycles. In healthcare and hygiene textiles, their antibacterial and anti-odor properties make them suitable for scrubs, bedding, and wound-care materials, particularly when biocompatibility and safety are established [17, 38]. In sportswear and outdoor apparel, their ability to impart UV protection and antioxidant stability under conditions of sweat and heat can enhance both performance and comfort while also contributing to low-odor profiles during active use. Within the wellness and personal care sector, eco-carriers offer potential for imparting mild aromatherapeutic effects and maintaining skin-contact acceptability, provided that cytotoxicity and irritation testing confirm consumer safety [39]. Together, these application domains highlight the multifunctional potential of bio-based carriers to extend textile performance beyond coloration, positioning them as enablers of next-generation functional fabrics. The next section discusses the environmental and sustainability perspective.

ENVIRONMENTAL AND SUSTAINABILITY PERSPECTIVE

The environmental viability of carrier systems must be evaluated not only in terms of dyeing performance but also through a holistic sustainability lens. Conventional carriers such as chlorobenzenes and phenylphenols are well known to contribute significantly to wastewater pollution, raising COD, BOD, and ecotoxicity indices, while also generating volatile emissions during drying and thermofixation [39]. Their persistence and poor biodegradability make them increasingly incompatible with modern zero-discharge and circular economy objectives.

Eco-carriers derived from natural aromatics present a more favorable profile. They generally exhibit lower toxicity, higher partial biodegradability, and reduced volatility, thereby lowering the treatment burden on effluent management systems. For instance, Souissi et al. (2021, 2022, 2024) demonstrated that o-vanillin, p-vanillin, and coumarin substantially reduced the ecological impact compared to conventional carriers such as dichlorobenzene or o-phenylphenol while still improving dye kinetics and fastness. Their partial hydrophilicity ensures more efficient removal during reduction clearing, reducing recalcitrant residues in effluents [10, 20, 69]. Energy-related sustainability gains are another key dimension. By enabling effective dyeing at 90°C–100°C instead of 130°C–135°C, eco-carriers translate into measurable reductions in thermal energy consumption and, consequently, carbon footprint. This effect is further amplified when combined with process intensification. For example, Kalaycı [6] showed that microwave-assisted dyeing with turmeric extracts not only enhanced dye uptake but also dramatically reduced energy demand compared to conventional heating, directly supporting low-carbon textile processing strategies.

Alternative process platforms also contribute to sustainability outcomes. Broadbent highlighted the potential of supercritical CO_2 dyeing systems, where the utilization of aromatic acids such as benzoic and salicylic acid enhanced disperse dye uptake under milder conditions while eliminating wastewater discharge entirely. By avoiding aqueous effluents, $ScCO_2$ dyeing addresses COD/BOD at the source and aligns with near-zero liquid discharge (ZLD) goals [31]. Similarly, Atav reported that essential-oil-based auxiliaries (e.g., sandalwood oil and argan oil) not only improved color yield in polyester printing but also reduced VOC emissions and toxic residues relative to petrochemical carriers, providing both ecological and functional benefits [17]. Ultimately, the sustainability credentials of eco-carriers rest on their performance across interconnected axes. Key sustainability outcomes of eco-carriers (Figure 10.2):

FIGURE 10.2 Key sustainability outcomes of eco-carriers.

(This figure was created using free icons from Canva, no permission required under Canva Free Media License.)

- *Reduced Toxicity*: Natural aromatic carriers (e.g., vanillin, coumarin, and essential oils) exhibit lower toxicity compared to chlorobenzenes or phenylphenols. They reduce worker exposure risks and lower ecotoxicity indices in effluents [17, 37].
- *Reduced Water Impact*: Partial hydrophilicity of eco-carriers enables more efficient removal during reduction clearing, reducing recalcitrant residues and lowering COD/BOD load in wastewater. When combined with $ScCO_2$ or near-ZLD systems, they further minimize liquid discharge [6, 31, 37].
- *Reduced Energy Consumption*: Eco-carriers enable polyester dyeing at 90°C–100°C instead of 130–135°C. Integration with alternative heating methods (e.g., microwave-assisted dyeing) provides additional energy savings and lowers the carbon footprint [19, 39, 66, 68].
- *Waste Minimization*: By improving dye kinetics and color yield, eco-carriers decrease the need for excess auxiliaries, salts, and reduction agents. This results in less process waste and aligns with circular economy goals [31].
- *Biodegradability*: Plant-derived carriers possess partial biodegradability, ensuring effluent residues break down more readily in natural systems compared to persistent petrochemical carriers [6].

FUTURE PERSPECTIVES

The transition from petrochemical to bio-based carriers is still in its early stages, and future studies must address both scientific and industrial challenges to ensure broader adoption. Beyond vanillin and coumarin, a wider portfolio of phenolic and terpenoid compounds—such as thymol, eugenol, ferulic acid, catechins, and lignin-derived aromatics—deserves systematic exploration, and advanced screening tools including quantitative structure–property relationships, molecular dynamics simulations, and high-throughput experimental platforms can accelerate the identification of effective T_g depressants with favorable toxicological profiles [6, 18, 37–39]. Integration of synthetic biology and biorefinery concepts may further enable cost-efficient production of novel bio-carriers from agricultural waste streams. In parallel, eco-carriers are exclusively situated to contribute to the next generation of multifunctional textiles; their inherent antibacterial, antioxidant, UV-protective, and fragrance-retention properties offer the potential to diminish the necessity for separate finishing steps, streamlining processes while simultaneously adding value [17, 31, 38, 71, 72]. To realize this potential, future work should prioritize durability under repeated laundering and compatibility with diverse fiber systems such as aramids, polyimides, and bio-based polymers, ensuring that multifunctionality is both stable and scalable. Moreover, the deployment of eco-carriers can be further optimized through smart process control and digital manufacturing platforms. By coupling dye house operations with real-time sensors for K/S, pH, and temperature, alongside machine learning-based predictive models, overdosing can be minimized and reproducibility improved [73–77]. Emerging technologies such as microwave irradiation, ultrasound, and supercritical CO_2 dyeing provide synergistic pathways for reducing energy and water use, and within the framework of Industry 4.0, eco-carriers may form part of digitally integrated, low-impact dyeing ecosystems [6, 31, 78–80]. Finally, the long-term success of eco-carriers will depend on supportive policy frameworks and regulatory clarity. As global restrictions on toxic carriers and hazardous auxiliaries continue to tighten, eco-carriers stand to benefit from eco-labeling schemes, green chemistry certifications, and compliance with OEKO-TEX® (International Association for Research and Testing in the Field of Textile and Leather Ecology), REACH (Registration, Evaluation, Authorisation and Restriction of Chemicals), and ZDHC (Zero Discharge of Hazardous Chemicals) guidelines, while harmonized international standards would provide the textile industry with confidence to scale adoption and avoid regional discrepancies in compliance requirements [1, 39, 72, 81–84].

CONCLUSION

The exploration of bio-based carriers represents a critical step toward reconciling textile coloration with sustainability imperatives. Conventional carriers, while effective in facilitating dye diffusion in hydrophobic fibers, impose severe environmental and health burdens due to their toxicity, poor biodegradability, and volatile emissions. In contrast, eco-carriers derived from natural aromatics—most notably vanillin and coumarin—demonstrate the capacity to enhance dye uptake under milder conditions, thereby reducing both energy consumption and chemical load. Their structural features, combining aromaticity with polar functionalities, not only enable T_g depression and fiber swelling but also impart value-added functionalities such as antibacterial activity, UV protection, antioxidant behavior, and fragrance retention.

Recent literature and experimental evidence confirm that these carriers can perform comparably to, or even surpass, petrochemical counterparts in terms of color yield, dye exhaustion, and color fastness, while aligning with circular economy frameworks and near-zero-discharge goals. Moreover, alternative approaches—such as microwave-assisted dyeing and supercritical CO_2 processing—when combined with eco-carriers, amplify environmental benefits by minimizing water consumption and reducing carbon footprints.

Nevertheless, challenges remain. Variability in natural feedstocks, scalability of production, cost competitiveness, and incomplete elimination of wastewater concerns highlight the need for comprehensive life cycle assessments and standardization in performance testing. The alignment of eco-carriers with international regulatory frameworks and eco-labeling initiatives will be decisive for industrial adoption.

Overall, eco-carriers should not be viewed merely as substitutes for conventional systems but as enablers of a paradigm shift in textile wet processing. By coupling coloration efficiency with ecological responsibility and functional versatility, they pave the way for the development of a new generation of sustainable, high-performance textiles. In this light, the conclusion is clear: the future of diffusion is green.

ACKNOWLEDGMENTS

This chapter utilized ChatGPT for language refinement and grammar checking. The authors reviewed and validated all content to ensure its accuracy and originality.

REFERENCES

1. Kalayci, E., O. Avinc, A. Yavas, and S. Coskun, Responsible textile design and manufacturing: environmentally conscious material selection, in *Responsible Manufacturing: Issues Pertaining to Sustainability*, A.Y. Alqahtani, et al., Editors. 2019, Taylor & Francis Group. https://doi.org/10.1201/9781351239141
2. Kalayci, E. and O. Avinc, Life cycle assessment (LCA) of biocomposites and its impact, in *Biocomposites—Bio-Based Fibers and Polymers From Renewable Resources*, M.R.S.R.A. Ramnath, S.M. Gorbatyuk, S. Fischer, S. Siengchin, Editor. 2024, Elsevier. https://doi.org/10.1016/B978-0-323-97282-6.00002-5
3. Kumartasli, S. and O. Avinc, Recycled thermoplastics: Textile fiber production, scientific and recent commercial developments, in *Recent Developments in Plastic Recycling*. 2021, Springer. pp. 169–192. https://doi.org/10.1007/978-981-16-3627-1_8
4. Periyasamy, A.P. and J. Militky, Sustainability in textile dyeing: Recent developments, in *Sustainability in the Textile and Apparel Industries: Sustainable Textiles: Production, Processing, Manufacturing & Chemistry*, S. Muthu, Gardetti, M., Editor. 2020, Springer. pp. 37–79. https://doi.org/10.1007/978-3-030-38545-3_2
5. Pandit, P., T. Gayatri, and S. Maiti, Green and sustainable textile materials using natural resources, in *Green and Sustainable Advanced Materials: Processing and Characterization (Volume 1)*, S.A.C.M. Hussain, Editor. 2018. pp. 213–261.
6. Kalaycı, E., Polieterimid liflerinin boyanma özelliklerinin araştirilmasi/investigation of dyeing properties of polyetherimide fibers, in *Textile Engineering*. 2024, Pamukkale University: Denizli.
7. Li, M.-Z., J.-X. Song, J.-M. Sun, and T.-T. Dong, Study on carrier dyeing for high strength high modulus PVA fiber. *The Journal of The Textile Institute*, 2022. **113**(12): pp. 2798–2807. https://doi.org/10.1080/00405000.2021.2018780
8. Kumartasli, S. and O. Avinc, Recycling of marine litter and ocean plastics: A vital sustainable solution for increasing ecology and health problem. *Sustainability in the Textile and Apparel Industries*, 2020: p. 117. https://doi.org/10.1007/978-3-030-38013-7_6
9. Kumartasli, S. and O. Avinc, Important step in sustainability: Polyethylene terephthalate recycling and the recent developments. *Sustainability in the Textile and Apparel Industries*, 2020: p. 1. https://doi.org/10.1007/978-3-030-38013-7_1
10. Souissi, M., R. Khiari, M. Abdelwaheb, M. Zaag, N. Meksi, and H. Dhaouadi, Kinetics study of dyeing bicomponent polyester textiles (PET/PTT) using environmentally friendly carriers. *RSC Advances*, 2022. **12**(4): pp. 2361–2374. https://doi.org/10.1039/D1RA08416J
11. Lima, C., K. Silva, J. Nascimento, C. Rossi, M. Granato, and F. Oliveira, Optimization of the polyester fabric dyeing process using coumarin as a green carrier. *Textile Research Journal*, 2022. **92**(19–20): pp. 3875–3888. https://doi.org/10.1177/00405175211026539
12. Kalayci, E. and O. Avinc, The disperse dyeing of polyetherimide fibers using orthogonal array (Taguchi) design. *Fibers and Polymers*, 2024. https://doi.org/10.1007/s12221-024-00641-1
13. Arcoria, A., A. Cerniani, R. De Giorgi, M. Longo, and R. Toscano, Carrier dyeing of polyester fibre with some disperse azo dyes. *Dyes and Pigments*, 1989. **11**(4): pp. 269–276. https://doi.org/10.1016/0143-7208(89)85045-4
14. Ingamells, W. and R. Peters, The carrier dyeing of synthetic fibers. *Polymer Engineering & Science*, 1980. **20**(4): pp. 276–281. https://doi.org/10.1002/pen.760200409
15. Ingamells, W., R. Peters, and S. Thornton, The mechanism of carrier dyeing. *Journal of Applied Polymer Science*, 1973. **17**(12): pp. 3733–3746. https://doi.org/10.1002/app.1973.070171215

16. Murray, A. and K. Mortimer, Carrier dyeing. *Review of Progress in Coloration and Related Topics*, 1971. **2**(1): pp. 67–72. https://doi.org/10.1111/j.1478-4408.1971.tb00182.x
17. Atav, R., B. Yıldırım, Ö. Çolakoğlu, F.N. Kuğu, and D. Kara, A sustainable innovative approach for disperse printing of polyester fabrics: Utilisation of essential oils extracted from plants as carrier. *Coloration Technology*, 2025: pp. 1–11. https://doi.org/10.1111/cote.70001
18. Jalali, S., R. Rezaei, M.G. Afjeh, and N. Eslahi, Effect of vanilla as a natural alternative to traditional carriers in polyester dyeing with disperse dyes. *Fibers and Polymers*, 2019. **20**: pp. 86–92. https://doi.org/10.1007/s12221-019-8482-2
19. Pasquet, V., A. Perwuelz, N. Behary, and J. Isaad, Vanillin, a potential carrier for low temperature dyeing of polyester fabrics. *Journal of Cleaner Production*, 2013. **43**: pp. 20–26. https://doi.org/10.1016/j.jclepro.2012.12.032
20. Souissi, M., R. Khiari, M. Zaag, N. Meksi, and H. Dhaouadi, Ecological and cleaner process for dyeing bicomponent polyester filaments (PET/PTT) using ecological carriers: Analysis of dyeing performance. *RSC Advances*, 2021. **11**(42): pp. 25830–25840. https://doi.org/10.1039/D1RA04771J
21. Wang, X., J. Wang, J. He, X. Gu, and K. Qin, A comparative study on dyeing properties of iterative polyester and ordinary polyester dyed with disperse dyes. *The Journal of The Textile Institute*, 2023. **114**(5): pp. 774–781. https://doi.org/10.1080/00405000.2022.2079821
22. Parmar, M., D.R. Maurya, and N. Sisodia, Developing faded denim using surface-modified disperse dye dyeable cotton yarn. 2023. https://doi.org/10.56042/ijftr.v48i4.7647
23. Vigo, T.L., *Textile Processing and Properties*. 1997, New Orleans, LA, USA: Elsevier Science B.V.
24. Vigo, T.L., *Textile processing and properties: Preparation, dyeing, finishing and performance*. 2013, New Orleans, LA, USA: Elsevier.
25. Burkinshaw, S.M., *Chemical Principles of Synthetic Fibre Dyeing*. 1995, Cornwall: Chapman & Hall.
26. Glenz, O., W. Beckmann, and W. Wulder, The mechanism of the dyeing of polyester fibres with disperse dyes. *Journal of the Society of Dyers and Colourists*, 1959. **75**(3): pp. 141–147. https://doi.org/10.1111/j.1478-4408.1959.tb02315.x
27. Koh, J., Dyeing with disperse dyes. *Textile Dyeing*, 2011. **10**: pp. 195–220.
28. Murray, A. and K. Mortimer, Dye auxiliaries in the application of disperse dyes to man-made fibres. *Journal of the Society of Dyers and Colourists*, 1971. **87**(6): pp. 173–181. https://doi.org/10.1111/j.1478-4408.1971.tb03018.x
29. Mao, Y.H., Y. Guan, Q.K. Zheng, Q.S. Liu, X.N. Feng, and X.X. Wang, Carrier dyeing of polyphenylene sulphide fabric with disperse dye. *Coloration Technology*, 2013. **129**(1): pp. 39–48. https://doi.org/10.1111/cote.12000
30. Xu, X.F., Q.K. Zheng, and J.Q. Zheng, Effect of carrier on the structure and performance of polyphenylene sulfide fiber. *Advanced Materials Research*, 2012. **554**: pp. 147–152. https://doi.org/10.4028/www.scientific.net/AMR.554-556.147
31. Broadbent, P.J., C.M. Carr, D.M. Lewis, M.L. Rigout, E.J. Siewers, M.S. Kaveh, and P. Desai, Investigation into the aqueous and supercritical carbon dioxide dyeing of polyester and cellulose acetate: The influence of temperature and carriers on fabric coloration and dye levelling. *Coloration Technology*, 2025. https://doi.org/10.1111/cote.12821
32. Sharma, V. and S.W. Ali, Functionalization of cellulosic and polyester textiles using reduced Schiff base (RSB) of eco-friendly vanillin. *Cellulose*, 2023. **30**(5): pp. 3317–3338. https://doi.org/10.1007/s10570-023-05085-z

33. Sharma, V. and S.W. Ali, A greener approach to impart multiple functionalities on polyester fabric using Schiff base of vanillin and benzyl amine. *Sustainable Chemistry and Pharmacy*, 2022. **27**: p. 100645. DOI: https://doi.org/10.1016/j.scp.2022.100645
34. Tai, A., T. Sawano, F. Yazama, and H. Ito, Evaluation of antioxidant activity of vanillin by using multiple antioxidant assays. *Biochimica et Biophysica Acta (BBA)-General Subjects*, 2011. **1810**(2): pp. 170–177. https://doi.org/10.1016/j.bbagen.2010.11.004
35. Gressier, P., N. Behary, and C. Campagne. Functionalisation of Polyester Fabric With Bio-Based Antibacterial Ortho-Vanillin Using Microwaves, in *Proceedings of the 19th World Textile Conference-Autex 2019*. 2019. https://doi.org/10.21825/autex.63864
36. Namita, P., R. Mukesh, and K.J. Vijay, Camellia sinensis (green tea): A review. *Global Journal of Pharmacology*, 2012. **6**(2): pp. 52–59.
37. Rehman, A., M. Naveed, and M.I. Siyal, Low-temperature dyeing of recycled polyester fibers using COUMARIN as eco-friendly carrier. *Chemical Papers*, 2025: pp. 1–13. https://doi.org/10.1007/s11696-025-04177-3
38. Ibrahim, N.A., E.M. El-Zairy, H.M. Khalil, and S.E.A. Almaksoud, A green sustainable strategy for eco-friendly sublimation transfer printing and multifunctional finishing of cotton/polyester blended fabrics using eco-functional additives. *Coloration Technology*, 2025. https://doi.org/10.1111/cote.70009
39. Yadav, S., K.S. Tiwari, M.K. Sharma, P. Meena, N. Usmani, and L. Agnihotri, Micelle Dyeing: Green Chemistry, in *Sustainable Coloration Techniques in Textiles*, M.S.S. Maiti, R.V. Adivarekar, Editor. 2025. p. 267. https://doi.org/10.1007/978-981-96-4975-4_10
40. Walton, N.J., M.J. Mayer, and A. Narbad, Vanillin. *Phytochemistry*, 2003. **63**(5): pp. 505–515. https://doi.org/10.1016/S0031-9422(03)00149-3
41. Walton, N.J., A. Narbad, C. Faulds, and G. Williamson, Novel approaches to the biosynthesis of vanillin. *Current Opinion in Biotechnology*, 2000. **11**(5): pp. 490–496. https://doi.org/10.1016/S0958-1669(00)00125-7
42. Cicchetti, E. and A. Chaintreau, Comparison of extraction techniques and modeling of accelerated solvent extraction for the authentication of natural vanilla flavors. *Journal of Separation Science*, 2009. **32**(11): pp. 1957–1964. https://doi.org/10.1002/jssc.200800650
43. de Jager, L.S., G.A. Perfetti, and G.W. Diachenko, Determination of coumarin, vanillin, and ethyl vanillin in vanilla extract products: liquid chromatography mass spectrometry method development and validation studies. *Journal of Chromatography A*, 2007. **1145**(1–2): pp. 83–88. https://doi.org/10.1016/j.chroma.2007.01.039
44. Sinha, A.K., U.K. Sharma, and N. Sharma, A comprehensive review on vanilla flavor: extraction, isolation and quantification of vanillin and others constituents. *International Journal of Food Sciences and Nutrition*, 2008. **59**(4): pp. 299–326. https://doi.org/10.1080/09687630701539350
45. Sinha, M.K., A review of processing technology for the utilisation of agro-waste fibres. *Agricultural Wastes*, 1982. **4**(6): pp. 461–475. https://doi.org/10.1016/0141-4607(82)90041-5
46. Ramachandra Rao, S. and G.A. Ravishankar, Vanilla flavour: production by conventional and biotechnological routes. *Journal of the Science of Food and Agriculture*, 2000. **80**(3): pp. 289–304. https://doi.org/10.1002/1097-0010(200002)80:3%3C289::AID-JSFA543%3E3.0.CO;2-2
47. Lavine, B.K., D.T. Corona, and U.D.N.T. Perera, Analysis of vanilla extract by reversed phase liquid chromatography using water rich mobile phases. *Microchemical Journal*, 2012. **103**: pp. 49–61. https://doi.org/10.1016/j.microc.2012.01.004
48. McShan, D. and I. Shah, Heurstic search for metabolic engineering: de novo synthesis of vanillin. *Computers & Chemical Engineering*, 2005. **29**(3): pp. 499–507. https://doi.org/10.1016/j.compchemeng.2004.08.038

49. Boisde, P.M. and W.C. Meuly, Coumarin. kirk-othmer encyclopedia of chemical technology, 2000. https://doi.org/10.1002/0471238961.0315211302150919.a01
50. Lake, B., Coumarin metabolism, toxicity and carcinogenicity: relevance for human risk assessment. *Food and Chemical Toxicology*, 1999. **37**(4): pp. 423–453. https://doi.org/10.1016/S0278-6915(99)00010-1
51. Ballin, N.Z. and A.T. Sørensen, Coumarin content in cinnamon containing food products on the Danish market. *Food Control*, 2014. **38**: pp. 198–203. https://doi.org/10.1016/j.foodcont.2013.10.014
52. Jose, A., N. Leela, T. Zachariah, and J. Rema, Evaluation of coumarin content and essential oil constituents in Cinnamomum cassia (Nees & T. Nees) *J. Presl*. 2019. https://doi.org/10.25081/josac.2019.v28.i1.5743
53. Woehrlin, F., H. Fry, K. Abraham, and A. Preiss-Weigert, Quantification of flavoring constituents in cinnamon: high variation of coumarin in cassia bark from the German retail market and in authentic samples from Indonesia. *Journal of Agricultural and Food Chemistry*, 2010. **58**(19): pp. 10568–10575. https://doi.org/10.1021/jf102112p
54. Austin, D. and M. Meyers, The formation of 7-oxygenated coumarins in hydrangea and lavender. *Phytochemistry*, 1965. **4**(2): pp. 245–254. https://doi.org/10.1016/S0031-9422(00)86170-1
55. Kunhachan, P., C. Banchonglikitkul, T. Kajsongkram, A. Khayungarnnawee, and W. Leelamanit, Chemical composition, toxicity and vasodilatation effect of the flowers extract of Jasminum sambac (L.) Ait."G. Duke of Tuscany". *Evidence-Based Complementary and Alternative Medicine*, 2012. https://doi.org/10.1155/2012/471312
56. Reshma, D., C. Anitha, and T.T. Sheeja, Phytochemical and pharmacological properties of five different species of Jasminum. *Plant Archives*, 2021. https://doi.org/10.51470/PLANTARCHIVES.2021.v21.no2.022
57. Nair, R., A. Whittall, S. Hughes, A. Craig, D. Revell, S. Miller, T. Powell, and G. Auricht, Variation in coumarin content of Melilotus species grown in South Australia. *New Zealand Journal of Agricultural Research*, 2010. **53**(3): pp. 201–213. https://doi.org/10.1080/00288233.2010.495743
58. Wu, F., Z. Duan, P. Xu, Q. Yan, M. Meng, M. Cao, C.S. Jones, X. Zong, P. Zhou, and Y. Wang, Genome and systems biology of Melilotus albus provides insights into coumarins biosynthesis. *Plant Biotechnology Journal*, 2022. **20**(3): pp. 592–609. https://doi.org/10.1111/pbi.13742
59. Zhang, J., H. Di, K. Luo, Z. Jahufer, F. Wu, Z. Duan, A. Stewart, Z. Yan, and Y. Wang, Coumarin content, morphological variation, and molecular phylogenetics of Melilotus. *Molecules*, 2018. **23**(4): p. 810. https://doi.org/10.3390/molecules23040810
60. Alihosseini, F., Plant-based compounds for antimicrobial textiles, in *Antimicrobial Textiles*, G. Sun, Editor. 2016, Elsevier. pp. 155–195. https://doi.org/10.1016/B978-0-08-100576-7.00010-9
61. Mishra, S., A. Pandey, and S. Manvati, Coumarin: An emerging antiviral agent. *Heliyon*, 2020. **6**(1). https://doi.org/10.1016/j.heliyon.2020.e03217
62. Qin, H.-L., Z.-W. Zhang, L. Ravindar, and K. Rakesh, Antibacterial activities with the structure-activity relationship of coumarin derivatives. *European Journal of Medicinal Chemistry*, 2020. **207**: p. 112832. https://doi.org/10.1016/j.ejmech.2020.112832
63. Abraham, K., F. Wöhrlin, O. Lindtner, G. Heinemeyer, and A. Lampen, Toxicology and risk assessment of coumarin: focus on human data. *Molecular Nutrition & Food Research*, 2010. **54**(2): pp. 228–239. https://doi.org/10.1002/mnfr.200900281
64. Lunagariya, J. and D. Bhavasar, Coumarin as a potential pharmacophore in medicinal chemistry. *International Journal for Research in Management and Pharmacy* 2014. **3**(1): pp. 44–51.

65. Elgemeie, G., K. Ahmed, E. Ahmed, M. Helal, and D. Masoud, Microwave synthesis, photophysical properties of novel fluorescent iminocoumarins and their application in textile printing. *Pigment & Resin Technology*, 2015. **44**(2): pp. 87–93. https://doi.org/10.1108/PRT-04-2014-0029
66. Carrion-Fite, F.J. and S. Radei. Development auxiliaries for dyeing polyester with disperse dyes at low temperatures, in *IOP Conference Series: Materials Science and Engineering*. 2017. IOP Publishing. https://doi.org/10.1088/1757-899X/254/8/082006
67. Radei, S., F.J. Carrión-Fité, M. Ardanuy, and J.M. Canal, Kinetics of low temperature polyester dyeing with high molecular weight disperse dyes by solvent microemulsion and agrosourced auxiliaries. *Polymers*, 2018. **10**(2): p. 200. https://doi.org/10.3390/polym10020200
68. Radei, S., F.J. Carrión-Fité, M. Ardanuy, and J.M. Canal, Thermodynamic and kinetic parameters of polyester dyeing with Disperse Blue 56 using bio-based auxiliaries and co-solvent microemulsion. *Textile Research Journal*, 2020. **90**(5–6): pp. 523–536. DOI: https://doi.org/10.1177/0040517519871264
69. Souissi, M., A. Moussa, and H. Dhaouadi, Valorization of natural dye extracted from date palm pits (Phoenix dactylifera) for ecofriendly dyeing of bicomponent polyester filaments. *Fibers and Polymers*, 2024: pp. 1–11. https://doi.org/10.1016/j.jclepro.2018.08.115
70. Atav, R., N. Kugu, and Ö. Çolakoğlu, A sustainable innovative approach for low-temperature dyeing of polyester fabrics: Using essential oils extracted from plants. *Orclever Proceedings of Research and Development*, 2024. **5**(1): pp. 253–260. https://doi.org/10.56038/oprd.v5i1.538
71. Eren, S., M. Öztürk, H.A. Eren, and S.T. Günç, Bleaching terry fabric in supercritical carbon dioxide medium: a sustainable solution. *Clean Technologies and Environmental Policy*, 2025. https://doi.org/10.1007/s10098-025-03296-5
72. Kalayci, E., O. Avinc, and H.A. Eren, Bleaching processes of pineapple fibers assisted with sustainable technologies: Microwave, ultrasound and ozonation. *Fibers and Polymers*, 2025. https://doi.org/10.1007/s12221-025-01155-0
73. Pervez, M.N., W.S. Yeo, L. Lin, X. Xiong, V. Naddeo, and Y. Cai, Optimization and prediction of the cotton fabric dyeing process using Taguchi design-integrated machine learning approach. *Scientific Reports*, 2023. **13**(1): p. 12363. https://doi.org/10.1038/s41598-023-39528-1
74. Jong, W.-R., Y.-M. Huang, Y.-Z. Lin, S.-C. Chen, and Y.-W. Chen, Integrating Taguchi method and artificial neural network to explore machine learning of computer aided engineering. *Journal of the Chinese Institute of Engineers*, 2020. **43**(4): pp. 346–356. https://doi.org/10.1080/02533839.2019.1708804
75. Ingle, N. and W.J. Jasper, A review of the evolution and concepts of deep learning and AI in the textile industry. *Textile Research Journal*, 2025. **95**(13–14): pp. 1709–1737. https://doi.org/10.1177/00405175241310632
76. Busari, M., *AI and Machine Learning Algorithms for Predictive Supply Chain Management in Textile Manufacturing*. 2025. Available: https://www.researchgate.net/profile/MuhammedBusari/publication/390141733_AI_and_Machine_Learning_Algorithms_for_Predictive_Supply_Chain_Management_in_Textile_Manufacturing/links/67e214ac3ad6d174c4bdb899/AI-and-MachineLearning-Algorithms-for-Predictive-SupplyChain-Management-in-Textile-Manufacturing.pdf
77. Haji, A., M. Vadood, M. Öztürk, I. Yigit, S. Eren, and H.A. Eren, Prediction of colour strength in environmentally-friendly dyeing of polyester fabric with madder using supercritical carbon dioxide. *Coloration Technology*, 2025. **141**(1): pp. 44–53. https://doi.org/10.1111/cote.12757
78. Malashin, I., D. Martysyuk, V. Tynchenko, A. Gantimurov, V. Nelyub, A. Borodulin, and A. Galinovsky, Machine learning in polymeric technical textiles: A review. *Polymers*, 2025. **17**(9): p. 1172. https://doi.org/10.3390/polym17091172

79. Wang, Y., A. Brahmia, A. Shahbaz, H. Sahramaneshi, T. Alkhalifah, and J. Yang, A novel machine learning model for innovative microencapsulation techniques and applications in advanced materials, textiles, and food industries. *Renewable and Sustainable Energy Reviews*, 2025. **224**: p. 116082. https://doi.org/10.1016/j.rser.2025.116082
80. Iannacchero, M., J. Löfgren, M. Mohan, P. Rinke, and J. Vapaavuori, Machine learning-assisted development of polypyrrole-grafted yarns for e-textiles. *Materials & Design*, 2025. **249**: p. 113528. https://doi.org/10.1016/j.matdes.2024.113528
81. Amutha, K., Sustainable chemical management and zero discharges, in *Sustainable Fibres and Textiles*, S.S. Muthu, Editor. 2017, Elsevier. p. 347–366. https://doi.org/10.1016/B978-0-08-102041-8.00012-3
82. Prasannamedha, G. and P. Senthilkumar, Chemical compliance and regulations in textiles and fashion, in *Chemical Management in Textiles and Fashion*, S.S. Muthu, Editor. 2021. pp. 135–154. https://doi.org/10.1016/B978-0-12-820494-8.00007-1
83. Naikwadi, S., Role of Governments and Consumers, in *Climate Action Through Eco-Friendly Textiles*. 2024, Springer. pp. 163–178. https://doi.org/10.1007/978-981-99-9856-2_11
84. Sharma, S., M. Jahanzaib, J. Lee, and D. Park, Global legislation, schemes and standards to control environmental concern for textile-generated waste. *Nanotechnology-Assisted Recycling of Textile Waste: Sustainable Tools for Textiles*, 2025: pp. 435–459. https://doi.org/10.1002/9781394175017.ch15

Index

Pages in *italics* refer to figures and pages in **bold** refer to tables.

D

E

F

G

L

M

N

O

P

Q

R

S

X

Y

Z

For Product Safety Concerns and Information please contact our EU representative GPSR@taylorandfrancis.com Taylor & Francis Verlag GmbH, Kaufingerstraße 24, 80331 München, Germany

Batch number: 10755284

Printed by Printforce, the Netherlands